Palgrave Studies in the History of Science and Technology

Designed to bridge the gap between the history of science and the history of technology, this series publishes the best new work by promising and accomplished authors in both areas. In particular, it offers historical perspectives on issues of current and ongoing concern, provides international and global perspectives on scientific issues, and encourages productive communication between historians and practicing scientists.

Olga Dubrovina

The Soviet Space Program and International Collaboration during Perestroika, 1985–1991

palgrave
macmillan

Olga Dubrovina
Carpi, Italy

ISSN 2730-972X ISSN 2730-9738 (electronic)
Palgrave Studies in the History of Science and Technology
ISBN 978-3-032-03156-3 ISBN 978-3-032-03157-0 (eBook)
https://doi.org/10.1007/978-3-032-03157-0

This Palgrave Macmillan imprint is published by the registered company Springer Nature Switzerland AG.
The registered company address is: Gewerbestrasse 11, 6330 Cham, Switzerland

To my parents,
Elena and Vladimir

Foreword

Is it ignorance? Or is it amnesia? As the war in Ukraine dragged on in 2025, time and again the press and some public intellectuals were surprised by the cooperative posture struck by Donald Trump towards Vladimir Putin, as if the US and Russia were always enemies. They were not of course. The collapse of the Soviet Union and its fragmentation into a patchwork of separate states in 1991 was a process that was not just welcomed by Washington, but seized as an opportunity to help transform the Soviet command economy into a capitalist free market system. The Soviet military and civil space programs were necessarily engaged in this process, a process that was already under way in the late 1980s during the period of *perestroika* and its demand for transparency (*glasnost*) undertaken by Mikhail Gorbachev. This fine study uses international collaboration with (mostly) NASA and ESA and Western Europe before the 1990s as an instrument to explore the challenges faced by the Soviet system as it navigated the tensions between a reform-driven intelligentsia sympathetic to *perestroika* and an administrative and party elite with a vested interest in preserving a militarised national space program shrouded in secrecy. In their eyes the technological and scientific success of that program was a source of national pride and national security as well as a tribute to the communist system. International collaboration called for openness and reciprocity, undermined Soviet military capabilities, and exposed the system's failures and weaknesses, further discrediting the command economy, and eventually contributing to the collapse of the Soviet system itself—or so they feared.

Gorbachev's first steps advocating *perestroika* in outer space combined concerns about the weakening economy and the need to react to Ronald Reagan's SDI (Strategic Defense Initiative, or "Star Wars"). Both skeptical and fearful of the USA president's bellicose agenda, he advocated international collaboration in outer space for peaceful purposes. He was willing to share the country's extensive scientific and technological hardware and knowhow to continue the exploration of space in cost-sharing ventures that would help preserve a national space sector threatened by major financial cuts and industrial stagnation. Collaborative international space ventures would help transform key sectors of the Soviet economy that needed to be modernised by abolishing secrecy and encouraging competitive innovation. Coupled with *glasnost*, *perestroika* in space opened the black box that shrouded the space industry in secrecy. It facilitated open communication with scientists and engineers in the west (and enabled some of them to emigrate). It was essential to the commercialisation of some sectors of the space program, notably the provision of launch services. The demand for transparency also allowed the people and the world watch Soviet space endeavors as they happened—even when they failed, undermining confidence in the entire space engineering and production system.

Collaboration with Western Europe and with ESA was impacted by geopolitical tensions with the Soviet Union and by internal tensions between national priorities (notably in France) and international obligations in ESA. The Soviet invasion of Afghanistan in 1979, the declaration of martial law in Poland, and President Reagan's characterisation of the Soviet Union as an "evil empire" in March 1983 all called for prudence in dealings with the Soviet space system. Ambiguity on the Soviet side over the financing of joint scientific missions—did it involve clean interfaces and no exchange of funds, as with NASA, or were the foreign partners expected to contribute financially to Soviet costs and to share sensitive technology—added to the confusion. This did not stop France, the UK, the Federal Republic of Germany, Italy, and Austria from working with their Soviet homologues. It simply exposed them to transformations under way in the economic, industrial, and institutional restructuring of the Soviet space program in the late 1980s, that they legitimated by their persistence and gradually normalised by brokering agreements that were mutually acceptable.

The collaborative international effort to observe Halley's comet as it passed by the earth in 1986 on its regular 76-year journey around the sun remains an unforgettable and outstanding success, even eclipsing the Bion life sciences experiments that got under way in the 1970s. Two Soviet Vega spacecraft tracked the comet on its approach, sharing data with probes launched by ESA and by NASDA (the Japanese space agency) using NASA's Deep Space Network, to facilitate their rendezvous in outer space. It showed that a sound scientific project could maintain its momentum even as geopolitical tensions intensified in the early 1980s, when planning got underway. It dispelled doubts over the possibility of the Soviet space system to build and deliver two functioning spacecraft under pressure of time and to share information with Western colleagues. And it threw light on the meaning of the extension of *perestroika* and *glasnost* into space for both Soviet and Western space communities. It was exceptional in another way. Each national space agency was responsible for separate spacecraft in the armada that met Halley in space, maintaining clean technological interfaces and no exchange of funds. Only the technical data needed by ESA's and NASDA's probes to successfully rendezvous with the comet were transmitted through NASA's network of tracking stations. Above all the project escaped the chilling effect of the export controls that, as explained in Part IV of the book, were one omnipresent obstacle to the commercialisation of space technology by the Soviets in the 1990s.

This book deftly describes the twists and turns in international collaboration in space with the Soviet Union and Russia during the last two decades of the twentieth century. It is a reminder of the important role that science and technology can play in sustaining foreign relations between governments, even when they support very different modes of political organisation. The shared desire to know more and to know better, as well as to build a dynamic international epistemic community, enables cooperative ventures to persist even when the political climate is strained. It can also serve as a stimulus to social change, transparency, and mutual tolerance, as Olga Dubrovina's insightful investigation of the extension of *perestroika* and *glasnost* into space shows. Whether or not science (and technology) diplomacy will serve any purpose in the particularly confrontational and anti-science political climate that is taking hold as I write, is another question.

Kranzberg Professor Emeritus, Georgia Tech — John Krige
Paris, France

Acknowledgments

In many ways, this book is a collective work, shaped by the written and oral testimonies of numerous participants in collaborative projects. I am deeply grateful to all those who generously dedicated their time, effort, and willingness to share their memories with me. These individuals include (in alphabetical order): Valery Abrashkin, Mark Belakovsky, Karl Bergquist, Thierry Courvoisier, Fabio Favata, Viktor Nikolaev, Mikhail Mogilevsky, and David Southwood.

I extend my special thanks to Genrikh Avanesov, Mike Gruntman, and Lev Zelenyi for their trust, invaluable suggestions, and insightful comments on IKI activities and the Soviet space program as a whole. René Demets, Leonid Gurvits, and Antonio Verga read an early draft of the book and provided invaluable feedback. I sincerely appreciate their time, assistance, moral support, and unwavering belief in my project. I deeply regret that Antonio did not live to see the completion of this book, to which he contributed so much.

I would also like to express my heartfelt gratitude to Evgeny Ilyin, who responded to my request to share his memories with remarkable kindness and enthusiasm. Sadly, in September 2023, Evgeny Aleksandrovich passed away, as did another key figure in my book, Wim Jansen, in February 2024. Together, they laid the foundations for years of international cooperation.

I am especially grateful to William P. Barry, Yuri Baturin, Davide Casali, Slava Gerovitch, John Krige, and Pierre-Bruno Ruffini for their discussions on various book chapters and their valuable insights, which helped me navigate the complex developments and transformations explored in this work. Brian Harvey provided invaluable support throughout the years of writing this book, offering pioneering examples and patiently answering my many questions.

The book would have lost an essential part of its source material without the generous assistance of Tatiana Agaptseva (IBMP), Piero Messina (ESA), and Olga Zakutnyaya (IKI). I also owe special thanks to the archival staff who kindly assisted me in researching and working with historical materials: Gherardo Bonini and Andreja Casar (HAEU) and Sarah Jenkins and Sarah LeClaire (NASA HQ Archives).

I would like to express my sincere gratitude to my anonymous reviewer and series editor, whose valuable suggestions and constructive criticism significantly contributed to improving my research.

Finally, I must acknowledge my close friend and mentor, Andrea Panaccione, who accompanied me through every step of this journey, sharing his ideas and comments on the drafts they reviewed. My heartfelt and boundless gratitude goes to my friend Carolyn Green, who generously proofread the entire book and greatly improved my English. Special thanks also go to Natalia Dubrovina for her artistic contributions. Above all, this book could not have become a reality without the patience, understanding, and moral support of the two people closest to me: Nicola and Nadia Negri.

Music runs throughout this book, connecting voices and memories into a single symphony. I see my role as that of a humble conductor, bringing together talented musicians—each with their own unique perspective—to build something coherent and meaningful.

Competing Interests This book is the result of my continuous involvement in several scientific projects. Its compilation began within the framework of the project "*A Mission for All Mankind? Imaging Russian Science Diplomacy in the Space Race from Cold War Cooperation to Global Competition*", which was part of the *InsSciDE Horizon 2020* project. I continued writing the book within the *PRIN 2017* project, "*Science, Technology, and International Relations: Case Studies in Italian Foreign Policy*", and later as part of the *PRIN 2022* project, "*Europe's Research for Strategic Autonomy: From the Late Cold War to the New Millennium*". All these projects were led by Professor David Burigana. The Department of Political Science, Law, and International Studies (SPGI) at the University of Padua provided the necessary funding to complete this book.

Praise for *The Soviet Space Program and International Collaboration During Perestroika, 1985–1991*

"'Blessed is he who visited this world in its fateful moments,' wrote Pushkin. Olga Dubrovina's book evokes that same blend of nostalgia, urgency, and regret. She captures a turbulent era and the complex ESA-USSR space partnership. As a reader, I strongly recommend this book and would welcome a continuation exploring Russia's dramatic post-Perestroika space journey."

—Lev M. Zelenyi, *Scientific Director, Space Research Institute of the Russian Academy of Sciences*

"Having lived through these turbulent years of international space cooperation, I saw firsthand how perestroika reshaped Soviet space science. This book is an important historical work, capturing the challenges faced at that time, and how determination from scientists kept international collaboration alive."

—Roger Bonnet, *European Space Agency Director of Science, 1983–2001*

"Olga Dubrovina has produced an extraordinarily important book about an understudied aspect of the Soviet space program, the years during perestroika. As Dubrovina shows, science was an important lever of the larger political changes in the Soviet system during this crucial time. Her work is indispensable for anyone seeking to understand the source of the challenges faced by the current Russian space program."

—Asif Siddiqi, *Professor, History Department, Fordham University*

"Olga Dubrovina sheds light on a fascinating chapter in the history of cooperation in space between previously adverse partners during the perestroika period, 1985–1990. With great detail and profound insight, she explores the political and practical conditions which enabled a wave of short- and mid-term opportunities—paving the way for the long-term collaboration that would ultimately culminate in the creation of the largest object ever built in space: the International Space Station."

—René Pischel, *Head of the European Space Agency Permanent Mission in Russia, 2009–2024*

Contents

About the Author

Olga Dubrovina is a historian of international relations with a particular focus on the foreign policy of the USSR and Russia. She has lectured at the Universities of Milan and Bologna and conducted research at the University of Padua. She is also a member of the History Committee of the International Academy of Astronautics (IAA). Dubrovina has authored numerous essays on the history of international relations and Soviet foreign policy, including *Space Politics in Cold War Central and Eastern Europe* (2025), *INTEGRAL Observatory: Rescue at All Costs* (2023), *Russia's Space Diplomacy: Why We Should Look Back to the Soviet Years* (2022), and *Scientists in the Soviet Aerospace Sector: Revealing a Complicated Co-Existence* (2021).

Abbreviations

CAST	China Academy of Space Technology
CMEA	Council for Mutual Economic Assistance
CNES	Centre National d'Études Spatiales
CNRS	Centre National de la Recherche Scientifique
CoCom	Coordinating Committee for Multilateral Export Controls
COPUOS	Committee on the peaceful uses of Outer Space
CPSU	Communist Party of the Soviet Union
DFVLR	German Research and Test Institute for Aviation and Space Flight, 1969–1989
DLR	Deutsches Zentrum für Luft- und Raumfahrt (German Centre for Aviation and Space Flight)
DSN	Deep Space Network
EEC	European Economic Community
EO	Earth Observation
ESA	European Space Agency
ESOC	European Space Operations Centre, Darmstadt (GER)
ESRO	European Space Research Organisation
ESTEC	European Space Research and Technology Center, Noordwijk (NL)
IACG	Inter-Agency Consultative Group
IAF	International Astronautical Federation
IBMP	Institute of Biomedical Problems
IKI	Space Research Institute
INMARSAT	International Maritime Satellite Organisation
IRAC	International Relations Advisory Committee
IRAG	International Relations Advisory Group
ISRO	Indian Space Research Organisation

ISTP	International Solar-Terrestrial Physics
KNA	Scientific Equipment Container
MGU	Lomonosov Moscow State University
MNTS KI	Interdepartmental Scientific and Technical Council on Space Research
MOM	Ministry of General Machine Building
NASA	National Aeronautics and Space Administration
NASDA	National Space Development Agency of Japan
NII	Scientific and Research Institute
NPO	Scientific-Production Association
NPP	Research & Development Production Enterprise
NRC	American Nuclear Regulatory Commission
OKB	Experimental Design Bureau
PTT	Postal Telegraph and Telephone
SAR	Synthetic Aperture Radar
SDI	Strategic Defence Initiative
SDIO	Strategic Defence Initiative Organisation
SSD	Space Science Department of ESA
TsNIIMash	Central Research Institute of Machine Building
TsSKB	Central Specialised Design Bureau in former Kuybyshev
TsUP	Mission Control Centre
VDNKh	Exhibition of Achievements of National Economy in Moscow
VLBI	Very Long Baseline Interferometry
VNIIPAS	All-Union Scientific Research Institute for Applied Automated Systems
VPK	Military-industrial complex
WSO	World Space Organisation
Yuzhmash	Southern Machine-Building Plant in former Dnepropetrovsk, Ukraine

List of Figures

Introduction

Historians and political analysts often seek to explain contemporary Russia and its relations with the West, tracing the origin of today's dynamics in the so-called dashing nineties, a time of profound upheaval in Russia, marked by sweeping internal social, political, and economic transformations, and the redefinition of Moscow's role on the global stage. However, the roots of these changes should be traced back to the earlier era of *perestroika*—a unique period in history that simultaneously marked the peak of Soviet power and set the stage for the collapse of the USSR, the end of the Cold War, and the beginning of a new chapter for Russia. In many ways, analysing this period is essential to understanding the developments of the decades that followed.

The choice of cosmonautics as the subject of this book is no coincidence. By its very nature, cosmonautics serves as a clear and sensitive indicator—a kind of litmus test—of the social, political, and economic transformations of the late Soviet era. The space field reflects both universal characteristics shared by many countries and features unique to the Soviet system of that time. It is a highly visible and crucial sector, both domestically and at the international level. Closely tied to advanced technologies, the space industry occupied a leading position in the USSR and the USA economic development. As Walter McDougall aptly described it, the space program was a "subsystem of the state", deeply embedded in the broader political, economic, and military structures of the USA and the Soviet Union.[1]

In the Soviet Union, cosmonautics was an inseparable part of the military-industrial complex (VPK), which conferred upon it an exceptional significance not only in military and strategic terms but also in social terms—as evidenced by the hundreds of thousands of people employed in the field. The physical and intellectual resources of Soviet society were largely concentrated on sustaining the growth of the VPK, which was seen as essential for national security but consumed vast human and economic capital. Moreover, the cultural aspect of cosmonautics was central to Soviet self-identity.[2] The end of the 1980s marked a decisive turning point when the Soviet planned economy, strained by the efforts of multiple generations, finally collapsed—bringing cosmonautics, one of its key pillars, down with it.

The end of the confrontation between the two space superpowers during the Cold War has, somewhat surprisingly, attracted far less scholarly attention than its initial and intermediate phases. Both Russian and Western historians have rightly focused on the beginning of the space race,[3] its uneven progression, and its key milestones, such as the launch of the Sputnik and its multifaceted effects,[4] the first spectacular manned missions, and the rivalry of the Moon programs. The 1970s and early 1980s are justifiably considered the most dynamic period,[5] marked by groundbreaking international space projects, joint missions of great symbolic significance, and major scientific achievements. Projects such as the Soyuz-Apollo mission, the development of reusable space transportation

[1] Mc.Dougall, W. (1997) …*The Heavens and the Earth. A political history of the Space Age.* The John Hopkins University Press, 445.

[2] See more in Andrews, J. T., & Siddiqi, A. A. (Eds.). (2011). *Into the Cosmos: Space Exploration and Soviet Culture.* University of Pittsburgh Press.

[3] E.g. Siddiqi, A. (2000). *Challenge to Apollo: The Soviet Union and the Space Race, 1945-1974.* NASA History Office; Hardesty, V. and Eisman, G. (2007). *Epic Rivalry: The Inside Story of the Soviet and American Space Race.* Washington, DC: National Geographic.

[4] E.g. R. Lanius, R. Smith, J. Logsdon. (Eds.). (2000). *Reconsidering Sputnik: Forty Years Since the Soviet Satellite.* Routledge; D. Cozzoli (Ed.) *The Sputnik Effect in Global Perspective. Competition, Collaboration, and the Circulation of Knowledge.* Forthcoming with Palgrave Macmillan.

[5] For a detailed discussion on the significance of this period, see A. C. T. Geppert. (Ed.). (2018). *Limiting Outer Space: Astroculture after Apollo.* Palgrave Macmillan.

systems (the Space Shuttle and the Buran[6]), and the Strategic Defence Initiative (SDI) have deservedly occupied a central place in historiography. It is no coincidence that most studies focus on Soviet-American[7] and Soviet-French[8] space relations, as well as the USSR's cooperation with countries within the socialist bloc.[9] Within this historiographical framework, Brian Harvey's exploration of Soviet-European cooperation in space[10] and Asif Siddiqi's study of Soviet-African cooperative programs[11] stand out as a pioneering contribution.

Despite the richness of existing literature, the era of *perestroika*—which marked the end of Soviet-American rivalry, including in the space sector—has not yet received the scholarly consideration it deserves. In the 1980s, intense focus on other international developments, combined with a noticeable decline in public interest in space affairs, contributed to the marginalisation of this topic within historiographical debates. However, this is not to say that the period has been entirely overlooked by historians of space exploration. Several specialised studies do address space aspects of the *perestroika* period. Notably, "space sleuths" of different generations Christian Lardier, Asif Siddiqi, and Bart Hendrickx have highlighted the significance of *perestroika* in opening the Soviet space program to Western

[6] E.g. Hendrix, B. and Vis, B. (2007). *Energiya-Buran: The Soviet Space Shuttle*. Springer-Praxis.

[7] Karash, Yu. (1999). *The superpower Odyssey: A Russian Perspective on Space Cooperation*. AIAA; Long Callahan, A. (2013). NASA and Eastern Europe. In Krige, J., Long Callahan, A., Maharaj, A. (Eds.), *NASA in the world. Fifty years of international collaboration in space*, Palgrave Macmillan.

[8] E.g. Lamy J. (2011). What Management Does to Space Projects: The Franco-Soviet Project ARCAD 3 in the Late 1970s. *Science in Context*, 24(4); Ammar-Israël, A. (Ed.) (2015). *50 ans de coopération spatiale France-Urss/Russie : Genèse et évolutions 1966-2016*. Éditions Tessier et Ashpool ; Gouarné, I. (2016). Dépasser les tensions Est-Ouest pour la conquête de l'espace. La coopération franco-soviétique au temps de la guerre froide. *Les Cahiers Sirice*, 2 (16), 49-67. Roche Nye, L., (2017) *Coopération spatiale franco-soviétique et réseaux scientifiques en temps de guerre froide (1966-1988) : Transferts, circulations, pouvoirs*. Doctoral thesis;

[9] Burgess, C., Vis, B. (2015). *Interkosmos: The Eastern Bloc's Early Space Program*. Springer.

[10] Harvey, B. (2021). *European-Russian Space Cooperation*, Springer.

[11] Siddiqi, A. (2021). Shaping the World: Soviet-African Technologies from the Sahel to the Cosmos. *Comparative Studies of South Asia, Africa and the Middle East*. 41 (1), 41–55.

understanding.[12] Brian Harvey captured the "splendours and miseries" of the Soviet space program in the late 1980s.[13]

The evolution of scientific relations between the USSR and Western countries during the *perestroika* period has been mentioned by Roger M. Bonnet and Vittorio Manno,[14] as well as by Olga Zhdanovich in her study of international scientific cooperation in the USSR.[15] The impact of *perestroika* on manned spaceflights is addressed in one of the chapters of Robert Zimmerman's *Leaving Earth*,[16] as well as partially in the work of Luc van den Abeelen,[17] who examines the Soviet Union's involvement in the European Hermes shuttle project.

The initial stages of commercialisation of Soviet manned spaceflights are discussed by Chris Dubbs and Emeline Paat-Dahlstrom, particularly the first commercial flights of foreign citizens aboard Soviet space stations, as well as the earliest USA efforts to establish commercial ventures within the USSR.[18] Similarly, the Soviet Union's first steps towards the development of a commercial space sector are explored by Christian Lardier and Stefan Barensky in their history of Soviet Soyuz[19] and Proton[20] launch vehicles. The authors underscore the significance of Gorbachev's new political direction also in the field of space, which resulted in the establishment of Glavkosmos and the initiation of an extensive campaign to commercialise the Soviet space infrastructure. They identify key trends in the evolution of Glavkosmos' commercial activities, including its strategic focus on the Indian market, unsuccessful attempts to penetrate Western

[12] D. Phelan. (Ed.). (2013). *Cold war space sleuths. The Untold secrets of the Soviet space program*, Springer.

[13] Harvey, B. (2007). *The rebirth of the Russian space program. 50 years after Sputnik, new frontiers.* Springer, 5-8.

[14] Bonnet, R.M and Manno, V. (1994). *International Cooperation in Space: The Example of the European Space Agency.* Harvard University Press

[15] Zhdanovich, O. (2012). International cooperation in Space, from U.S.S.R. to Russia. In L. Morris, K. J. Cox. (Eds.), *International Cooperation for the development of space.* Aerospace Technology Working Group.

[16] Zimmerman, R. (2003). *Leaving Earth. Space stations, rival superpowers, and the quest for interplanetary travel.* Joseph Henry Press.

[17] Van den Abeelen, L. (2017). *Spaceplane Hermes. Europe's dream of independent manned spaceflight.* Springer.

[18] Dubbs, Ch. and Paat-Dahlstrom, E. (2011). *Realizing tomorrow. The path to private spaceflight.* University of Nebraska press.

[19] Lardier, Ch. and Barensky, S. (2010). *Les Deux vies de Soyouz.* Editions E-dite.

[20] Lardier, Ch. and Barensky, S. (2018). *The Proton Launcher: History and Developments.* Wiley-ISTE.

markets, and the gradual decentralisation of the space industry's management structure, which had begun even before the dissolution of the USSR.

A recurring theme in studies on international space cooperation during the Cold War between the Soviet Union and Western countries is the persistent challenge of mutual understanding—or, more often, misunderstanding—and the divergent interpretation of shared phenomena. Reflecting on this divergent understanding from personal experience, only after moving to Italy from Russia in the early 2000s I came to realise the commonly accepted Western view that the USSR had lost the Cold War. For many post-Soviet citizens, the collapse of the USSR did not automatically equate to a definitive defeat in the nearly half-century-long confrontation with the United States. This discrepancy serves as yet another illustration of how internal and external perspectives frequently diverge.

Such ambiguity also characterises the history of the USSR's relations with the West in the realm of space programs. Information manipulation (including disinformation), deliberate concealment, and opportunistic exploitation of the adversary's perceived strengths and weaknesses created an ideal environment for the proliferation of myths, misconceptions, and misunderstandings in the field of space exploration. Throughout the Cold War, Western space experts sought to penetrate the veil of secrecy surrounding the Soviet space program, often relying on indirect sources to reconstruct the carefully guarded Soviet activities. Soviet intelligence services demonstrated a similar interest in Western space efforts.

In the West, an entire movement of "space sleuths" emerged, united by their determination to uncover the truth and circumvent Soviet information barriers. Among them, James Oberg stands out as one of the most reputable specialists on the Soviet space program. His extensive research activity provides a unique lens through which to examine the 1980s from the intertwined perspectives of past, present, and future. In his groundbreaking book on the Soviet space program *Red Star in Orbit*, published at the beginning of 1980s, Oberg endeavoured to forecast the future of Soviet space exploration. Drawing upon the information available at the time, he predicted steady advancement within the Soviet space industry, envisioning ambitious projects such as *Kosmograd*—a codename for "Space Colony Number One"—as well as manned lunar and Martian missions and the development of a powerful launch vehicle to pursue these objectives. Oberg further anticipated that by the end of the twentieth century, space travel would become routine, Soviet cosmonauts would initiate the colonisation of space, and a new type of "space human" would emerge.[21]

Two decades later, James Oberg—once an optimistic forecaster—had become a disillusioned observer after learning what he described as the true condition of the Soviet space industry. As he later admitted, his disappointment stemmed from the realisation that "there was a lot we all thought we knew that wasn't so".[22] This situation, shaped by severely limited access to reliable information, sustained an extended period of mutual misunderstanding and misinterpretation of intentions between the two space powers. Oberg noted that when "*glasnost* appeared in the USSR, a torrent of restricted information poured forth". Yet, in his view, neither the era of openness nor the collapse of the Soviet Union steered Russia to become a "normal country", and the misjudging in the domain of space persisted. According to Oberg, the primary obstacle to meaningful space cooperation was the persistent lack of transparency and sincerity on the side of the Soviet Union, and later Russia.

A somewhat different perspective is offered by Jeffrey Manber, a direct participant in the first Soviet-American commercial space ventures in the late 1980s. In recounting the history of these early commercial interactions, Manber also emphasises the prevalence of misunderstandings on both sides. However, he places the main responsibility on the American bureaucracy, which, in his view, contributed to the situation where "the working relationship between the Russians and Americans was born in confusion and misunderstanding". Manber believes that in the initial stages of the commercialisation of Soviet space technologies, the Soviet side fundamentally misjudged American intentions. Consequently, the Soviets naively assumed that "the US would welcome a commercial Russian space program, even if it competed against the US government space station, or the monopoly status of US launch vehicles, and the generally non-commercial orientation of NASA's space program."[23]

Space historian Andrew L. Jenks also emphasises the relative secrecy of the American space program, contending that the Soviet space industry began to engage in international cooperation during the period of *détente*

[21] Oberg, J. (1981). *Red Star in Orbit*. Random House Inc.

[22] Oberg, J. (2001). *Star-crossed orbits. Inside the U.S.-Russian space alliance*. McClaw-Hill, 2.

[23] Manber, J. (2000). Russian-American space miscommunication: a study in missed opportunities. *Space policy* 16, 3-6.

under Brezhnev. Whereas American policy on technology transfer remained shaped by a determination to safeguard its presumed technological superiority—even from its allies—the Soviet Union sought cooperation partly to assess the extent of its own technological lag. Jenks underscores the influence of Soviet internationalist traditions, which, in his view, "encouraged the Soviet leadership during the 1970s and 1980s to pursue collaborative opportunities and to gradually overcome the internal arguments for the continued securitisation of their space program".[24] Jenks argues that an opposite dynamic was unfolding in the United States. By the mid-1980s, he maintains, the United States had deepened its reliance on practices of secrecy, securitisation, and obfuscation—habits that American leaders, persistently fearful of the "communist bogeyman", routinely projected onto their ideological adversaries.

Christine Evans and Lars Lundgren, authors of *No Heavenly Bodies*, reach a similar conclusion regarding the Soviet Union's openness and willingness to coordinate activities in space. Their analysis focuses on the behaviour of the Soviet leadership in the development of satellite communications infrastructure, which they argue served as a testing ground for new patterns of international engagement. Specifically, they examine the commercialisation of satellite communication services through cooperation between the Soviet-led Intersputnik and the USA-led Intelsat, beginning in the 1970s. On the one hand, Intersputnik positioned itself as a competitor to Intelsat, particularly in the markets of the Global South. On the other hand, "the Soviet Union simultaneously pursued integration into Intelsat and sought to expand Intersputnik's global presence—despite its status as an asymmetrical actor, with significantly lower investment levels and less advanced technology, but offering lower prices and demonstrating a strong motivation to profit from its space technology". Evans and Lundgren convincingly illustrate how the global integration and commercialisation of Soviet infrastructure, even in a seemingly less visible area such as satellite communications, had already begun in the 1970s.[25]

If, in the context of international space cooperation, the principal historiographical concern at the end of the Soviet era was the degree of openness, mutual understanding, and trust between the parties, then

[24] Jenks, A. (2022). *Collaboration in space and the search for peace on Earth.* Anthem Press.

[25] Evans Ch. and Lundgren L. (2023). *No Heavenly Bodies: A History of Satellite Communications Infrastructure.* The MIT Press.

domestically, scholars were preoccupied with other issues. The interest of historians and cultural studies scholars was attracted by the very phenomenon of the diminishing significance of space exploration in late Soviet and post-Soviet society. One of the first scholars to address the theme of nostalgia for Soviet space achievements was Svetlana Boym. As she recognised, "by the late 1980s, cosmos had become devalued within Soviet youth culture. [...] The myth of cosmos came to an end together with the dissolution of the Soviet Union, or perhaps even earlier. [...] In post-Soviet Russia the exploration of space continues, but it is devoid of the mythical and ideological significance it vested in the past. The Soviet space program is no longer the model of the future but a souvenir of the past. By the end of the twentieth century, cosmos had turned from a futuristic utopia into a space for nostalgic remembrance".[26]

Later, Asif Siddiqi elaborated this idea, offering his own interpretation of the phenomenon, which he described as a "nostalgia for the future". As Siddiqi argues, "the unfulfilled expectations were at the heart of this nostalgia for the future; having grown into middle age in the 1980s, the Khrushchev generation felt a deep nostalgia for a time when the future was still ahead, while subsequent generations identified Soviet exploits of the 1980s with economic stagnation; for them, the past was as mysterious as the present was mundane".[27]

Having lost its mythical aura of romanticism by the end of the Soviet era, the space sector acquired new traits of business like efficiency and pure pragmatism. Previous researches make it clear that several phenomena typically associated with the *perestroika* period in international space cooperation had, in fact, already begun to emerge in the preceding decades. However, during the 1980s, under Gorbachev's *new thinking* policy and amidst shifting geopolitical conditions, these new modes of space activity—particularly the strategic pursuit of commercial and scientific opportunities—acquired greater significance. In this context, the deepening of relations with Western partners rendered issues such as transparency of intentions, mutual understanding, and knowledge of each other's internal realities increasingly critical.

[26] Boym, S. (2001). Kosmos: Remembrances of the Future. In Bartos, S. and Boym, S. (Eds.).

Kosmos: A Portrait of the Russian Space Age. Princeton Architectural Press.

[27] Siddiqi, A. (2011). From Cosmic Enthusiasm to Nostalgia for the Future. In E. Maurer, J. Richers, M. Rüthers, C. Scheide (Eds.). *Soviet Space Culture: Cosmic Enthusiasm in Socialist Societies*. Palgrave Macmillan.

This book seeks to shed light on the actual conditions of the Soviet space industry and science at the end of the Cold War and on the reasons behind the Soviet attempts to enter the international market for launch and satellite services. These developments are analysed against the background of a profound transformations of the Soviet space sector since the time of *perestroika*, which epitomise the country's dramatic transition, which led from a communist regime with a planned economy to a liberal democracy with a capitalist system in the 1990s.

To achieve these objectives, a comprehensive perspective on the Soviet scientific and commercial space program during the final years of the USSR was assembled. This research draws on a wide range of historical sources, including archival materials from the Historical Archives of the European Union (HAEU) in Florence, Italy, and the NASA Archives in Washington, D.C., USA. Additionally, contemporary press coverage and first-hand accounts from interviews with key witnesses have provided invaluable insights into this critical period.

This book explores themes of resilience in the face of adversity, the pursuit of ambitious goals despite setbacks, and the enduring tension between ideals and reality. It is dedicated to those pioneers who laid the foundation for Soviet/Russian-Western collaboration in space, individuals who, through perseverance and unwavering dedication, built not walls but bridges. Their contributions cannot be measured solely in monetary terms but in the immeasurable cost of time, energy, and personal sacrifice.

To those who shaped the course of history in ways both seen and unseen, this work stands as a tribute.

PART I

Gorbachev's Policy in the Space Sector: From Stars to Earth

1.1 Introduction

As part of the acceleration policy (*uskoreniye*), the beginning of which was announced by Gorbachev on April 23, 1985, at the Plenum of the Central Committee of the CPSU,[1] special importance was attributed to the scientific and technological progress and the implementation of its achievements in the development of the country's economy. One of the leading technology and science-intensive areas for the creation and execution of advanced knowledge was the space sector. However, there lay one large obstacle in the way of reform and the application of the successes achieved during the sector's entire existence: an excessive secrecy, due to its direct involvement in strategic security issues.

The activities of the USSR in space exploration and utilisation were not confined solely to military and strategic purposes; human space endeavours also aimed at advancing scientific knowledge and technologies that provided solutions to numerous "earthly" challenges. Moreover, it served as a powerful tool in the hands of the political leadership. Gorbachev inherited this ambivalent approach from former Soviet leaders, which simultaneously combined the desire to guarantee the Soviet Union's dominant role as a military power, relying on, among other things, space science and technology, and the use of space as a *soft power* in support of communist ideology.

[1] The Communist Party of the Soviet Union.

Gorbachev had to integrate the traditional Soviet attitudes towards space, inherited from his predecessors, with his own evolving perspective, shaped by the unique conditions of the *perestroika* era. The worsening state of the socialist command economy and the imperative to accelerate and expand liberalisation reforms necessitated a radical revaluation of his space policies to align with the broader economic context. Equally significant was the political dimension of his *new thinking* policy, which profoundly influenced reforms in the space sector. Domestically, the policy of *glasnost* fostered increased transparency and openness, while efforts to normalise relations with the West introduced dynamic changes to what had historically been a highly secretive and isolated domain.

As Anatoly Dobrynin, Soviet ambassador to the USA, remembered, Gorbachev did not have a clear program for the country's foreign policy, although he expressed dissatisfaction with the fact that it lacked dynamism and strategic room for wider manoeuvre to break the dangerous confrontation between the USSR and the USA.[2] New words entered the political lexicon of the Soviet leadership: such concepts as "balance of interests", "unity and interdependence of the world", "reasonable sufficiency" of the armed forces, the "multi-vector basis of international relations", "cooperation and co-development" and finally "the principle of freedom of choice" of political regime were phrases hitherto unheard of in the Soviet Union. Within the framework of this general shift in policy, it made sense to consider not only initiatives to stop the arms race and to advance peacekeeping proposals, but also to develop the space program itself.

Chapter 1 of this section will examine the space policies implemented by Gorbachev's administration. Utilising a range of surviving official documents and contemporaneous memoirs, we will endeavour to reconstruct the approaches adopted to reform cosmonautics, analyse Gorbachev's personal perspective in this field, and assess its significance within the broader political framework of *perestroika*. In Chap. 2, the practical implementation of these reformative ideas within the space industry and the scientific community will be explored.

[2] Dobrynin, A. (2016). *Sugubo doveritel'no*. Tsentropoligraf, 697.

CHAPTER 1

Gorbachev's "New Thinking Policy" in Space

The purpose of this chapter is to explore the main visions and approaches to reforming the space field during *perestroika*. By identifying the features of continuity in Gorbachev's policy in relation to previous periods, the aim is to analyse the changes and their connections within the general provisions of the "new thinking" policy. On the one hand, Gorbachev was tasked with continuing the legacy of his predecessors by actively addressing key issues such as space-based disarmament and the Reagan-era Strategic Defence Initiative (SDI) program. He also had to navigate the cultural and historical legacy of the space sector within the broader framework of state policy. On the other hand, the era of change introduced its own imperatives. Gorbachev began by advocating for enhanced international cooperation, notably proposing the establishment of the World Space Organisation. Domestically the liberalisation of the entire sector was envisioned by reformers as a pathway to increased industrialisation and commercialisation.

Starting Points

Before proceeding to the analysis of Gorbachev's vision of the space program, several starting points that in one way or another determined the space policy during *perestroika* are worth looking at. First, the issue of

O. Dubrovina, *The Soviet Space Program and International Collaboration during Perestroika, 1985–1991*, Palgrave Studies in the History of Science and Technology,
https://doi.org/10.1007/978-3-032-03157-0_1

disarmament, inherited from previous Soviet diplomacy, acquired a new and more relevant significance during the *perestroika* period, due to the increased risks of turning outer space into a location for nuclear weapons and the new vision of international relations promoted by Gorbachev. The struggle for disarmament was part of the *soft power* policy traditionally pursued by the Soviet government and Gorbachev could rely on the already tested experience of using space as an instrument of propaganda. Secondly, his assessment of the danger posed by President Ronald Reagan's "Star Wars" program played a considerable role in his attitude to the evolution of both military and civilian space programs. Finally, the concept of weakness hung over him like the sword of Damocles. He had to step through a figurative minefield in search of a compromise between the need to ease the pressure on the economy from the Military-industrial complex (VPK) and accusations of weakening national security through disarmament and cuts in funding military industries, including the space area. All of this was undertaken with the recognition that cosmonautics represented a significant aspect of the Soviet Union's cultural and historical legacy, serving as a source of national pride and a symbol of the success of the communist system.

Disarmament Issue

The interconnection between disarmament efforts and the military use of outer space was a key element of Gorbachev's narrative, rooted in the policies of previous Soviet leadership. With the renewed escalation of the Cold War in the background due to some major events (the entry of Soviet troops into Afghanistan in 1979, the introduction of martial law in Poland in 1981, the deployment of American medium-range missiles in Europe) the Soviet authorities attempted to resume dialogue on the use of outer space for peaceful purposes. Consequently, in August 1981, the USSR proposed to the United Nations a draft "Treaty on the Prohibition of the Deployment of Weapons of Any Kind in Outer Space". Reagan's announcement regarding the commencement of work on the SDI program provided new momentum for Soviet initiatives. In 1983 a new Soviet proposal followed: a draft "Treaty on the Prohibition of the Use of Force in Outer Space and from Outer Space against the Earth". A year later, this was supplemented by a recommendation to ban the use of force from the Earth against space objects. The Soviet Union had already assumed an obligation to not be the first country to launch any type of anti-satellite weapons into outer space if other states also refrained from doing so. This moratorium also covered test launches of anti-satellite weapons.[1]

During the spring and summer of 1984, the Soviet Union repeatedly put forward proposals to kick-start Soviet-American negotiations, aimed at preventing the militarisation of outer space. In response to the appeal of the American scientists Richard Garwin and Carl Sagan, the General Secretary of the Central Committee of the CPSU, Konstantin Chernenko, stated that: "In an effort to block the path of weapons to outer space, we proposed an agreement on the prohibition of the use of force in outer space and from outer space against the Earth. This agreement would imply the complete renunciation of anti-satellite weapons, including the destruction of already existing systems. To encourage international agreement, the USSR unilaterally announced a moratorium on deploying anti-satellite weapons in space, provided that other countries reciprocated".[2]

Thus, in the early 1980s, space issues once again became relevant as an instrument of Soviet diplomacy: firstly, it made it possible to find a common platform for negotiations with the USA to overcome the international crisis; secondly, to demonstrate to a wider audience the peaceful intentions of the USSR, and finally, to attempt to save the already clearly weakening economy from a new round in the arms race. In terms of a propagandistic use of space, the USSR could boast experience lasting many years. Starting from the late 1950s space was a convenient tool in the hands of the Soviet leadership that resorted to it to demonstrate the superiority of the socialist regime and readiness to begin the disarmament process. At the same time, it served as an opportunity to demonstrate the effectiveness of cooperation, provided there was goodwill on both sides (e.g., the Soyuz-Apollo project of 1975 and Soviet-French collaboration in space beginning in 1966).

In the first months after his appointment to the supreme position in March 1985, Gorbachev continued down the line of publicly advocating the peaceful use of space. He advocated for the urgent need for disarmament and the definitive elimination of international tensions through a series of bans on the development and use of all types of weapons, particularly in outer space. For the first time as General Secretary Gorbachev

[1] Antonov, A. (2012). International Legal Regulation of Military Outer Space Activities. *MGIMO Review of International Relations*, 4 (25), 190–197.

[2] Pravda, May 20, 1984, cit. in Batyuk, B. (2018). *Kholodnaya voyna mezhdu SShA i Rossiey (1945–1991)*. Ves' Mir, 235.

spoke at Plenum of the Central Committee of the CPSU on March 11. Gorbachev noted that "In the field of foreign policy, our course is clear and consistent. This is the course of peace and progress". He went on to say: "Never before has humanity been threatened by such a terrible threat as in our day. The only reasonable way out of the current situation is an agreement between the opposing forces on the immediate cessation of the arms race - primarily nuclear - on Earth and its prevention in outer space. An agreement on an honest and equal basis, without trying to 'outplay' the other side and dictate your terms to it".[3] The importance of outer space as a promising strand of foreign policy was also emphasised by the fact that the Department for the Peaceful Use of Nuclear Energy and Space was created in the Ministry of Foreign Affairs.[4]

Gorbachev emphasised the idea of the rejection of an aggressive space policy in a series of subsequent interviews. In response to an appeal from the American public organisation, the Union of Concerned Scientists, he affirmed that the Soviet Union would not be the first to deploy weapons in space and would make every effort to dissuade the United States from doing so as well.[5]

The leaders of *perestroika* used the United Nations arena also to convey to the international community their views about the further development of space exploration and the nature of human activity in outer space. A few months after Gorbachev's appointment to the post of General Secretary, the newly nominated Minister for Foreign Affairs, Eduard Shevardnadze, presented to the UN the Soviet proposal on "International cooperation in the peaceful exploration of outer space under conditions of its non-militarisation". This contained a call to halt the militarisation of space and direct the technical and intellectual potential of all participants in space exploration to a peaceful conclusion. Moreover, he expressed his concern about American intentions to transform outer space into a new battlefield. "Plans now being proclaimed and activities now being undertaken are aimed at developing and deploying space strike weapons to destroy objects in space and to launch attacks from space against objects in the

[3] Pravda, March 12, 1985.

[4] Barsenkov, A. (2012). "New thinking" in foreign policy U.S.S.R (1985–1991). *Vestnik Moskovskogo universiteta. Mezhdunarodnyye otnosheniya i mirovaya politika*, 1, 45.

[5] Aleksandrov, M. (1985). SOI – strategiya pervogo udara. *Aviazia i kosmonavtika*,10, 43.

atmosphere and on Earth, including the creation of a large-scale anti-missile system with space-based components".[6]

Gorbachev's peaceful rhetoric on the use of outer space harmoniously combined with his style of communication fitted perfectly into the general narrative of his domestic and foreign policy. Overall, Gorbachev's foreign policy was determined by dynamism and a noticeable improvement in its *soft power* skills. The space policy under Gorbachev was part of the same "public diplomacy" that Ronald Reagan appreciated himself. As Dobrynin recalled, Reagan's National security advisor, Robert McFarlane, said that "the president gave props to the new Soviet leader, who, unlike his predecessors, possessed the ability to use public diplomacy against his foreign opponents, setting the public opinion of his own countries against them".[7]

However, the peace-loving rhetoric was accompanied in the USSR with both the development of the most powerful rocket for that time (Energya) and reusable spacecraft (Buran) whose production was in full swing. This monumental space project was inherited by Gorbachev from the 1970s, a period when the decision was made to develop a "fitting reply" to the American Space Shuttle program. The Soviet reusable space transportation system Energya-Buran was first tested in automatic mode on November 18, 1988. In the best traditions of Soviet leadership, the date of the successful launch of Buran coincided (no matter how strange it may sound) with the speech of the leader of the USSR, Gorbachev, at the UN on… disarmament issues. According to the Soviet political elite of that time, this allowed him to strengthen his position and speak on an equal footing with the USA delegation.[8]

Even in the 1980s, the Soviet intellectual community remained deeply divided into two distinct camps: proponents and opponents of the Energya-Buran system. Gorbachev had to navigate these opposing viewpoints while considering the dynamic shifts in foreign policy and economic conditions. By the end of *perestroika*, the prioritised development of the new space complex was at risk. Despite the promising potential of this advanced transportation system, its developers observed a lack of interest from the new leadership in cutting-edge space technologies. They believed

[6] Soviet proposal "Main lines and principles of international cooperation in the peaceful exploration of outer space under condition of its non-militarisation", August 16, 1985. In ESA-9783, HAEU.

[7] Dobrynin. (2016), 715.

[8] Baturin, Yu. (2006). *Kosmicheskaya diplomatitiya i mezhdunarodnoe pravo.* Zvezdny gorodok, 73.

this disinterest resulted in insufficient funding for the project and a general underestimation of the contributions of domestic scientists and design engineers in maintaining the country's defence capabilities. Chief Designer of the Energya-Buran reusable space transport system, Boris Gubanov reminisced that: "Energya and Buran were developed for the advanced technology of military means of conducting 'Star Wars'. Later, in 1986, the concept changed. The thesis of so-called 'reasonable sufficiency' was voiced, which would have been more understandable in the times of knights' battles".[9]

Not everyone, however, shared the same disappointment as the developers concerning the lack of demand for Energya. Anatoly Adamishin, then Deputy Foreign Minister of the Soviet Union, pointed out the excessive costs paid for what he considered to be serious American "disinformation". "What about the famous Shuttle? We believed that we would not escape from this space killer if we did not create the same apparatus at home. The super-powerful rocket lifted off only twice. About a million people worked on it. And the rocket, by all accounts, turned out to be successful, just not for our pocket. Unlike the zealous Americans (the Congress' control hung over them), we spent immeasurable money right away in two expensive enterprises: both the orbital station and the Energya-Buran system".[10] Still, the Soviet Shuttle did not become the fateful challenge for Gorbachev's space policy. More credibly, the actual threat was represented by the SDI program.

The SDI Program's Impact

It is crucial to consider the highly significant factors that shaped Gorbachev's disarmament policy and, by extension, the entire space program. What was Gorbachev's understanding of the SDI program, launched by Reagan in 1983, and inherited by Gorbachev from his predecessors? How did he perceive its actual threat and the Soviet Union's capacity to counter it? These questions warrant careful and thorough consideration. In this context, we will attempt to outline the potential challenges.

[9] Gubanov, B. (1998). *Triumf i Tragediya "Energii". Razmyshleniya Glavnogo Konstruktora*. Vol. 3. NIER.

[10] Adamishin, A. (2005). Eti luchshie tri goda.... In V. Kuvaldin, A.Veber (Eds.), *Proryv k svobode: o perestroyke dvadtsat' let spustya*, Al'pina biznes buks, 219.

According to Dobrynin, Moscow could not immediately assess the military-technical consequences of SDI. He also gave a detailed picture of the perception of SDI in various circles of the Soviet leadership before Gorbachev came to power. Dobrynin asserted that military engineers believed that American technology might have made a breakthrough in the field of space weapons. They were supported by the Minister of Defence, Dmitry Ustinov, and the entire Soviet political leadership convinced of the potentially great technical capabilities of the United States. At the same time, theoretical physicists led by the vice-president of the Academy of Sciences of the Soviet Union, Evgeny Velikhov, expressed scepticism about the feasibility of SDI, but, as Dobrynin explained, their point of view did not find broad support from the leadership.[11]

Cosmonaut, scientist, and Boris Yeltsin's adviser on national security issues, Yuri Baturin, pointed out that the very idea of creating SDI came from American scientists who were more interested in obtaining government funding for their projects, while the level of their feasibility was not always considered when deciding on the distribution of the national budget.[12] Dobrynin adhered to a similar point of view: "the real authors of the program were American scientists led by the famous physicist Edward Teller, as well as the leadership of the Pentagon. They 'sold' this program to the USA president, who was fired up with the idea of finding such protection against Soviet missiles".[13]

Gorbachev and his team had to answer two questions simultaneously: how realistic was the SDI project and what should have been the reaction of the USSR. The answer to the first question was echoed in a 2005 interview with Shevardnadze. He recalled how he and Gorbachev made the decision to cooperate with the Americans in disarmament, under pressure from the possible consequences of SDI for Soviet national security. As Shevardnadze put it, such a decision was taken because of two meetings with leading physicists of that time, among whom were Evgeny Velikhov, a science advisor to Mikhail Gorbachev, and Roald Sagdeev, the Director of the Space Research Institute (IKI).[14] At the first meeting the scientists argued that: "such an initiative of Reagan was unrealistic. All this was

[11] Dobrynin. (2016), 641.

[12] Interview with Yu. Baturin, September 17, 2022.

[13] Dobrynin. (2016), 640.

[14] For an analysis of Andrei Sakharov's position on the SDI, see A. Pensias, Sakharov i SOI. (2002). In M. Sakharova-Liberman, L. Gurvits (Eds.), *A. D. Sakharov: grazhdanin Vselennoy*. 55–63.

nonsense, empty phraseology, and nothing more. The only reason was to scare the Soviet Union".[15] During a second meeting, which took place two months later, the same group of the scientific delegation, according to Shevardnadze's memoirs, radically changed its mind. "We thought it over again, at first it seemed that it was a fantasy, that it was unreal. But if the [American] economy allows it, and if they will not protect the entire territory [of America], but only individual objects and cities, for example, Washington, then this is feasible. It will be impossible to break through the American shield, but the Americans can hit the Kremlin".[16]

The question on the Soviet side was not only in assessing the level of danger that SDI posed to national security, but also the way they responded to it. Within the Soviet leadership, two approaches emerged: the first one focused on building up military potential to meet the new American challenges, while the second involved an emphasis on building peaceful relations. In his memoirs, published in the USA in 1994, Sagdeev, who was also a Gorbachev advisor on issues related to civil and military space matters, summarised the Soviet dilemma. Speaking to a seminar of political scientists and strategic analysts in Paris, he confessed that "if Americans oversold SDI, we Russians overbought it".[17] He pointed out that there were influential members of the military and industrial sectors in the Soviet Union that sought to involve Gorbachev's administration in a nonstop escalation of an SDI budget at the expense of the deterioration of the strategic stability and Soviet economy. One of these supporters of retaliatory actions from the USSR was the Minister of the Ministry of General Machine Building (MOM), Oleg Baklanov, appointed to this post in 1983. He strongly believed that the SDI "was another next step towards the conquest of space as a springboard for influencing a possible rival".[18]

To avoid the increasing costs of a new round of the arms race, Soviet scientists, politicians, and militaries worked on the reaction towards the SDI threat. Domestic scientists and specialists, primarily within the framework of the activities of the "Velikhov group", formulated the concept and specific programs of an "asymmetric response" to the American

[15] Shevardnadze, E. (2005). Ya 12 raz vstrechalsya s Reyganom odin na odin. *Mezhdunarodnaya zhizn'*, 12, 143.

[16] Ibid., 144.

[17] Sagdeev, R. (1994). *The Making of a Soviet Scientist: My Adventures in Nuclear Fusion and Space from Stalin to Star Wars.* John Wiley & Sons Inc., 272.

[18] Baklanov, O. (2012). *Kosmos - moya sud'ba. Zapiski iz "Matrosskoy tishiny"*. Obshchestvo sokhraneniya literaturnogo nslediya. Vol. 1, 364.

initiative. The concept of an "asymmetric response" was implemented through overcoming great obstacles because the USSR had a tradition of predominantly symmetric actions, of "point against point" actions. This tradition was fully manifested when the Soviet officials debated the question of how to respond to Reagan's "Star Wars". The essence of the "asymmetric response" was reduced primarily to the fact that under the most difficult conditions, when the United States deployed a multi-echelon missile defence with the use of a variety of missile defence means, it was possible for Soviet nuclear missile arsenal to inflict "unacceptable damage" on the aggressor in a retaliatory strike, thus convincing it to abandon a pre-emptive strike.[19] The developed concept was eventually accepted by the Soviet leadership, so it gave a sense of security to Gorbachev who, during private talks in 1986 with members of the Politburo, admitted that the Soviet Union should not be afraid of the SDI, at least in the short term. The USA were far from completing work in this area, and the Soviet Union has something to answer for, both asymmetrically and in terms of costs.[20]

After further consultations with academics and leading Soviet economists, Gorbachev and Shevardnadze were finally convinced of the SDI practicality; thus, they considered the compromise with the United States as a necessary tool to stop the arms race, which, according to Shevardnadze, the country was no longer able to withstand: "…And then Gorbachev and I finally decided: we must at all costs seek a common language with the Americans. Negotiate with them!"[21]

The decision made by the Soviet leadership, based on the calculations of some experts, did not convince everyone. Yuri Baturin believes that at that time, the top Soviet authorities did not have enough information to make the right decision. "This misconception about SDI has shaped both Gorbachev's foreign policy strategy and his attitude to space program that was conceived as a result of his ideas about the need to disarm. As a consequence, he reduced the military and civil space programs too".[22] I therefore argue that Gorbachev's steps in foreign policy were aimed at protecting national security (according to his own perceptions) and the struggling

[19] Oznobishchev, S.; Potapov, V; Skokov, V. (2008). *Kak gotovilsya «asimmetrichnyy otvet» na «strategicheskuyu oboronnuyu initsiativu» R. Reygana: Velikhov, Kokoshin e drugie*. Lenand.

[20] Barsenkov (2012), 58.

[21] Shevardnadze (2005), 144.

[22] Interview with Yu. Baturin, September 17, 2022.

economy against the threats posed by Reagan's SDI program. National security issues and economic shortcomings were expressions of the Soviet leadership's concerns about the inability of the Soviet Union to give a worthy response to the challenges set by the American side. However, what Gorbachev and his team could not afford was public recognition both in the international arena and in the eyes of their compatriots, of Soviet Union weakness and the insufficiency of its military potential. This principle became the third starting point of his space policy.

No Country for Weak Men

Recognising how the Soviet political and military elite, as well as ordinary citizens, associated power with the build-up and demonstration of military force (especially with the heightened American challenge looming), Gorbachev had to emphasise that his intention was not to weaken the country's space sector, even if in his public speeches the message of the superiority of moral strength over the military began to be heard more often. He continued to stress the will to preserve the Soviet Union military capacity despite his course towards the disarmament process. In 1986, during his visit to Togliatti, Gorbachev sought to justify his peaceful policy in the frame of the "new thinking" program and to commit to protect previous Soviet positions: "... It is time to refuse to build relations with the USSR on erroneous ideas, on illusions. One of the most dangerous such illusions is that the peaceful intentions and calls of the Soviet Union are considered as a sign of weakness. Hence, it is impossible to run us into the ground by an arms race, to throw us out from outer space, to overtake us in technology. Nothing good will come of these attempts".[23] Allegedly addressing an invisible interlocutor on the other side of the Atlantic, Gorbachev sought to convince his own audience of the safety of his disarmament course and to drop accusations of weakening the country's defence capability.

Gorbachev's visit to Baikonur on May 11, 1987, was a landmark event in this context. According to Boris Gubanov, during his meeting with engineers and workers, Gorbachev stated: "Our course towards peaceful space is not a sign of weakness... Any speech about protection from nuclear weapons is the greatest deception of the people... The Strategic Defence Initiative is not only a robbery of taxpayers, but also a senseless

[23] Pravda, April 9, 1986.

undertaking from the point of view of defence... The arms race is moving into space".[24] In the Leninist manner, he sought to reach out to American public opinion, bypassing official authorities: "We do not intend to weaken our efforts and lose our avant-garde positions in space exploration... Our course towards peaceful space... Our interests here coincide with the interests of the American people and other people of the world. They do not coincide with the interests of those who do business in the arms race and who want to achieve military superiority through space... The task is not to miss the opportunity that arises thanks to our proactive policy".[25]

Gorbachev's peace-oriented rhetoric, however, concealed a reality well understood by those involved in space activities: the disarmament policy inevitably led to a significant decline in funding for the space industry. Gorbachev's remarks during his visit to Baikonur made his intentions regarding the future of the space industry unmistakably clear. As Boris Gubanov noted, "Gorbachev's last thesis answered all our questions about our future. It became clear what awaits Energya".[26] Cutting-edge military space technology was no longer on the Soviet agenda. Instead, the focus shifted towards fulfilling international disarmament commitments and reaping economic benefits, including cost savings in the VPK and fostering market relations within the space sector. Despite numerous speeches extolling the achievements of scientific and technological progress in space, Gorbachev's actions reflected his view of space development as a secondary priority compared to other pressing policy concerns.

The visit to Baikonur on May 11, 1987, became a pivotal moment in the history of the final phase of Soviet cosmonautics, marking the beginning of the decline for the state military space program. Yuri Baturin recalls that the space technologies Gorbachev witnessed during his visit so impressed him that he expressed regret about the preliminary agreements already made with the United States to reduce strategic arms during the Reykjavik Summit with Reagan in October 1986.[27] Nevertheless, the gradual dismantling of the Soviet space programs had become irreversible. Gorbachev faced the dual challenge of aligning his space policy with the evolving domestic and international landscape while also considering the symbolic significance of space activities in the USSR Cosmonautics

[24] Gubanov (1998).

[25] Ibid.

[26] Ibid.

[27] Interview with Yu. Baturin, September 17, 2022.

remained an indispensable attribute of superpower status in the eyes of segments of the political and intellectual elite, as well as a significant portion of the population. However, under the constraints of the times, the political leadership was compelled to prioritise other objectives.

"New Thinking Policy" in Space

The challenges Gorbachev inherited from his predecessors, which he was obliged to address, were combined with the new approaches introduced by *perestroika* reformers to integrate the space sector into their broader policy framework. Although the prominence of the space sector steadily declined within the *perestroika* agenda, its strategic value, particularly as a tool of *soft power*, remained significant. To realise Gorbachev's new vision for the future of space exploration, the Soviet Union presented the global community with the "Star Peace" program, signalling a markedly different approach to the subject compared to that of the United States. In the realm of international politics, the apex of the program was the proposal to establish a World Space Organisation. Domestically, the economic turmoil of the time was compelling Gorbachev to leverage the potential of space to address pressing national challenges.

Cooperation for Peaceful Purposes

Perestroika brought a new image of the Soviet Union's role in space exploration and set new goals for the country, in line with the spirit of international de-escalation of tensions and domestic political and economic reforms. The demilitarisation of space was considered as the first step towards large-scale international cooperation in outer space becoming the key concept of the follow-up Soviet program of future space exploration and development. By 1985, in an address to the UN, Shevardnadze announced: "The Soviet Union is proposing the implementation of a set of practical measures that would help unite the efforts of states in the peaceful exploration of space and the use of space technology for the benefit of all states, including the provision of comprehensive assistance in this area to the developing countries. It is obvious that this task, the common task of all mankind, can be successfully accomplished if agreements for effectively ensuring the non-militarisation of outer space are reached".[28]

The principal ideas of Gorbachev's appeal to cooperation in outer space were continued and presented in more detail one year later (in June 1986), in a letter from Nikolay Ryzhkov, the Chairman of the Council of Ministers of the USSR, where he addressed the United Nations Secretary General. In essence, Ryzhkov's proposals touched upon three main issues: a ban on the spread of weapons in outer space and on using outer space for military purposes; the use of outer space exclusively for peaceful purposes for the benefit of the development of all mankind; and, finally, the creation of an international organisation for the use and exploration of outer space. As for the issue of demilitarisation, Ryzhkov's proposal paid particular attention to artificial satellite protection and to the prohibition of the development and manufacturing of anti-satellite systems. Later, in an interview with the TASS agency, given in January 1987, Ryzhkov emphasised the feasibility of the Soviet program for the development of international cooperation in space. The possibility of its implementation, as Ryzhkov put it, was indicated by the gradual realisation and the thoughtful approach of the Soviet leadership to the very idea of international cooperation in space.

World Space Organisation

Among various proposals for the application of the international cooperation in space aimed at preventing it from becoming a military ground, the Gorbachev administration put forward the idea of creating a World Space Organisation (WSO). They pointed out that this "World Space Organisation would deal with international cooperation in the peaceful exploration and use of outer space under conditions of its non-militarisation".

Ryzhkov believed that it would become unreasonable "to scatter and duplicate the efforts of states in space exploration". "The WSO would promote a possibility for all the states to use achievements of space science and technology for purposes and needs of their social and economic development and would help developing countries become direct participants in the great process of space exploration". Therefore, the leading space

[28] Shevardnadze, E. (1985). *Letter dated 15 August 1985 from the Minister for Foreign Affairs of the U.S.S.R. addressed to the Secretary-General.* In ESA-9783, HAEU.

powers were supposed to assist developing countries in accessing space achievements. It is no coincidence that Gorbachev first voiced the idea of creating an international organisation in front of the Indian parliament. The realisation of this program concerning the creation of WSO and the collaboration between leading space powers and developing countries "could be a landmark in laying the foundation of star peace".[29]

The idea of the WSO received a more detailed description in the follow-up publications of Soviet space experts. The cosmonaut, Vitaly Sevastyanov, and diplomat Vladimir Pryakhin, delivered further information on the Organisation's project and the prospective for its implementation in their popular-scientific brochure published in 1989. First, they acknowledged that the Soviet side proceeded from the positive experience of the International Atomic Energy Agency (IAEA), in whose practice the functions of cooperation and control purposes were uniquely combined. The authors contended that the WSO should be a highly effective, dynamic, universal organisation of a new type—free from bureaucratic structures and academicism. The WSO should assume the role of a "think tank" and "global sponsor" of international cooperation in the exploration and use of outer space for peaceful purposes. It could coordinate the national space exploration plans of the member states, facilitate the exchange of space research results, provide practical assistance in involving various states in space activities, and serve as a centre for organising commercially profitable and scientifically and technically promising joint space projects.[30]

The World Space Organisation was conceived not only as a coordinator of international scientific and economic activities in outer space, but also a necessary instrument for maintaining political stability in the world. In the opinion of the Gorbachev administration, the creation of an international organisation regulating the use of outer space for peaceful purposes would lead to relaxation in international relationships. In this manner, it would be up to the international legal system to monitor compliance with agreements on preventing the spread of an arms race to outer space and on limiting and reducing armed forces and armaments on Earth.

Among the main goals of cooperation were the prevention of a global nuclear war and economic concerns, such as raising the standard of living

[29] Interview of N. Ryzhkov to TASS, January 1987. In ESA-28130, HAEU.

[30] Sevastianov, V., Pryakhin, V. (1989). *Kosmos, chelovek, politika: (Kosmicheskomu veku – kosmicheskoye myshleniye)*. Znaniye.

on the entire planet. Gorbachev's team had also foreseen cooperation on scientific matters regarding the exploration of the Moon and the development of its territory. "Joint construction of spaceborne infrastructure, including orbital stations and platforms for scientific and commercial undertakings, interplanetary manned spaceships for practical exploration of the Moon in the first decades of the twenty-first century, which could be also used as a base for flights to other planets".[31]

Mars was also considered as an attractive and rewarding object of cooperation. On June 8, 1988, Shevardnadze spoke at an unscheduled session of the UN General Assembly with an address on the Soviet disarmament program. Among the numerous areas of cooperation included in the disarmament process, the Soviet Minister of Foreign Affairs also proposed a joint expedition of American astronauts and Soviet cosmonauts to Mars that could serve, in his opinion, as "an example of cooperation" of leading space nations.[32] Shevardnadze mentioned again the need to create the World Space Organisation and expressed the requirement to take measures shielding space from pollution. Moreover, he raised from the international platform the question of protecting the environment by means of space technology.

Gorbachev expanded on these arguments in his renowned speech delivered from the UN podium on December 7, 1988. He called for international cooperation in space to prevent an environmental catastrophe. "The Soviet Union is ready to cooperate in the creation of an international space laboratory or manned orbital station, which would be engaged exclusively in monitoring the state of nature". In this same speech, he also drew attention to the lack of an updated legal framework for the peaceful exploration of outer space. In his opinion, the Treaty of 1967 had to be supplemented with new articles corresponding to the current state of space exploration and the international situation: "However, there is already a need to develop a comprehensive regime for peaceful work in outer space". Once again, Gorbachev turned to the necessity of creating a World Space

[31] United Nations: Soviet proposal for international cooperation in the peaceful exploration of outer space. In ESA-10448, HAEU.

[32] Intervention de E.A. Shevardnadze au débat général à la session extraordinaire de l'Assemblée Générale consacrée au désarmement, 8 juin 1988, New York, 13. In ESA-28130, HAEU.

Organisation, which would "verify the compliance [of national space activities] to the regime for peaceful work in outer space".[33]

Consequently, there were three main aspects of the future international organisation, as Gorbachev and his supporters had planned: a scientific centre for coordinating space research, a trade and economic platform for evaluating commercial projects and their enforcement, as well as a political regulator designed to legally control the actions of all participants in the organisation. Such a bold and universal proposal inevitably elicited reactions from potential partners expected to participate in its implementation. Were these responses commensurate with the grandeur of the proposal?

International Reaction

In response to the ESA inquiry—addressed to the NASA leadership regarding the agency's position on Gorbachev's proposal to create an international space laboratory and a manned orbital station for environmental monitoring—NASA remained non-committal. The agency did not provide a clear or definitive answer. They instead limited themselves to expressing an unspecific interest in the ideas of the Soviet leader and expressed their readiness to consider a more detailed plan. As for the attitude of the American side to the WSO it was suggested to refer "any questions to the State Department".[34] The restrained reaction of the USA to the Soviet suggestion was probably explained, firstly, by the fact that they had suspicions of purely political overtones and detected just another Gorbachev populist gesture; secondly, by the will to preserve their own leadership in space exploration. This was clearly evidenced by the episode connected with the Space Future Forum held in Moscow on October 2–4, 1987 (see Chap. 9).

The Forum was organised by the IKI of the Academy of Sciences on the 30th anniversary of the 1st Sputnik.[35] With approximately one thousand scientists, engineers and managers attending, the Forum was one of the largest international meetings on space ever to have taken place. Despite the efforts of the organisers, whose task was to demonstrate the

[33] Gorbachev M. (1988), *Speech at the United Nations*, December 7, 1988. Retrieved January 17, 2025, from https://www.gorby.ru/userfiles/file/vystuplenie_v_oon.pdf.

[34] James V. Zimmerman. *Nasa European Office*, 9 December 1988. In ESA-28130, HAEU.

[35] Sagdeev, R. Telegram to ESA's authorities, 21 August 1987. In ESA-28130, HAEU.

achievements of Soviet space science and strengthen relations with foreign colleagues, the mood of the Forum remained delicate. The World Space Organisation, promoted by the Soviet side, was at the centre of the potential conflict. Soviet officials were expected by their international colleagues to repeat earlier calls for the foundation of the new organisation, but the USA did not support the WSO proposal because, as journalist Craig Covault argued, the USSR used the concept to counter the USA Strategic Defence Initiative. To reassure his American guests and to not cause political tension during the sessions Roald Sagdeev guaranteed that the Space Future Forum would not be used as a platform to criticise SDI.[36]

However, the issue of the creation of the WSO was raised in the Space Future Forum. The Soviet speakers, among other things, raised the possibility of a special development fund within the context of the WSO to organise major international projects, primarily with a view to assist developing countries.[37] It also became apparent that the Soviets decided not to completely abandon the idea of the WSO project, once again mentioning it in front of the international scientific and space community. However, to avoid heightening tensions during the Forum, the question of establishing the Organisation was raised without reference to the USA military program and the final joint statement (signed by the participants of the Forum) did not include any statements calling for the creation of the WSO. The participants limited themselves to the classic declarations of intent to carry out the peaceful exploration of outer space in the interests of, among other things, developing countries. In fact, the main goals of the planned Organisation were announced without even mentioning its formulation: "The cumulative effect of coordinated efforts in space will benefit all countries, including the developing ones, which will be able to utilise its results to address their social and economic challenges".[38]

Gorbachev's universal and seemingly harmless proposal to create a World Space Organisation did not withstand the test of viability in the prevailing international environment. On one hand, his actions were still interpreted through the lens of the Cold War paradigm, viewed as an attempt to promote pro-Soviet propaganda. On the other hand, the

[36] Covault, C. (1987). Soviet Space Records Mark. Opening of International Forum, *Aviation week & space technology*, October 5.

[37] Sevastianov, Pryakhin (1989).

[38] Statement by the participants in the International Forum "Cooperation in space for peace on Earth". In ESA-28130, HAEU.

emerging reality of a unipolar world, dominated by a single remaining post-Cold War superpower, was gaining increasing influence.[39]

Space in the Context of Economic Development

In April 1985, at the Plenum of the Central Committee of the CPSU, Gorbachev proposed the concept of economic development, based on the intensification and acceleration of scientific and technological progress, "upgraded capabilities" of all sectors of the national economy "applying the cutting-edge achievements in science and technology" and giving priority to the development of machinery manufacturing. It was envisaged "in the twelfth five-year plan to double the growth rate of the industry". "Revolutionary shifts are needed—a transition to fundamentally new technological systems, to the technology of the latest generations, giving the highest efficiency ... The main task is to quickly switch to the production of new generations of machines and equipment ... Priority should be given to the improvement of machine tools, accelerating the development of computer technology, instrumentation, electrical engineering, and electronics as impetus for scientific and technological progress... We need to boldly move forward on the path of empowerment enterprises, their independence, to introduce economic accounting, and on this basis to increase the responsibility and interest of labour collectives in the results of work".[40]

Of course, due to its nature, the space sector was directly affected by this new economic policy. In addition to the purely political aspect of space activities, the question of the profitability of the space domain remained. During the *perestroika* period, the space field began to play a major role in the country's civilian economy, in contrast to previous periods when it was associated almost exclusively with the activities of the VPK. This shift was also driven by public opinion, which increasingly viewed the space industry as excessively costly, consuming significant budgetary resources while providing minimal returns for the civilian economy and the broader national interest. As a means of "saving" domestic cosmonautics and, to some extent, justifying its existence in the eyes of the public, the Gorbachev government sought to expand the space sector's

[39] For more information on the emergence of the monopolar order in Brands, H. (2016). *Making the unipolar moment. U.S. Foreign policy and the rise of the post-cold war order*. Ithaca.

[40] Materials of the Plenum of the Central Committee of the CPSU April 23, Moscow, 1985. Retrieved January 17, 2025, from https://mpck.kpss.su/XXVI/XII.pdf.

involvement in civilian industries and facilitate its transition to a market economy.

Industrialisation

The ideas of space industrialisation which arose at the beginning of *perestroika* were nothing new. Soviet scientists expressed interest in industrial space exploration long before Gorbachev came to power and advocated for the expansion of space technologies within the civilian economy.[41] Soviet scientists, who closely followed these achievements in the West, were well aware of the possibilities of modern astronautics and its benefits for the development of national industry. They analysed the global scientific press, where various projects for the practical application of space technology were being increasingly discussed. International space scientists conjectured that new space systems would be created in the near future, making a tangible contribution to solving the problems of energy, informatics, and industrial production important for mankind.

The Soviet scientists emphasised the fact that the very task of industrial space exploration was first formulated by the founder of the Russian cosmonautics, Konstantin Tsiolkovsky. Tsiolkovsky believed that the main goal of the development of the space industry was the use of solar radiation in the interests of humanity. It is likely that at the dawn of *perestroika* space scientists intuitively felt climate change was on the horizon and that the greatest concern of the country's political leadership should be the application of space technologies in the interests of the national economy. Leonid Leskov, Doctor of Physical and Mathematical Sciences and designer, in an article written in early 1985, clearly called upon the new leadership of the USSR to revise the program for the development of outer space in the direction of its "humanisation". He stated that the forecast for future development of the world space industry could "be considered realistic only if humanity finds the strength and means to stop the arms race that diverted huge human and material resources. The released resources could be instead directed to the creation of the space industry,

[41] It suffices to reference the articles of K. Tsiolkovsky, compiled in Tsiolkovsky, K. (1989). *Promyshlennoye osvoyeniye kosmosa*. Mashinostroyeniye; Grishin S., Leskov L., Savichev, V. (1978). *Kosmicheskaya technologiya i proizvodstvo*. Znaniye.

to the solution of the vital peaceful tasks of the development of our civilisation".[42]

The idea of industrial use of space technologies received strong political support in the *perestroika* years. Gorbachev formulated his proposals for the industrialisation of space in 1985 at the fourth session of the Supreme Soviet of the USSR on November 27, 1985. Expressing concern about the state of the world ecology, Gorbachev moved on to specific proposals for the use of outer space in solving global world problems by common efforts.

> [...] Our country has submitted to the UN an extensive program of peaceful cooperation in outer space, the creation of a World Space Organisation that would coordinate the efforts of countries in research and space exploration. The possibilities for this are truly endless. These are fundamental scientific research and the application of their results in the fields of geology, medicine, materials science, climate, and environmental studies.
>
> This includes the creation of satellite global communication systems, remote sensing of the Earth. This, finally, is the creation and use of new space technology in the interests of all people, including large orbital scientific stations, various manned spacecraft, and in the future, the industrialisation of near-Earth space. This is a real alternative to the plans of "Star Wars", it is aimed at the peaceful future of all mankind.[43]

Thanks to the internal policy of *glasnost* and the external strategy aimed at fostering peaceful relations with international partners, the need to maintain a high level of secrecy regarding space technologies largely diminished, in line with the logic of Gorbachev's *perestroika*. In fact, it was the inflexibility of the command economy, secrecy, and a fear of losing technological superiority that had prevented the introduction of advanced technologies into civilian sectors of the Soviet economy in the preceding period. Numerous discussions on various topics related to the domestic space program appeared in the pages of Soviet magazines and newspapers.

Among specialised popular science publications, information on the application of space technologies to address various challenges in national

[42] Leskov, L. (1985). Industrializatsiya kosmosa: blizhaysheye tyshchyacheletiye. *Nauka i zhizn'*, 6, 28.

[43] From the report at the fourth session of the Supreme Soviet of the USSR of the eleventh convocation, November 27, 1985. In Veber A. (2010). *Gorbachev, Perestroyka i imperativy XXI veka*. Retrieved January 20, 2025, from https://www.gorby.ru/userfiles/imperativy2.pdf.

economic development was increasingly circulated.[44] Authors competed in describing the benefits of using space technologies in various areas of the economy and did not shy away from optimistic forecasts. "The implementation of space industrialisation programs was already in the 1980s of great interest for the development of national economies. Information space systems had been widely developed. Artificial satellites provided an overview of large areas of the earth's surface and atmosphere. Communications satellites made it possible to quickly transmit messages over vast distances. Space systems provided information about storms that arose in the ocean, sudden changes in the weather and helped predict future harvests. Using satellite information, geologists explored natural resources, and environmentalists were able to plan conservation activities in an optimal way. In the future, moving along this path, new technologies have been developed that can bring the technical development of mankind to a new level of mastering the information field of the planet. But space information systems were only one of the practical areas of space industrialisation. In the 1980s, the production of new unique organic and inorganic materials in space seemed topical, which, as expected, would find application in the national economy and healthcare".[45] At the same time, conferences were held dedicated to the introduction of scientific projects for solving problems of industrial development. At a scientific conference held in Gomel (in the former Byelorussian Soviet Republic) in 1988, Soviet scientists discussed the possibility of industrialising space without relying on military rockets.[46]

Thus, during *perestroika*, the theoretical visions of the scientific and political elite regarding the future directions of space research and technology development aligned. However, in practical terms, the realisation of this rare synergy was impeded by the economic challenges that both sectors faced towards the end of the *perestroika* era.

[44] For example: Avduevsky V., Leskov L. (1986). Industrialiazia kosmosa – sleduyushchiy logichesky. *Zemlya i Vselennaya*, 2; Avduevsky V., Leskov L. (1987). *Rabotayet nevesomost'*. Molodaya gvardiya; Grishin S. (1987). *Industrializatsiya kosmosa. Problemy i perspektivy*. Nauka.

[45] Leskov, L. (1988). Stanovlenie i razvitie ucheniya o neosfere. *Zemlya i Vselennaya*, 6, 48–49.

[46] Laurinavichus, I. (1988). Zvezdnyy mir v Gomele i mire. *Narodnoe khozyaystvo Belorussii*, 8, 24–28.

Commercialisation

In parallel with the industrialisation of the space sector, a closely related process of commercialisation started also intended to solve the problems of the national economy in general and of the space sector in particular. Commercialisation both fit into a broader process of economic reform, aimed at the decentralisation of the Soviet economy and stimulation of the growth, and into a series of reforms intended to change the USSR's foreign trade system. Driven by a desire to significantly increase exports of manufactured goods and the importation of advanced technologies, the reforms granted various ministries the authority to engage in direct trade with foreign countries, including capitalist ones. The Soviet government also opened the doors to foreign investment inviting the USA and other technologically advanced partners to participate in joint ventures with Soviet enterprises.[47]

Nikolay Ryzhkov made it clear during his interview to TASS given on January 5, 1987, that the Soviet Union was seeking to profit from space ventures. The Soviet Union sought to win a share of international launch business, to offer launchers and even the Soviet shuttle (Buran). Moreover, the Soviet Union's need for foreign currency was particularly acute at that time, as its primary export—oil—remained undervalued on the global market.[48] Gorbachev's policy of maximising all sources of foreign exchange to support his economic modernisation program added new urgency to the Soviet push for international sales.[49] Offering discounts to developing nations, exemptions from customs regulations, sealed freight containers to safeguard technical secrets, and insurance against failures, the Soviet partners provided several types of launchers with varying capacities for commercial use.[50]

However, the commercialisation of the Soviet space sector encountered a series of challenges, both domestically and in the international market. The emergence of a new player in the global space services arena elicited mixed reactions from its participants. Thus, the unprecedented openness of the Soviet space sector and the readiness of the political authorities to

[47] Interview with Sarah Carey (1987). *The Multinational Monitor*, 8 (11–12).

[48] Barringer, F. (1987). Soviet, with discounts, goes into space business. *The New York Times*, 8 January.

[49] D'Anastasio, M. (1987). Soviets step up efforts to win share in launching of commercial satellites. *The Wall Street Journal*, 20 January.

[50] Barringer (1987).

enter the international market faced fierce competition, primarily in the market of launch vehicles and satellites. These issues will be explored in greater detail in the final section of this book.

Conclusions

To summarise Gorbachev's vision for the development of the space sector, it is important to highlight several key aspects. First, Gorbachev "inherited" a powerful, highly advanced industrial and scientific field, the formation and growth of which had taken place over the previous three decades. As noted by Yuri Baturin, the development of the space program in the second half of the 1980s progressed due to the previously accumulated potential. He argues that "the highest level of design, engineering and technical culture of work, the interaction of many research teams, scientific and technological achievements of Soviet cosmonautics - all this had a positive effect on the development of the space sector in the late Soviet Union Buran, Energya, Halley's Comet scientific mission, Mir orbital station - all these successful projects were implemented not thanks to Gorbachev and his space policy, but simply because he did not have time to slow them down".[51] Hence, the space sector at the end of the Soviet period developed "by inertia", based on the impulses received back in the "golden age" of Soviet cosmonautics.

Secondly, this phenomenon could be explained by the fact that Gorbachev paid his principal attention to the problems of domestic and foreign policy both of which were becoming urgent by the mid-1980s. Economic difficulties and the gradual disintegration of the socialist bloc either distracted Gorbachev from cosmonautics, or forced him to dismiss its needs to other more prevailing tasks. The intensive strategy towards the industrialisation and commercialisation of space capabilities was an endeavour taken precisely to resolve these difficulties. Thirdly, unlike Nikita Khrushchev, who was personally involved in cosmonautics and extremely interested in its development, or Leonid Brezhnev for whom space was one of the most effective instruments of foreign policy, Gorbachev had vague ideas about the space program and knew little about its condition and *modus operandi*. The absence of a major international space project during the political rapprochement of the *perestroika* period—similar to the Soyuz-Apollo project culminated in docking in orbit in 1975—also

[51] Interview with Yu. Baturin, September 17, 2022.

testifies to the considerably lower priority Gorbachev placed on space exploration compared to his predecessors. The preparations for the international scientific Halley's Comet mission had commenced several years prior to his assumption of power, and its socio-political significance cannot be compared to the impact of the 1975 Apollo-Soyuz handshake in space on public opinion. To some extent, Gorbachev's personal lack of involvement in the space sector "freed his hands" and allowed for the relaxation of state control over this domain, which, throughout the entire history of the USSR prior to his leadership, had been viewed as an unassailable stronghold.

The final consideration is strictly connected with the above: *perestroika* leaders assigned to the space sector the priority of the propagandistic aspect of space against its strategic value. The importance of the demilitarisation of space and cooperation for peaceful purposes took precedence over the idea of space as a tactically important component of national security. Buran, which remained stationary for years in the Baikonur hangar and was displayed as an attraction in Moscow's Gorky Park, became one of the most striking symbols of the shift in state priorities and perceptions of national prestige.

CHAPTER 2

Impact of Perestroika on the Space Sector

A pivotal aspect of the *perestroika* reform was the imperative realignment of economic performance. Gorbachev and his proponents asserted that the restructuring of production relations, property dynamics, the establishment of a socialist market, pricing reforms, the advancement of cost accounting, and the steadfast implementation of the principle of distribution based on contribution were essential prerequisites. Furthermore, they emphasised that the transition of enterprises to self-management and self-financing was indispensable for propelling the Soviet economy into a qualitatively new state. In this chapter, our objective is to comprehend the repercussions of Gorbachev's economic reforms on the space industry and science. Specifically, we will focus on examining the influence exerted by the predominantly political and economic facet of *perestroika* on the execution of the space program.

Shaking The Foundations

Before delving into an analysis of the successes and failures of the *perestroika* reforms within the space industry and scientific research, it is crucial to comprehend the organisational system that evolved in the space sector and the innovations introduced by Gorbachev's reforms.

O. Dubrovina, *The Soviet Space Program and International Collaboration during Perestroika, 1985–1991*, Palgrave Studies in the History of Science and Technology,
https://doi.org/10.1007/978-3-032-03157-0_2

If we adopt the framework proposed by Mikhail Marov, a Soviet and Russian planetary scientist, for distribution of power within the Soviet Union as a basis for understanding the structure of the space sector, it becomes easier to identify the transformations that took place in this domain during the *perestroika* era.[1] He identified three key actors that not only governed the country but also wielded a significant influence on the space sector: the Communist Party, the government, and the military. Given that three civil-military institutions in the Soviet Union were concurrently responsible for executing the space program, a distinct demarcation between civilian and military space programs was notably absent.

The Soviet government consisted principally of the Council of Ministers, with its executive branch being the Presidium. Within the Presidium a Military-Industrial Commission coordinated the military-industrial activities of which space was one of the components. Fourteen ministries were responsible for the exploration and utilisation of space resources for various purposes. The most important of them and mostly involved in space activities, the Ministry of General Machine Building (MOM), was engaged not only in cosmonautics, but also in many other fields,[2] mainly in the development of military industry. The MOM exercised authority over the planning and budgetary aspects of the Soviet space program. Thus, the MOM played a pivotal role in establishing, overseeing, and financing various design bureaus (OKBs) responsible for the development of rocket and space systems, as well as scientific research institutes (NIIs) and scientific-production associations (NPOs) that furnished the necessary scientific and technical support for space projects.[3] It is important to emphasise that in overseeing the activities of the research centres, the MOM functioned in a dual capacity: as a customer, responsible for distributing military orders among research and technology institutes, and as an executor, ensuring compliance with and oversight of the execution of these orders. MOM specifically was tasked with the inception and advancement of space rocket

[1] For a more detailed account of the history of the organisation of the space sector, see Lardier, Ch. (2017). Organization of the governance of the rocket industry in Soviet Union and Russia (1946–2016). In *Institut istorii yestestvoznaniya i tekhniki im. S.i. Vavilova. Godichnaya nauchnaya konferentsiya*. IIET RAN.

[2] Kotlyar, P. (2022). "Yura, uderzhi otrasl'": s chego nachinalsya Roskosmos. Kosmonavt Zhukov rasskazal, kak sozdavalos' Rossiyskoye kosmicheskoye agentstvo. *Gazeta.ru*. Retrieved January 13, 2025, from https://www.gazeta.ru/science/2022/02/25/14574145.shtml

[3] Huntress, Marov. (2011), 21.

technology, including launch vehicles, manned and specialised unmanned spacecraft, ground infrastructure, as well as the execution of space activities for the benefit of national security, science, and economy, and subsequently, international cooperation.

During its operations, MOM collaborated closely with representatives from various industrial ministries and departments. This collaboration was particularly extensive with the group commonly referred to as the "defence nine" ministries. In addition to the MOM, this group included the Ministry of Aviation Industry, the Ministry of Defence, the Ministry of Medium Machine-Building Industry, the Ministry of Radio Technology, the Ministry of Shipbuilding Industry, the Ministry of Communications Equipment Industry, the Ministry of Machine-Building Industry, and the Ministry of Electronic Industry.[4] Moreover, other important organisations (research and development bureau, production centres, scientific institutes, etc.) were affiliated with the State Committee on Hydrometeorology, the State Committee on Education, and the Main Administration for Geodesy and Cartography.[5] Thus, MOM served as the central pillar of the formidable Soviet VPK, overseeing an extensive industrial system characterised by substantial funding and operating under conditions of utmost secrecy.[6] With regard to space activities, MOM served as the central coordinating body for all space-related endeavours in the USSR, encompassing both military and civilian domains.

Furthermore, MOM was involved in a partnership with the USSR Academy of Sciences, which traced its roots back to the inception of the space era. Organisationally, this collaboration was institutionalised in 1959 with the establishment of the Interdepartmental Scientific and Technical Council on Space Research (MNTS KI), which was later restructured in the post-Soviet period as the Space Council of the Presidium of the Academy of Sciences.[7] The Council and the Academy determined the objectives for the scientific space program, advised the government and recommended individual projects, provided expertise in space navigation,

[4] Baklanov, O., Rogozin O. (2013). *Otechestvennyy Voyenno-Promyshlennyy Kompleks i Yego Istoricheskoye Razvitiye*. OSLN, 357.

[5] Johnson, N., Rodvold D. (1991). *Europe and Asia in Space, 1991–1992*. Kaman Sciences Corporation, 1. Retrieved January 13, 2025, from https://apps.dtic.mil/sti/tr/pdf/ADA338582.pdf.

[6] Huntress, Marov. (2011), 21.

[7] Baklanov, Rogozin. (2013), 367.

and supplied scientific investigations for flight missions.[8] Through its many, widely dispersed institutes, the Academy of Sciences directed a few special scientific programs as well as supported a variety of other applications projects with both civilian and military objectives.[9] Being the Soviet Academy of Sciences subordinated to the Department of Sciences at the Secretariat of Central Committee of the Communist Party[10] and according to an unwritten rule, it was obligatory for the leading staff of the Academy to be members of the CPSU, the Academy was a Party organisation.[11] Therefore, the Party wielded control over the space program through two institutions integrated into the Secretariat of the Central Committee: the Department of Sciences and the Secretary of Defence Industry.

The intricate organisational structure of the Soviet space endeavours must also incorporate a third component: the military. The Armed Services' Strategic Missile Forces emerged as a significant contender for funding from MOM, with the military asserting control over the launch pads, launch ranges, and vast ground-based command and control infrastructure, including global tracking stations network.[12] In any case, the military constituted the primary consumers of rocket and space products, the manufacturing of which sustained the livelihoods of millions within the population of the USSR.

Drawing from the Resolution of the State Council of the USSR dated November 14, 1991, the Ministry of General Machine Building was officially dissolved on December 1 of the same year. However, this event marked the culmination of a protracted process that had commenced well before the Ministry's formal abolition. According to Yuri Mozzhorin, the director of the Central Research Institute of Machine Building (TsNIIMash), the structural integrity of MOM began to falter as early as 1989, "propelled by the winds of *perestroika* and evolving economic paradigms".[13]

By the late 1980s, the previously well-established organisational structure began to falter. According to one perspective, the absence of

[8] Huntress, Marov. (2011), 8.

[9] Johnson, Rodvold. (1991), 1.

[10] Huntress, Marov. (2011), 21.

[11] This fact, however, did not preclude a small number of prominent Soviet scientists, who were not members of the CPSU, from actively participating as members of the Academy of Sciences.

[12] Huntress, Marov. (2011), 22.

[13] Mozzhorin, Y. *Tak eto bylo.* (2000). Mezhdunarodnaya programma obrazovaniya, 338.

comprehensive planning and a strategic approach to the development of the space industry contributed to its internal fragmentation, including disintegration at the organisational level. According to the members of the Moscow Space Club, established in 1991, by the end of the Soviet era within the complex array of organisations, structures, and institutions in the USSR, there existed no efficient system for the distribution of space governance functions. According to observers, the management of Soviet cosmonautics evolved somewhat haphazardly, characterised by both an extreme level of departmental disunity and the monopolistic control of each department within its designated sphere. There was a notable absence of precise delineation of competencies and management functions among these entities. The existing coordination bodies, such as the MNTS KI, had lost control of the situation. Furthermore, the space program became contingent on the broader situation, with its susceptibility to political dynamics, chance occurrences, and subjective factors.[14] As observed by Sergey Zhukov, a test cosmonaut, and a full member of the K. E. Tsiolkovsky Russian Academy of Cosmonautics, "Cosmonautics grew up on the basis of military-political confrontation as a by-product of the arms race. Therefore, during the *perestroika* period, it remained a top-managed industry without a scientific approach to strategic planning of its development".[15]

According to another perspective, the primary cause of disorganisation in the management of the entire space system was the erosion of its foundational structure. The reformers' most significant impact was on both the Communist Party and the military—previously established as the foundational pillars of the space sector. Consequently, critics of *perestroika* argue that the reforms undermined the control mechanisms across the entire field, ultimately leading to its gradual decline. Thus, opponents of the *perestroika* reforms and advocates for the inviolability of the entire defence sector reacted with profound discomfort at the organisational changes introduced in the industry. They attributed the sector's challenges to the loss of control over what they perceive as a previously smoothly functioning system: "In 1987, the commencement of poorly conceived *perestroika* initiatives unfolded. The most detrimental factor was the erosion of control over industry, including the defence sector. Ostensibly presented as a reform to counter command-based management methods, it was coupled

[14] Virko, I. (Ed.). (1991). *Kosmonavtika – predlozheno vyzhit'*, Znaniye, 9.
[15] Kotlyar. (2022).

with demagoguery in the appointment of enterprise directors through local elections—allegedly under the guise of democracy. This approach infiltrated the established management structure of defence industries and enterprises, disrupting the proven system of personnel training. As a result, there was a noticeable decline in controllability".[16]

From another perspective, all organisational structures retained their original functions; however, their activities, both in the short and long term, were constrained by a reduction in funding compared to pre-*perestroika* levels. This financial limitation led to the disorientation of the entire space sector. Regardless of whether the primary cause of the difficulties in the space sector during the late 1980s was financial or political, a new phenomenon emerged during this period: the process of liberalisation.

Liberalisation in the Space Industry

The start of the liberalisation process in the Soviet economy within the socialist framework commenced with the June Plenum of the CPSU Central Committee of 1987, during which Gorbachev announced a series of economic reforms cantered around the shift towards a market economy within the context of socialism. The theoretical underpinning for this new phase of reforms was rooted in the ideology of "democratic socialism", which allowed for the prospect of a mixed economy and private ownership under the umbrella of socialism. The practical guidelines for implementing the tenets of "democratic socialism" in the economy were encapsulated in the Law on State Enterprise and the Law on Cooperation in the USSR. In practical terms, the realisation of "democratic socialism" necessitated the operationalisation of economic entities based on the principles of self-accounting. This involved the incorporation of self-management, self-payment, and self-financing as foundational elements in the functioning of economic entities.[17] As an integral component of the national economic system, the space sector found itself entangled in the *perestroika* initiative, undergoing substantial economic changes during the late 1980s.

The liberalisation of the space industry in the second half of the 1980s was driven by the leadership's desire to reform the stagnant economy

[16] Zabelin, L. (2013). Promyshlennost' mashinostroyeniya. In Baklanov, Rogozin. (2013), 351

[17] Eliseeva, N. (2017). *Istoria perestroiki v SSSR, 1985–1991*. RGGU, 214–216.

inherited from the previous era. This was pursued through the application of economic management mechanisms within the space sector, whose maintenance was becoming an increasing burden on the already strained state budget. Indeed, the liberalisation of the space sector was designed to bolster the industry by furnishing it with novel development opportunities, capitalising on the broader evolution of the domestic economy.

Among other *perestroika* measures, the Soviet government authorised some of the largest enterprises within MOM to independently negotiate space cooperation agreements with organisations in foreign countries. These measures aimed to enhance the participation of the USSR in the international space market.[18] Consequently, there were also sporadic initiatives at enterprises and scientific institutions within the rocket and space complex to establish commercial structures (for example, IBMP's (Institute of Biomedical Problems) Department of Application, Implementation and Promotion of Achievements of Space Biology and Medicine, TsSKB's Design Bureau Foton at Kuybyshev's plant).

Nonetheless, several impediments hindered the liberalisation of the space sector. Firstly, the individual components of the industry have faced challenges in achieving a state of free-floating autonomy. During this period, the predominant economic entity in the Soviet space sector became the NPO, a colossal economic conglomerate encompassing the entire specialised production cycle. Typically, a design bureau (e.g., NPO Energya, NPO Lavochkin, NPO Molnya) occupied the central position within an NPO, while peripheral components included scientific research institutes, testing facilities, and experimental high-tech production units. According to space experts, such an expansive economic structure, often referred to as an "economic monster", could not sustain itself through autonomous financing and remained heavily reliant on injections from the state budget. Such bulky space NPOs with highly specialised production could not, as Soviet specialists insisted, adapt to changing market conditions.[19]

Furthermore, shortcomings inherent in the organisational structure of the space research and production complex were emerging more and more. Departmental barriers arose between the space services of the MOM and enterprises within the national economic sectors that served as consumers of space information. In several instances, these enterprises were not technologically prepared to effectively integrate space

[18] Avduevsky, V., Leskov, L. (1990). *Kuda idet sovetskaya kosmonavtika?* Znaniye, 18.
[19] Virko. (1991), 9.

information into their operations. Regarding the dissemination of new materials, technologies, and equipment developed during space programs to other sectors of the national economy, several impediments were identified. These barriers included excessive secrecy, a tendency towards limited disclosure, a lack of information, a dearth of direct interest from MOM in facilitating knowledge transfer, and a lack of enthusiasm from other agencies to adopt and apply these advancements.[20]

It is essential to highlight the issue, closely linked to the preceding point, of the lack of communication amongst agencies (ministries, departments, etc.) rendering it impossible to accurately assess the returns from the space sector. In accordance with the established order in the Soviet Union the achievements in cosmonautics were not only transferred to other areas free of charge, but were also left unaccounted for. According to Anatoly Evich, a leading researcher at TsNIIMash, "the extent to which space resources were utilised and whether they yielded practical benefits, whether all images obtained from orbit became the property of consumers or remained unused in storage, what technological innovations were derived from cosmonautics and the resulting impact - none of these questions could be answered. Proper documentation was not systematically prepared, remained undisclosed, and comprehensive analyses were not conducted. Many details have been forgotten by living witnesses, and some are no longer with us".[21] Konstantin Feoktistov, Soviet engineer and cosmonaut, even gave an example of a case when non-use of space technologies led to direct economic losses. "For instance, in the United States, Japan, Australia, and Western Europe, satellite communications support millions of telephone channels. In our country, however, only a few thousand are available. It is more straightforward to shout to a neighbouring town than to establish a connection. This deficiency is detrimental to the national economy. The expenses incurred in creating a satellite communication infrastructure are outweighed by the losses caused by its practical absence".[22]

The process of liberalisation encountered difficulties as property relations remained unregulated, resulting in an inequitable distribution of income generated from the commercialisation of space products. In many cases, the primary profits from commercial operations in space activities

[20] Avduevsky, Leskov. (1990), 12.

[21] Evich, A. (1991). Vygodna li kosmonavtika. *Nauka v SSSR*, 1.

[22] Feoktistov, K. (1991). Zato my delaem rakety. *Znanie-sila*, 4.

did not accrue to those who had conducted the primary work but to those assigned with the task of selling the outcomes. According to experts, this situation led to a decline in initiatives and impeded the growth of profitable branches within the space sector.[23]

The notion of private property in a crucial sector like the space industry did not garner support within the ranks of the Soviet nomenklatura. Despite the simultaneous allowance of certain freedoms in industrial and commercial activities, manufacturing enterprises and research institutes persisted as exclusively state-owned entities. The possibility of privatising certain types of space activities was discussed in February 1986, at a meeting of the Scientific and Technical Subcommittee of the United Nations. "The Soviet Union, supported by some of its allies, spoke out against the privatisation of space activities as being harmful, especially to developing countries".[24] The Soviet Union's reluctance to engage in discussions about the privatisation of space activities further underscores the attitudinal Soviet leadership's unpreparedness for a shift towards trade relations within the industry. It also highlighted the inherent limitations of the command economy system, which struggled to embrace the principles of the free market. Nevertheless, this did not preclude the emergence of what were termed as "empires" in the late Soviet era, which evolved into independent, sizable private enterprises during the 1990s, such as NPO Energya near Moscow and TsSKB Progress based in Kuybyshev (modern Samara).

Finally, the process of liberalisation, involving the gradual opening up of the Soviet space infrastructure to foreign colleagues and partners, revealed the carefully concealed shortcomings of the entire sector. In this context, let us cite the observations of the Swiss astrophysicist Thierry Courvoisier, who visited the Moscow Khrunichev plant in the late 1980s. "It is a huge factory. The work was very different from European standards. But it was much cheaper. *Babushki*[25] came, took drawings, and used them to connect parts and fasten screws. That is, all the work was based on trust in the technicians who did the work. In the West, everything was more protocolised. A clear order was established, and quality control was implemented. Every action was anticipated, planned, and monitored in

[23] Virko. (1991), 9.

[24] Nations Unies: rapport sommaire sur la 23ème session du Sous-Comité scientifique et technique, 24 Février 1986. In ESA-10434, HAEU.

[25] Elderly women.

advance. In Russia, there was a sense that individuals took full responsibility for everything, whereas in the West, the entire system was accountable for the quality of work. People in Russia built rockets themselves. In the west the technicians had to follow exactly the instructions given to them by the engineers. And in Russia, the technicians took responsibility. That was the difference. This trust implied a simplification of the scheme of work. And it worked!"[26]

Here, however, two points must be considered. Firstly, the situation at the Soviet space enterprises in their later years, as observed by foreign delegations, did not fully reflect the actual state of affairs. According to Leonid Gurvits, an astrophysicist and a former IKI researcher, the members of Western delegations who visited secret production facilities in the USSR were presented with what could be described as typical Potemkin villages. "The Khrunichev Enterprise was the most advanced Soviet space plant, where the latest technologies were applied, the most skilled specialists were employed, and the most sophisticated rockets and spacecraft were produced. Its operations were even more secretive than those at NPO Energya or NPO Lavochkin. I am convinced that carefully prepared displays were prepared in advance for foreign visitors".[27] Secondly, while the scientific and technological achievements of the Soviet Union remained relevant at the end of *perestroika*, they did not prevent the Soviet space industry from falling behind in some other sectors of advanced technology.

Summarising the results of Gorbachev's reform of the space industry, Viktor Nikolaev, a former official at Glavkosmos and later a leading engineer at the Russian Space Agency, stated: "I was grateful to him that some processes had started. He contributed to lifting the taboo on declaring the space sector as an industry. Thanks to him, the space industry could now generate revenue based on its competencies, technologies, and the sale of services, including launch services. However, in my opinion, he became an impediment to the process he initiated due to his fear of losing party support. He saw prospects only if the party continued to lead the country. In fact, he was afraid of what he had started. Moreover, there was no clearly thought-out program. Everything was developing without direction, without preliminary calculations and plans; everything was carried out as the wind took it".[28]

[26] Interview with Th. Courvoisier, July 27, 2023.
[27] Interview with L. Gurvits, January 10, 2025.
[28] Interview with V. Nikolaev, June 30, 2024.

Liberalisation in the Scientific Field

In the scientific space community, *perestroika* and the subsequent liberalisation, which impacted scientists both politically and economically, elicited mixed repercussions and reactions. It is imperative to acknowledge that transformations within IKI, the main scientific entity for space research, commenced well before Gorbachev ascended to power. The Institute sensed the dawn of a new era as early as 1973, upon the appointment of Roald Sagdeev as its director. Reflecting on his arrival, Tamara Breus, science secretary at the IKI, recounted: "It is noteworthy that *Time*, in an article on space research in the USSR during the Gorbachev era, stated that Sagdeev initiated *perestroika* within space research long before its official course in the nation.[29] He exhibited remarkable political acumen and foresaw favourable trajectories well in advance of Gorbachev's public speeches. Operating with the demeanour of a shrewd politician, he employed assertive and unyielding tactics akin to those found not typically within the scientific realm but more so within the sphere of business competition, thereby invigorating the stagnant dynamics of the domestic scientific community".[30]

However, no less important for Sagdeev was the extension of international contacts and reaching a higher level of interaction with foreign partners. The Military-Industrial Commission, which oversaw the entire space program, stood in the way of achieving this goal. Its members believed that: "in no case should we co-operate with the West, so as not to give away our own secrets and not to undermine the dwindling prestige. The timing of spacecraft launches and future plans were carefully concealed. Scientific programs were made public only after they were completed, and even then, not always".[31]

In addition to maintaining the secrecy of information, Soviet governing bodies also exercised vigilance over scientific contacts. This vigilance arose from concerns over the transfer of scientific and technical knowledge, as well as apprehensions about undesirable emigration—referred to as

[29] "[soviet] planetary-science program did not really take off until shortly after the appointment in 1973 of a 40- year-old scientist named Roald Z. Sagdeev as head of IKI. From the outset, Sagdeev started to shake things up. He took physicists out of their labs and put them on production lines to watch their experiments being built". Lemonick, M. (1987). Surging ahead. The soviets overtake the U.S. as the no. 1 spacefaring nation. *Time*.

[30] Breus, T. (2002). *Institut. Nepridumannye istorii.* Kont, 57.

[31] Ibid.,74.

nevozvrashchentsy, or non-returning Soviet citizens—during foreign business trips, along with other potential consequences that could adversely affect the country's reputation. Yakov Alpert, a Soviet radio and space plasma physics scientist, who, after 1968, switched to dissident positions and became a *refusenik* emigrating to the United States in 1987, testified to how this approach of the political and military leadership affected science. "The progress of science in both Russia and the West suffered enormously because the Soviet regime allowed only a few scientists, and often not the best ones, to travel to the West or even to correspond and exchange ideas with scientists from other countries".[32]

According to Mike Gruntman, a Soviet-American professor of astronautics who emigrated to California in 1990, Sagdeev successfully persuaded the Soviet political leadership of the imperative for international cooperation in space by advancing two primary arguments in addition to getting access to Western technology: propaganda and national prestige. Gruntman recalls, "He effectively convinced the Central Committee that initiating joint missions for propaganda purposes was paramount... The USSR also sought to bolster the recognition of Soviet scientific prowess in the Western sphere. Consequently, he obtained authorisation and initiated the execution of these missions".[33]

Other contemporaries confirm the statement that Sagdeev's organisational and political talent manifested itself in his ability to convince opponents of close scientific contacts with foreigners of the need for cooperation in the interests of Soviet science, which could in turn ensure geopolitical prestige. For this purpose, he established close personal contacts with leaders of foreign space agencies and organisations, such as NASA, ESA, CNES and other entities. As Tamara Breus argued, he saw the meaning of international cooperation "in the optimal division of labour and its fruits. The labour was to be done by Soviet industry, which built space research vehicles with rich funding from the military budget. It also provided the Academy of Sciences with the launches of these vehicles". The contribution of foreign specialists laid in their financial resources and modern component base, which enabled the development of advanced scientific flight instruments. Then this equipment was installed on board by Soviet engineers from IKI; measurement data were also to be received by IKI from Soviet ground network of tracking and data acquisition stations and then

[32] Alpert, Ya. (2000). *Making Waves: Stories from My Life*. Yale University Press, 28.
[33] Interview with M. Gruntman, March 19, 2024.

distributed to co-investigators. Thus, as Breus recalled, "poor representatives of the Academy of Sciences, who had practically no money to develop unique scientific equipment for space, acted as a kind of intermediaries and could reap the benefits of modern scientific research, becoming co-authors of these experiments".[34]

Another motivation for Sagdeev's collaboration with foreign colleagues was the assurance of state support, as noted by Soviet and Russian physicist Lev Zelenyi, who served as Director of the IKI from 2002 to 2018. Specifically, commitments made to foreign counterparts regarding project implementation guaranteed their backing from the government.[35] As Leonid Gurvits explained: "Failing to keep a promise or fulfil a commitment to a domestic partner was common practice in the USSR. However, displaying the same behaviour toward a foreign partner would not only harm the professional reputation of the Soviet party but also diminish the overall status of the Soviet Union—an outcome that would be highly undesirable to the top Soviet authorities".[36] This further underscored Sagdeev and Vyacheslav Kovtunenko's (General Designer and General Director of NPO Lavochkin) efforts to involve foreign specialists in collaborative endeavours, recognising the unequivocal benefits of experience exchange and the potential for qualitative improvements in scientific equipment.[37]

For a thorough evaluation of Sagdeev's efforts in elevating the Institute's international collaboration to its apex, it is imperative to contextualise the unique position of IKI within the space sector's framework. Mike Gruntman explicated this notion succinctly: "Every entity experienced *perestroika* differently. IKI did not align with the conventional archetype within the Academy. This divergence was primarily due to the large size of the engineering and scientific teams involved. Unlike conventional scientific institutes which typically engage in smaller-scale projects, space missions overseen by IKI involved large teams comprising dozens or even hundreds of individuals, as well as considerable financial outlays reaching millions of dollars. Consequently, the administrative hierarchy at IKI, particularly the director and his deputy, had to navigate high-level

[34] Breus. (2002), 75–76.

[35] Zelenyi, L. *Boleye chetyrekh desyatiletiy partnerstva na zemle i v kosmose*. Paper presented by L. Zelenyi and kindly delivered to the author.

[36] Interview with L. Gurvits, December 20, 2024.

[37] Zelenyi, L. *Boleye chetyrekh desyatiletiy partnerstva na zemle i v kosmose*.

bureaucratic channels, including interfacing with the CPSU Central Committee, to secure funding for such monumental endeavours. Additionally, collaboration with industrial entities was imperative". Gruntman further elucidated: "Engaging in laboratory work and scrutinising data under a microscope is undoubtedly important, albeit it entails interactions with a distinctly different cohort of individuals".[38]

Lev Zelenyi also testifies to the special status of IKI, emphasising its difference from NPO Lavochkin, with which the Institute has closely cooperated since the 1970s. "For IKI it was easier—the staff of the academic institute co-operated quite freely and actively with their foreign colleagues. At that time, the NPO Lavochkin was still an 'off-limits organisation'".[39]

As Leonid Gurvits recalls: "Not a single major space science experiment in the USSR during the 1980s was executed as a purely national endeavour. Examples include all planetary science missions, plasma physics, and astrophysical missions and experiments".[40] This extensive network of international collaborations, including the invitation of foreign scholars, scientific exchanges, and joint projects, inevitably had a positive impact on the development of domestic science. Mike Gruntman affirmed the fundamental importance of establishing international cooperation for the development of his field of research—Energetic Neutral Atom (ENA) imaging technology. "We took full advantage of the possibility to invite foreign scientists as guests of the USSR Academy of Sciences which also aligned with Sagdeev's promotion of international cooperation". It is thanks to the international cooperation that Mike Gruntman believes his field of study "survived" and laid the foundations for further research. "Working relations with several colleagues in West Germany and Poland and later in the United States enabled our progress in the direct detection of neutral atoms. Much was achieved in an essentially joint effort. It was vitally important that we had established this mutually beneficial collaboration by the early 1980s".[41]

One of the initial indications of the new era was the increased communication with colleagues from foreign countries, particularly those labelled

[38] Interview with M. Gruntman, March 19, 2024.

[39] Zelenyi, L. *Boleye chetyrekh desyatiletiy partnerstva na zemle i v kosmose*.

[40] Interview with L. Gurvits, December 20, 2024.

[41] Gruntman, M. (2022). *My Fifteen Years at IKI, the Space Research Institute*. Interstellar Trail Press, 210.

as capitalist nations. First and foremost, on a purely technical level, it is important to highlight the role of facsimile machines, which had recently been introduced in the USSR and offered new opportunities for maintaining more consistent communication with Western colleagues. As Leonid Gurvits recalls: "Fax machines were a great rarity in the USSR, including within the Academy. At the onset of *perestroika*, sending and receiving fax messages required navigating a complex and elaborate process of obtaining permission from the 'competent authorities' (Secret Services Department). By the end of *perestroika*, however, the management of IKI had appointed one or two employees from each department, referred to as 'trusted persons,' to handle fax communications".[42]

Thus, the era of *perestroika* marked a significant shift towards unrestricted correspondence with international colleagues and the ability to publish scientific papers in foreign journals. As recalled by Mike Gruntman, "during the 1970s and early 1980s, all incoming letters to IKI from abroad underwent scrutiny by an institute employee proficient in a foreign language. Details of the sender, recipient, and contents of the letter were meticulously recorded in a dedicated log". In the second half of the 1980s, Gruntman took the initiative to submit one of his articles to the USA journal *Review of Scientific Instruments* without seeking authorisation from the IKI or the Academy of Sciences. As he noted: "I simply disregarded the rules mandating special authorisations, as the stringent Soviet system was already showing signs of faltering amidst the burgeoning *glasnost* and *perestroika*".[43] The control over foreign correspondence became less stringent and Gruntman began forwarding his letters, along with editorial responses and revised manuscripts, through trusted Western colleagues he had established connections with.

Among other notable benefits brought by *perestroika* was the facilitation of scientific collaboration through the opening of borders in 1985, immediately following Gorbachev's rise to the top position in the state. According to Lev Zelenyi: "Consequently, we began spending several months each year in Los Angeles. The strong computational capacities at UCLA (University of California) enabled us to assess theoretical models of

[42] Interview with L. Gurvits, December 20, 2024.

[43] Gruntman, M. (2015). Pyatnadtsat' let v IKI: mikrokanal'nyye plastiny, pozitsionnochuvstvitel'nyye detektory i Energichnyye neytral'nyye atomy. In Yu. Zaytsev (Eds.), *50 let Institutu Kosmicheskikh Issledovaniy Akademii Nauk. Obratnyy otchet... 3.* IKI RAN, 253.

Fig. 2.1 IKI. Source: G. Avanesov

chaotic and linear particle dynamics, leading to the revelation of a novel phenomenon—the structurisation of accelerated plasma flows".[44]

The intensified international collaboration resulted in an unforeseen outcome: the IKI underwent a process of substantial liberalisation and gradually disengaged from the oversight of secret intelligence services (Fig. 2.1). Due to the Institute's focus on topics closely or directly linked to national security, it was subject to heightened scrutiny by both military experts and security agencies. Nonetheless, the expansion of international engagements ultimately eroded the efficacy of surveillance mechanisms. Recollecting this transformation, Mike Gruntman remarked, "Sagdeev initiated the opening of IKI to international cooperation, which rapidly rendered the Institute's security apparatus incapable of monitoring its activities due to the extensive interactions with foreign entities. Consequently, breaches of protocol became commonplace, as I personally engaged in unauthorised meetings with foreign counterparts sometimes on a daily basis".[45]

Moreover, the onset of *perestroika* facilitated the Soviet scientific community's access to advanced foreign technologies, particularly in the realm of space instrumentation. Prior to Gorbachev's reforms, information about planned scientific projects in the USSR was typically not disclosed in the free press and only surfaced after successful implementation. As recounted by Georgy Managadze, who served as the Head of the Laboratory of Mass Spectrometry and Active Diagnostics at the IKI for many years, Sagdeev's *perestroika* innovations within the Institute played a pivotal role in overcoming this historical reticence. Managadze highlighted that the attainment of publicity had a transformative effect, enabling the advance publication of scientific objectives and project

[44] Zelenyi, L. (2015). Predislovie k poslesloviyu. In Yu. Zaytsev (Eds.), 18.
[45] Interview with M. Gruntman, March 19, 2024.

timelines. This, in turn, facilitated the active participation of scientists and specialists from other spacefaring nations in these projects. Their involvement extended beyond the contribution of ideas, encompassing the provision of specialised instruments. "Thanks to these reforms, the Institute was able to break through its traditional closedness and organise collaborative initiatives with numerous universities and institutes in foreign countries that possessed cutting-edge space technologies. This collaborative effort extended to prominent space centres situated in various developed countries. By that time, scientists in these countries had accumulated substantial experience in space instrumentation and had access to unique electronic components that were subject to export embargoes".[46]

As a result, a novel and unconventional framework for the formation, development, and realisation of space projects emerged. This framework allowed for: "open discussions not only with various organisations within the socialist camp countries but also with individual scientists and scientific institutions representing the world of capitalism". This collaborative interaction, according to Managadze, significantly contributed to the evolution of new space technologies, particularly in the realm of analytical instrumentation. "The exchange of ideas and expertise between scientists from diverse backgrounds played a pivotal role in advancing analytical instrumentation within the field of space exploration".[47]

Tamara Breus stressed a fundamental role of foreign contribution to the development of the Soviet scientific space sector: "In our country, characterised by chronic deficits and limited funding for scientific endeavours, the realisation of intricate and grandiose projects, as well as the development of cutting-edge equipment on a global scale, would have been unattainable without fruitful collaboration with foreign counterparts from diverse nations".[48]

Access to foreign technology also brought about certain drawbacks, exemplified by the case of the Information Processing and Display System: "Self-contained Video Information Terminal" (SVIT) developed by a group of scientists led by Lev Chesalin. According to the recollections of Tatyana Balakhovskaya, an IKI researcher during that period, the SVIT,

[46] Managadze, G. (2015). *Ekzoticheskaya mass-spektrometriya. Vozniknoveniye, stanovleniye i razvitiye mass-spektrometrii dlya issledovaniy v kosmose i soputstvuyushchiye sobytiya*. In Yu. Zaytsev (Eds.), 109.

[47] Ibid., 110.

[48] Breus. (2002), 226.

developed in the late 1980s, represented a “unique project, not only in its own right but also due to its capacity to produce not just one or two, but an entire series of devices. It appears that the total number of SVITs exceeded a hundred. Consequently, we found ourselves at the forefront of the burgeoning industry of computer image processing at that time. However, *perestroika* did not allow us to maintain this cutting-edge position. With the transformation of the country, the opening of floodgates, the lifting of embargoes, and the influx of imported video systems, notably spearheaded by Silicon Graphics, the landscape changed. Simultaneously, the advertising industry emerged and experienced rapid growth. Advertising and computer graphics found synergies: rotating letters and words dynamically entered television screens, objects seamlessly transformed into one another, and the visual landscape evolved. Advertising infused capital, leading to the proliferation of computer graphics studios, meeting the escalating demands of the advertising sector. Faced with these changes, what options remained for us, the humble scientists, with our cherished SVIT?...”.[49] Soviet technologies that had barely appeared were swept out of the way by foreign giants before they had a chance to gain a foothold and Soviet scientists still had to learn to live under the new laws of the capitalist economy.

Moreover, there existed divergent opinions within the scientific community concerning the implications of increased international contacts, travel opportunities, and the prospect of emigration. Tamara Breus also voiced a critical perspective on the outcomes of *perestroika*. She remarked that: “It was owing to *perestroika* that not only did entry into the orbit of international cooperation become possible, but the rotation on it reached such cosmic speeds that, due to centrifugal acceleration, almost all theorists departed the country and found themselves abroad. The prolonged anticipation for the chance to collaborate with professionally aligned colleagues abroad naturally steered them away from fostering close ties with colleagues domestically. *Perestroika* dispersed the principal potential of theoretical physics and astrophysics to other countries”.[50] Thus, brain-drain and the lack of favourable economic and social conditions were, in the first instance, unintended consequences of *perestroika*.

[49] Balakhovaskaya, T. (2016). SVIT I oskolki pamyati. In S. Vinogradova (Ed.), *Obratny otchet*...4. IKI RAN, 142.

[50] Breus, T. (2015). Nezabyvaemaya lichnost. In Yu. Zaytsev. (Eds.), 109.

It is important to note that, as Leonid Gurvits states, "For some, emigration was a goal, while for others it was an opportunity to work abroad temporarily as an intern, without necessarily intending to remain abroad permanently".[51] For some former Soviet scholars, the decision to stay abroad was influenced by the political and social consequences of the modernisation process in post-Soviet Russia.

It is also necessary to mention another outcome caused by the application of the new *perestroika* thinking in practice at IKI. In 1988, Roald Sagdeev, the Institute's director since 1973, decided to leave his post and give way to a new successor. This independent decision to leave such a prestigious post and the proposal to limit the powers of the director to two terms was a signal of serious change in personnel policy due to *perestroika*. His departure marked the beginning of a new era for IKI scientists signed with innovations in the very system of functioning of the space field. As Albert Galeev, Sagdeev's successor as director of IKI since 1988, argued: "IKI managers and scientists would face a number of new challenges as *perestroika* was implemented". *Aviation week & Space technology* quoted an anonymous Soviet scientist who not only stated the inevitability of imminent changes, but also uncertainty about the readiness to accept them and the ability to continue working under the new conditions: "There may be fundamental changes in the way IKI will operate. For example, we may end up having to buy the spacecraft we use for our missions instead of having them supplied by the government, as was the practice in the past. This means we must take a new look at the way we handle our budget—and we are not so sure how this will affect us".[52] In the 1990s, IKI faced the challenge of addressing a range of new phenomena and trends, many of which had their origins in the later stages of *perestroika*. Among Gorbachev's reforms, liberalisation had a substantial impact on the space sector. However, another aspect of these reforms, conversion, proved even more divisive, sparking significant differences in opinion.

[51] Interview with L. Gurvits, December 20, 2024.

[52] Sagdeev resigns from post at Soviet Space Research Institute. *Aviation week & Space technology*, July 18, 1988.

Conversion: Between the Hammer and the Anvil

One facet of the purported liberalisation involved the transformation of space enterprises, specifically, the relocation of a portion of their production capacity towards the manufacturing of consumer goods. In spearheading the reform of the VPK, the proponents of *perestroika* in the Soviet Union directed their efforts, among other objectives, towards the augmentation of civilian and consumer goods production within the defence industry. The Military-Industrial Commission, with the active involvement and cooperation of Oleg Baklanov, who was a minister of MOM until 1988, was assigned various responsibilities to coordinate and execute efforts towards the conversion of military production for the advancement of the civilian sector in the national economy.[53] For this purpose, enterprises that were previously part of other national ministries were transferred to the MOM. The endeavour to directly engage the military-industrial leadership in the implementation of the conversion policy constituted a strategic manoeuvre by Gorbachev. This move aimed to garner support for his reforms in areas where such endorsement was challenging to anticipate due to the inherent nature of the transformations.

Conversion was also one of the ways to switch to self-financing proclaimed by Gorbachev. The policy of conversion was aimed solving national economic problems through the reduction of funding for space activities, from scientific research to the space industry. In addition, pro-*perestroika* specialists suggested that the only correct approach would be the use of the accumulated scientific, technical, and organisational potential, as well as the rapid and effective promotion of advanced achievements in everyday economic practice. In the view of specialists, conversion should have become a tool for equalising production and technical potential, pulling up lagging areas to an advanced level achieved in cosmonautics.[54]

Consequently, both factions expressed dissatisfaction with the outcomes of conversion—both its advocates and adversaries were discontent. According to the opinion of the supporters of Gorbachev's policy aimed at reducing expenditures on the VPK, the conversion of defence factories to civilian production had little evidence that this reduced military spending. Pro-*perestroika* intellectuals argued that conversion programs actually

[53] Bystrova, I. (2006). *Sovetskiy voyenno-promyshlennyy kompleks: problemy stanovleniya i razvitiya (1930–1980-ye gody)*. IRI RAN, 299.

[54] Avduevsky, Leskov. (1990), 14.

increased spending on the military, because it gave factories a simple excuse for further state-funded investment, ostensibly for retooling production lines. As historian Chris Miller stated, Gorbachev may understand that defence conversion did not save money, but nonetheless supported it in the hope that it would give the VPK a greater incentive to back market reforms that they never did.[55]

Some sceptical Soviet experts warned about the risks of a rash and simplistic approach to conversion. According to them, the conversion of part of the space industry's capacities could lead to the loss of the accumulated potential of the whole industry.[56] The critics of Gorbachev's policies contended that neither the concepts nor the programs for conversion were adequately developed, and furthermore, the primary processes were not modelled. However, as rocket engineers argued, such a radical transformation of production capabilities required thorough preparation and reorganisation of the production process.[57]

Moreover, in the midst of making a conversion, they presupposed that an expansion of the range of consumer goods produced by airspace plants and enterprises, firstly, would bring a modest return from this kind of strategy, which would not fully exploit all the available opportunities. Secondly, it could even lead to the degradation of the rocket and space complex. "It is possible, of course, at enterprises specialising in the building of rockets, to master the production of pots and pans, but, firstly, their cost will most likely be prohibitive, and secondly, the quantities produced will still be insufficient".[58]

Conversion was also perceived by some as Gorbachev's strategy to combat the powerful leadership of the VPK. As Gubanov put it: "having narrowed the concept of conversion down to a simple struggle with the defence industry, naturally, they also came to a narrow path of methods for its implementation: to cut off oxygen through stopping funding".[59] The affiliates of the VPK lamented that in the pursuit of the objective to reduce military spending, the central government endorsed a resolute decision to

[55] Miller, Ch. (2016). *The Struggle to Save the Soviet Economy: Mikhail Gorbachev and the Collapse of the U.S.S.R.* The University of North Carolina Press.

[56] Avduevsky, V., Leskov L. (1990). O kosmicheskoy programme, konversii e kommercii. *Zemlya i Vselennaya*, 5, 13.

[57] Gubanov. (1998).

[58] Avduevsky, Leskov. (1990). *O kosmicheskoy programme, konversii e kommercii*, 13.

[59] Gubanov. (1998).

curtail funding for several enterprises and "freeze" numerous promising programs.

According to opponents of Gorbachev's reforms, the process of "conversion on command" had taken on the features of a political campaign. Reports were sent to the political leadership from the relevant ministries about the deployment of the conversion and positive results. Gubanov acerbically observed the circumstance wherein the MOM, which, when considering the discontent expressed by several of its subordinate enterprises, ultimately acquiesced to the government's decision on conversion: "The leadership of the MOM which was responsible for overseeing the development and organisation of work in the rocket and space industry, reoriented its focus despite resistance from enterprise heads. By 1990, the share of civilian production had increased to 60% of total output, and at some enterprises, it reached up to 80%".[60] The figures cited by Baklanov were less pessimistic for supporters of the development of the military-industrial sector: "During 1988–1989, the allocation was distributed as follows: 40% for the production of national economic output and 60 percent for arms production. By 1990, the ratio had been equalised to 50/50. In 1991–1992, there existed the potential to establish a 60–40 ratio, thereby prioritising peaceful production".[61]

While generally endorsing the concept of conversion, Baklanov cautioned against an overly rapid and forceful execution of the process. Reflecting on the outcomes of Gorbachev's conversion efforts in the early 1990s, Baklanov articulated: "Conversion is not an end in itself. The conversion plan was adopted with great effort in 1987. All of this necessitated in-depth scientific study and planning, addressing not only technical but also social issues. However, the fervour generated by the «democrats» and the breakdown of the state management structure distorted the conversion. High-tech, knowledge-intensive production concentrated in the VPK has already surpassed the threshold of self-preservation and is being dismantled. It is neither in demand for peaceful nor military production".[62] Gubanov summed up the disappointing result of the conversion in the late Soviet Union: "The current conversion is the collapse of the VPK and of the space industry".

[60] Ibid.

[61] Baklanov. (2012). Vol. 2, 661.

[62] Ibid., 685.

The example of NPO (RKK) Energya illustrates the evolution of conversion efforts within the Soviet and Russian space industry and provides an opportunity to assess their specific outcomes. It is important to note, first of all, that conversion was not a new phenomenon in the Soviet industry. Its origins can be traced back to the post-war period, gradually gaining momentum towards the end of the Soviet era.

The development and production of civilian goods at NPO Energya can be divided into three distinct stages. From 1946 to 1970, the enterprise primarily engaged in technical assistance to the medical sector and other industries. This work was typically carried out either on the instructions of party and state authorities or on the personal initiative of the enterprise's leadership. The production volumes during this period were minimal, often measured in units, and accounted for only a fraction of a per cent of the plant's total output.

Between 1970 and 1988, the enterprise received assignments through government decrees, ministerial decisions, and directives from party and state bodies at both the Moscow regional and Kaliningrad city levels. During this period, projects typically advanced to testing and the introduction of goods into serial production. The share of civilian production in the plant's total output became more substantial.

Since 1988, conversion efforts were simultaneously driven by top-down government initiatives—particularly through state decrees up to 1992—and bottom-up efforts to identify the most profitable civilian markets. In some years, the volume of civilian production reached as much as 39% of the plant's annual output. However, on average, the share of civilian products in the enterprise's total output did not exceed 5%.

Throughout all periods, the plant's civilian production was characterised by the sporadic manufacture of a wide range of products, generally in small batches. It is also notable that production initiated by government orders was, as a rule, economically unprofitable.[63]

It is evident that conversion within the VPK, which included the space sector, was critically necessary. To address pressing challenges and effectively reform the economy, it was imperative to adjust the balance of funding and prioritise the development of the civilian sector over the military sector. This imbalance had been a key factor contributing to the stagnation and eventual collapse of the socialist economy. The issue laid not in

[63] Semenov, Yu. (Ed.) (2011). *Raketno-kosmicheskaya korporatsiya Energiya imeni S.P.Koroleva*. Tom 1: 1946–1996. Energya.

the direction chosen by the *perestroika* reformers but in the methods they employed to achieve their objectives. Paving the way for diversification of their capabilities and entry into domestic and foreign markets, the *perestroika* reformers found it necessary to reassess another longstanding postulate inherited from earlier times—space propaganda.

Propaganda Burden

Propaganda had played an important role in the Soviet space program since the golden space age. But the sustained implementation and upkeep of this propaganda, at its erstwhile elevated standard, had exerted escalating pressure on the state budget. Within this context, a renewed examination of the interplay between propaganda and material interests in the evolution of the space industry has emerged. Specifically, discussions have focused on the use of space for propaganda purposes to maintain the political influence of the USSR within socialist bloc countries and the so-called Third World. In these settings, propaganda remained a fundamental factor in shaping international cooperation in space activities.

As some Soviet observers argued, the space program in the USSR, unlike its main rival, the United States, did not include the gain of material benefits. Space activities were used by the leadership for propaganda and international diplomacy, without first considering any possible profit from national and international projects. In the Soviet press, two opposing approaches concerning cooperation with developing countries were analysed: the philanthropic Soviet approach and the material American one. At the same time, the interests of the American side, according to Soviet space experts, were not limited solely to the appropriation of profits received from the sale of goods and services to developing countries. "The United States seeks to turn this kind of 'cooperation' into another channel for strengthening the technological dependence of the states of Asia, Africa, Latin America on imperialism in order to obtain unilateral political, economic, propaganda benefits. Their international space programs are inextricably linked with selfish foreign policy goals".[64]

On the contrary, according to these experts, the USSR did expand international cooperation which was aimed at assisting developing countries to obtain sufficient knowledge for their own space programs, so developing their national economies and raising the standard of living of

[64] Osvoenie kosmosa i razvivaushchiesya strany, *Zemlya i Vselennaya*, 4, 1985.

those populations. Among the implemented Soviet joint projects with these countries were the famous "free tickets" to the Soviet orbital stations, within the framework of cooperation under the Intercosmos program.[65] "On equal and mutually beneficial terms", Moscow also provided everyone who wished to participate in the international communications organisation Intersputnik,[66] and seminars for specialists from developing countries. In addition, Moscow issued annual scholarships to students to study in Soviet universities through the program of cooperation between the USSR Academy of Sciences and the corresponding organisations of Third World countries. It is noteworthy that, at the outset of the *perestroika* era, such a generous approach came under criticism from participants in the space sector, as it no longer aligned with the evolving needs of the Soviet Union. Subsequently, as we will explore further, the model shifted from free access to space to a system where access became a privileged commodity. Fees were introduced for satellite launches aboard Soviet rockets, which had previously been offered as a complimentary and goodwill-based service.

Certainly, beneath the veneer of peace-loving rhetoric, there existed a component of geopolitical calculation and overt propaganda. However, until a certain juncture, collaboration with developing and socialist bloc countries was not distinctly oriented towards reaping economic rewards. Though this did not imply that the USSR's cooperation with countries within the socialist bloc, for instance, operated solely within the ambit of a charitable program. Scientists from Czechoslovakia, the German

[65] See more in Burgess, Vis. (2016).

[66] For a more comprehensive exploration of the history of this organisation in the international context, refer to Downing. J. D. H. (1985). The USSR and satellite communications: Competition and cooperation. *Space Policy*, 1 (3), 250–262; Evans, Ch., Lundgren L. (2022). Dividing the Cosmos? INTELSAT, Intersputnik, and the Development of Transnational Satellite Communications Infrastructures during the Cold War, in A. Lovejoy, M. Pajala (Eds.), *Remapping Cold War Media: Institutions, Infrastructures, Translations* (pp. 260–276). Indiana University Press; Borisov, V. (2003). Iz Istorii Sozdaniya Otechestvennogo Kosmicheskogo Televideniya. *Upravlenie naukoy: teoriya i praktika*, 5 (4) 166–173. DOI: https://doi.org/10.19181/smtp.2023.5.4.9 EDN: RPCUKQ; Stovbun, V. (2009). Osobennosti mezhdunarodnoy organizatsii kosmicheskoy svyazi «intersputnik» v kachestve operativnoy (ekspluatatsionnoy) organizatsii. *Istoricheskiye, filosofskiye, politicheskiye i yuridicheskiye nauki, kul'turologiya i iskusstvovedeniye. Voprosy teorii i praktiki*, 3 (4), 192–196; Wagner, H. (2024). *Space enthusiasm in pre-and interwar times between national policy, aspirations of internationalism, and the open question of Europe.* In T. Nanz, H. Wagner (Eds.), *Cold War Europe: A Space of Communication*, (pp. 61–86). De Gruyter.

Democratic Republic (GDR), Poland, Bulgaria, and Hungary indisputably contributed to the advancement of Soviet space activities.[67]

It is worth bearing in mind that on the other side of the Iron Curtain, the space field played an equally important political and ideological role. As historian of science and technology John Krige stated, NASA's collaboration with European space engineers and scientists "was also an instrument of soft power that provided a counterweight to the attraction that some French scientists felt working with the highly successful Soviet program".[68] One of the negative aspects of cooperation with Soviet colleagues for Europeans was the extreme secrecy of the Soviet space program. In this sense, the American alternative therefore seemed much more attractive.

The Conquests of Glasnost

This secrecy, prevalent within the space sector, not only constituted a substantial hindrance to the promotion of scientific collaborations, but also impeded the establishment and growth of trade and commercial relations. Indeed, a significant impediment to the space industry's transition to a market economy was the pervasive secrecy that defined the preceding era. For instance, the inability to openly disclose research findings precluded the opportunity to secure patents and, consequently, the financial benefits derived from one's own scientific discoveries.

With the policy of *glasnost*, the veil of secrecy has begun to gradually disappear from such a tightly regulated and closed field as cosmonautics. The burgeoning occurrence of manifestations hinted at this phenomenon.

[67] See Gubarev, V., Klimuk, P., Germashevsky, M. (1987). *Interkosmos – eto sotrudnichestvo*, Mashinostroeniye; Foody, P. W. (1994). Detailed case study for Hungary, *International Journal of Remote Sensing*, 15, 2995–3001; Sylwester, J. (2001). Impact of the Intercosmos Programme on the Past, Present and Future of Solar Space Research. *Space Research Center*, Polish Academy of Sciences.; Getsov, P. S. (2003). Bulgarian space studies - past, present and future. *International Conference on Recent Advances in Space Technologies*, Istanbul, 542–546; Bareš, P. (2001). Czechs in space. *Acta Astronautica*, 48, (5–12), 957–965; Manikowski, P. (2013). Developments in space activities in Poland. *Space Policy*, 29 (1), 35–39.

Sagath, D., Adriaensen, M., Giannopapa, Ch. (2018). Past and present engagement in space activities in Central and Eastern Europe. *Acta Astronautica*, 148, 132–140.

[68] Long Callahan, A. (2013). NASA and Eastern Europe. In Krige, J., Long Callahan, A., Maharaj, A. (Eds.), *NASA in the world. Fifty years of international collaboration in space*, Palgrave Macmillan, 31.

On March 13, 1986, for the first time in the history of the Soviet Union, the launch of a manned spacecraft was pre-announced and shown live on central television. The Soyuz T-15 crew, consisting of cosmonauts Leonid Kizim and Vladimir Solovyov, was to be the first to visit the new Mir orbital station. Prior to this historic event, everyone learned about a successful launch only after the rocket had already been launched into Earth's orbit. Instances of failures and delayed launches were systematically concealed, and catastrophes, with few exceptions, were shrouded in secrecy. The discernible positive transformations resulting from Gorbachev's *perestroika*, which laid the groundwork for *glasnost* and transparency in the space industry, were evident in this context.

As recounted by Brian Harvey, an expert in the history of space exploration, the primary impact of *glasnost* was reflected in the historical narrative of the Soviet space program. The Cosmonautics Day of 1986 witnessed the debut publication in *Izvestia* of a four-part series detailing the inaugural cosmonaut squad, disclosing their identities, and showcasing their complete portraits for the first time. Subsequently, a few years later, the authentic narrative of the Moon Race and the Soviet Union's fervent endeavours to outpace the Americans in reaching the Moon emerged, owing to interviews conducted by Western interviewers with the former chief designer, Vasily Mishin. No adverse consequences befell him; and consequently, others followed suit.[69]

Glasnost reached such an extent that, in August 1989, a televised broadcast featured an extensive meeting of the board of the Ministry of General Machine Building. During this public session, in the presence of correspondents, Yuri Mozzhorin discussed the draft program for the development of space equipment—intended for scientific and national economic purposes spanning the period up to 2000.[70]

Nonetheless, according to certain experts, *glasnost* had a dual impact on the space sector. According to the famous dissident, Yuri Orlov, "...what Gorbachev really should have been afraid of was *glasnost*. By introducing it so self-confidently, he unintentionally broke the network of global, pervasive lies that lay at the heart, at the foundation of this system".[71] In this sense, the effect of *glasnost* in the space sector did not

[69] Harvey, B. (2013). Hidden in plain view. In D. Phelan (Ed.), *Cold war space sleuths. The Untold Secrets of the Soviet Space Program (Space Exploration)*. Springer, 49.

[70] Avduevsky, Leskov. (1990). *Kuda idet sovetskaya kosmonavtika?*

[71] Orlov, Yu. (2008). *Opasnye mysli. Memuary russkoy zhizni*, Zakharov.

differ much from other areas of life in the USSR. On one hand, the adoption of openness and transparency in the space sector played a pivotal role in enhancing the industry's capabilities, rectifying erroneous decisions, and fostering its overall progression. On the other hand, an open discussion not only of achievements but also of the challenges within the space program significantly undermined the industry's image in the eyes of the public. As the country's overall financial situation deteriorated, a growing number of voices emerged in the public press advocating for reduced expenditures on the development of the space industry, deeming it "ineffective" for national welfare. In essence, *perestroika* created ample opportunities for public criticism of the space industry as unduly expensive and draining budgetary resources, with only minimal returns to the civilian economy for the benefit of the general consumer.

Individuals directly engaged in the space sector attributed this unwarranted criticism to the longstanding policy of intense secrecy that had been implemented across all space-related activities. This extreme secrecy ultimately backfired, adversely affecting the space industry itself. Even those of space activities that provided services to the national economy (i.e. meteorological, remote sensing, communication technologies, and Earth observation) had no public support. This circumstance arose from the absence of initiatives dedicated to elucidating to the public the essentiality of conducting and financing such research activities. In the minds of Soviet citizens, space exploration was primarily for military purposes and for the interests of maintaining national prestige. Against the backdrop of change in the political and economic situation at the end of the 1980s, both these *raisons d'être* gradually lost their significance in the eyes of public opinion and, eventually, in the view of Gorbachev himself.

Even according to the scientists, who were keen on securing continued funding for space research, there existed a lack of accurate public understanding regarding the benefits of space exploration for the economic development of the country. As recalled by Leonid Ksanfomaliti, a leading astrophysicist at IKI, the juxtaposition of "significant space achievements with the backdrop of modest living conditions in the USSR fostered the perception that 'the authorities spared no expense for space'. However, in reality, financial allocations were primarily directed towards defence needs, while space exploration, utilising similar defence technologies, constituted only a tip of the iceberg of the overall budget. For several years, deep-space science missions' cost, on a per capita basis, only a few Rubles for each citizen of the USSR".[72]

In her memoirs, Tamara Breus recounted a sentiment prevailing among the general public and numerous influential politicians. "There existed a prevailing belief that space exploration represented an exorbitant drain on national finances, prompting suggestions to eliminate it from the budgetary allocation in favour of bolstering the national economy. The dearth of public awareness contributed to a misunderstanding wherein the costs associated with all reconnaissance launches and manned stations, despite their portrayal as successful endeavours in scientific inquiry, were seemingly lumped into the science expenditure category by ordinary citizens".[73]

The root cause for this lack of awareness was identified as the absence of proactive promotion of space science and, according to space scientists, a limited interest from political leadership in the advancement of national cosmonautics. In 1989, the aforementioned national program for the development of space technology for scientific and national economic purposes presented by Yuri Mozzhorin was released in a succinct summary, primarily within a specialised promotional brochure titled "USSR in Space. 2005", published at the behest of Glavkosmos (see Chap. 9). Limited information about the program also appeared in select newspapers, such as *Krasnaya Zvezda* on August 23 and 25, 1989. As noted by Soviet space experts, the USSR did not publish a single scientific or popular science journal exclusively dedicated to practical issues in cosmonautics. Furthermore, the involvement of Soviet specialists in international forums and conferences on cosmonautics was still limited, with the USSR's contribution typically constituting only a small percentage of the total number of papers presented.[74]

The incapacity to effectively convey to the public the advantages and significance of space research for the development of the national economy emerged as a notable deficiency in space policy during the *perestroika* period. A particular inconsistency of the reformists is noteworthy in this context. On one hand, the political narrative of *perestroika* accentuated the material well-being of the population, creating an inverse relationship with the emphasis on preserving national pride in space achievements, as per

[72] Ksanfomaliti, L. (2010). Vspominaya 1990-e... In A. Pevzner (Ed.), *K 45-letiyu Instituta kosmicheskikh issledovaniy RAN. Obratnyy otschet... 2*, IKI RAN, 152–153.

[73] Breus. (2002), 156.

[74] Avduevsky, Leskov. (1990). *Kuda idet sovetskaya kosmonavtika?*

Soviet standards.[75] On the other hand, the potential of space research for meeting the demands of the civilian economy, thereby targeting the enhancement of the population's economic well-being, remained under-utilised during that era and had not garnered maximum publicity.

Glasnost proved to be a double-edged sword for the Soviet space sector. Open criticism from democratic politicians and the public was echoed by advocates within the field. Simultaneously, there existed a stark contrast between the content of their criticism and their respective objectives. While the former expressed dissatisfaction with government spending on the space sector's development, the latter aimed to draw attention to the critical state of cosmonautics and initiate a search for solutions to overcome the crisis. At the end of the *perestroika* period, Soviet specialists made no secret of the real state of the space field. They bitterly admitted that the United States was ahead of the Soviet Union, not only in the implementation of manned expeditions to the Moon and on the surface of the Moon, but also in the voyages to distant planets by automatic devices. As Sagdeev put it,

> The most we could do, with what might be given to us by our industry, was to reach and land on Venus, and survive for a while in an extremely hot and hostile environment. But to undertake trips to distant planets would require something different. The technology had to be able to work for a very long time, readapting to constantly varying conditions. Such a long survival time would require a different technological culture. It would require special dedication, which on top of it all would also cost a lot.[76]

In addition, the growth of the new-generation super-heavy launch vehicle Energya and the Buran orbital spaceplane, as well as the vast ground-based infrastructure of the Buran-Energya project, took much more time to complete in the USSR than in the USA. Moreover, it was no longer a secret to anyone that the Soviet Union at the end of the 1980s was inferior to the USA, Western Europe, and Japan in terms of digital electronics technologies, information processing tools, microminiaturisation of equipment and devices and other specialty fields. In turn, this affected the technical characteristics of radio-electronic equipment for space vehicles and orbital stations. Therefore, on-board systems suffered

[75] See the interview with V. Avduevsky (1989). Kosmos. Dayet li on pribyl. *Ogonyok*, 24 (3229).

[76] Sagdeev. (1994), 237.

from insufficient reliability and required redundancy which led to an inevitable increase in their production. Even this method did not always provide sufficient operational lifespans for space systems. For example, Soviet communications satellites operated for only 3–5 years, less than 50% of the lifespan of similar American satellites. As a result, the USSR was forced to launch 110–115 spacecraft annually, while the United States launched five or more times fewer. At the same time in 1990, according to the free press, there were 180–220 American operational satellites in orbit and only 150–180 Soviet satellites.[77]

The state of the space sector inevitably drew the attention of Gorbachev's political adversaries. Thus, Boris Yeltsin, who at the time served as a deputy in the Congress of People's Deputies, demonstrated awareness of public sentiment and addressed the issue of the future development of the space sector. He argued that: "At present, space programs are having a difficult time. There is need to shelve them for 5–7 years. This is a serious matter. In addition to nuclear-phobia, there has emerged a space-phobia. It is not just a question of economic difficulties. The fact is that there is no openness as far as spending on particular projects is concerned. In my election platform, I urge the re-establishment of the popular belief in space science and exploration through truthful information. I will press for the publication of the space budget and spending in particular projects and activities. Nothing short of open debates will convince the public of the need of such work. I am positive that we have already created it. But a number of projects may seem to be before their time. For instance, there appears to be nothing at this point to justify the outlays for reusable spacecraft Buran".[78] One cannot disregard the fact that Yeltsin criticised aspects that were intended for reform during *perestroika*, including budget coverage, civil space development, and transparency in all space sector activities. The failure to fulfil these promises enabled political opponents of the *perestroika* agenda, both from the liberal-radical and conservative opposition, to assume the role of advocates for the sector's interests.

The barrage of often non-constructive criticism directed towards the space field, originating from both external and internal sources, inevitably impacted the moral and material well-being of its participants. While we

[77] Avduevsky, V., Senkevich, V. (1990). Sovetskaya kosmonavtika: dostizheniya i perspektivy. *Zemlya i Vselennaya*, 4.

[78] APN, Phobos Loss – Spacecraft Designers Blamed. (1989). *Spaceflight*, July 31. In NASA HQ Archives.

have previously touched upon the latter aspect, it is pertinent to delve further into the former. When contemplating Soviet cosmonautics and its moral and emotional dimensions, which served as both a unifying force within society and a factor in identity formation, we must acknowledge the blow dealt to this very component during the era of *perestroika*. A particularly telling example in this regard is the criticism directed at the IBMP on an issue that seemed entirely unexpected at the time.

Glasnost and IBMP

In the years of late *perestroika*, when the entire Soviet space sector was subjected to spontaneous criticism for the absorption of financial resources and low returns, the IBMP could not avoid disclosure and blame for the secret pages of the past. In October 1991, the popular Soviet magazine *Ogonyok* (Spark) published an article written by journalist Nina Chugunova[79] dedicated to the "poster guys", secret participants in medical experiments conducted within the walls of the Institute. The article was based on interviews taken by the author with test participants and their wives. Despite the voluntary and commercial basis of the tests, the author blamed these scientists for inhumane treatment of the test subjects, disrespect for their rights, their social insecurity, and careless attitude towards "spent human material".

As Chugunova stated, space medicine, in contrast to disciplines studying the consequences of the influence of hostile environments like the sea on human beings, endeavoured to anticipate the challenges posed by a new element humanity had to confront—space. Consequently, the primary objective of space medicine became the exploration of the limits of human tolerance. Beyond the extreme stresses of take-off and landing, a myriad of issues emerged concerning the extended duration of human presence in space. According to Chugunova, "the Western powers conducting research in the field of space medicine did not have at their disposal scientific and military centres of professional subjects". Meanwhile, in the USSR, "there were always such groups, the most numerous of them could be seen in the 1970s. These were the most disenfranchised conquerors of space". In the 1970s in Moscow, within a "paramilitary scientific institution under the general designation of the Institute of Biomedical

[79] Chugunova, N. (1991). Bez ogranicheniy k voprosu ob eksperimentakh nad zhivymi lyud'mi. *Ogonyok*, 43–44.

Problems, dedicated to addressing the challenges of human habitation in space, there existed a group of professional subjects who were officially listed as mechanics or laboratory assistants, receiving a meagre salary even by the standards of that time. In the records of 'tests', they were designated as volunteers".

As Chugunova further asserted, "not a single space or military power (with the exception of states with a totalitarian socialist regime) has employed its citizens as test subjects for military, medical, and other purposes. These individuals, whose official responsibilities solely involved exposing themselves to hostile environments or harmful factors in exchange for monetary compensation—their only source of income, which had to support both themselves and their families—dedicated significant time and effort to these activities. As a result, the pursuit of alternative forms of self-expression, personal development, and creativity became nearly impossible for them".

This article did not go unnoticed in the West. *The European* newspaper published a material with a brief retelling of Chugunova's article and the author's comments. The latter drew attention to the fact that the article was published "at a time when the Soviet program lost much of its glory and was under threat from chronic shortage of money". Moreover, the author added that "there has been no reaction to the allegations from authorities" and "from public hardened by revelations of past abuses and more concerned with the day-to-day problems of survival".[80]

Hence, the very myth propagated by the Soviet official narrative concerning the pioneers of space, the heroism of cosmonauts,[81] and the selfless dedication of scientists to the collective efforts came under scrutiny, eroding the foundational legitimacy of cosmonautics. Formerly revered as heroes, those involved in space endeavours were now portrayed either as victims or as perpetrators, leading to the displacement of the mythical superstructure on a more grounded base. Such a societal trend was frequently not embraced by individuals directly engaged in both research and production endeavours. Among the myriads of reasons for this divergence were attitudinal inertia, preservation of professional identity, adherence to specific values, ideological convictions, and more. However, changing times have presented the space sector with new challenges that must be

[80] Conradi, P. (1991). Untitled. *The European*, 15 November.

[81] See more in Gerovitch S. (2015). *Soviet Space Mythologies: Public Images, Private Memories, and the Making of a Cultural Identity*, University of Pittsburgh Press.

addressed. Among these challenges was the opportunity to engage with both domestic and international markets. The introduction of market-oriented elements—the central focus of the *perestroika* era—into the space sector exposed its internal tensions.

Market In Space: Pros And Cons

The second half of the 1980s was a period when various hitherto classified research centres, institutes, and space equipment manufacturers were looking for ways to enter the international market. However, each institution responded uniquely to the opportunities and the new reality, which starkly differed from what generations of scientists and engineers had grown accustomed to. One of the objectives established by the front line was to create specialised departments to engage with foreign partners and promote internationally the services and products of their institutions. The leadership of these enterprises and their designated managers perceived the new conditions as a challenge and promptly responded to them.

To promote its international and commercial activities in 1987, the Foton Design Bureau was established at the TsSKB Progress in Kuybyshev. This small structural unit was responsible for the international cooperation of the still closed space enterprise. One year later, in 1988, within the framework of the IBMP the Department for the Implementation and Promotion of Scientific Achievements was established under the leadership of Mark Belakovsky. One of the main competences of the department was (and still is) international cooperation.

Belakovsky recalls that at that time, there was no such concept as "commercialisation". The director of the Institute, Anatoly Grigoriev, set before him both a simple and difficult task: to earn money.[82] The competence of the new Department included the conclusion of contracts on a commercial basis, the introduction into medicine of the achievements and results of the scientific activities of the IBMP, informing the public about the activities of the Institute, as well as the development and implementation of international projects on earth and in space. Despite the substantial overlap in the spheres of operation between the Department of IBMP and Glavkosmos, established three years prior to the foundation of Belakovsky's Department, both organisations did not engage in cooperation and lacked shared points of contact. In reality, the IBMP was a component of the 3rd

[82] Interview with M. Belakovsky, February 2, 2023.

Main Directorate under the USSR Ministry of Health. The 3rd Main Directorate, serving as the central administrative body for the Ministry of Health of the USSR, operated as a classified structure in the Soviet Union. Its primary objective was to conduct research associated with life support for crews during space flights, as well as the prevention and treatment of personnel from enterprises within the Ministry of Medium Machine Building of the USSR. Additionally, it was also responsible for addressing the health needs of employees in research institutes linked to nuclear topics and the development of new types of nuclear weapons.

According to Belakovsky, the international cooperation, led by the IBMP under his guidance, brought in an additional 15–20% of funding to the Institute's budget. The experiments were commissioned both by the European Space Agency, with corresponding payments from the European side, and through direct funding in cases where the Soviet side was also interested in conducting specific experiments. Belakovsky highlighted the mutual interest of both parties in enhancing the quality of experiments, utilising state-of-the-art European equipment, and introducing new and innovative ideas into collaborative projects.[83] Evgeny Ilyin, who was responsible within IBMP for overseeing international cooperation on the Bion project, also harboured positive recollections of the final years of *perestroika*. He reminisced about the turn of the 1980s and 1990s in the following manner: "We did not feel any sudden changes because there were such strong, stable, and trusting relations between us and TsSKB, the manufacturer of Bion capsule. As the program on Bion-10 was planned, so it went. We did not feel any difficulties [organisationally—O.D.]. As it was planned, so it proceeded".[84]

Institutions and individuals adapted to the changes brought about by *perestroika* in accordance with their capacities, skills, and circumstances. For instance, as per Ilyin's recollections, medical and biological scientists from IBMP faced challenges in securing positions in foreign research institutes, which in his opinion somewhat compelled them to remain in the USSR and explore opportunities to sustain their professional pursuits. "The financial collapse resulted in a decline in currency value, leading to reduced wages. Many individuals ventured into other business, and some chose to relocate abroad. Within our Institute, some specialists departed, but the core team was preserved. Enthusiastic and dedicated individuals,

[83] Ibid.
[84] Interview with E. Ilyn, 13 March, 2023.

passionate about their work, continued their efforts. In my team, most members remained intact, with minimal loss of personnel. The impact varied widely across disciplines; physicists, being in higher demand, found opportunities in commercial firms, while biologists and medics persisted in their positions. Although funding constraints affected scientific equipment acquisition and travel, the Institute continued its operations, and experiments were conducted at the Mir station".[85]

A similar situation occurred at another scientific centre, the IKI, where several researchers relocated abroad and continued their work in foreign scientific institutions. As Genrikh Avanesov, leading IKI engineer and head of the Optical and Physical Research Department (1988–2002), recalls, "during the *perestroika* out of 1500 employees working in the late 1980s, 20–30 leading specialists emigrated abroad, while others moved into private business. In general, by the end of *perestroika* and the beginning of the 1990s, about 900 people remained".[86] This figure aligns with the data provided by Aleksandr Zakharov, Scientific Secretary of the Scientific Council on Solar System Physics of the Russian Academy of Sciences: "During the years of *perestroika*, the total number of the Institute's staff decreased by more than a third".[87] Zakharov, however, underscored the significance of the brain-drain that had subsequently transpired: "The Institute emerged as a benefactor for several universities and scientific organisations in Western Europe, the USA, and even Brazil. This outcome was a consequence of *perestroika*, as many individuals chose to pursue work abroad for various reasons".[88]

In this regard, Leonid Gurvits adds that "the issue was not about quantity or numbers, but about the 'quality' of those who left IKI—whether temporarily or permanently". The researchers, who represented the pride of Soviet space science, numbered at about 300 at IKI. Consequently, the emigration of even 30 of them had a significant impact on the future development of science in the USSR. These individuals were typically internationally renowned scientists who had received personal invitations to prestigious positions in Western scientific institutions. Their positions in the late Soviet Union were often filled by less accomplished individuals.

[85] Ibid.

[86] Interview with G. Avanesov, April 15, 2023.

[87] Zakharov, A. (2006). Vmesto predisloviya. In A. Pevzner (Ed.), *Obratny otschet vremeni*. IKI RAN, 10.

[88] Ibid.

While the numerical composition of IKI's scientific staff remained largely unchanged, its overall professional potential experienced a substantial decline. Furthermore, he emphasises that it is crucial to recognise that "*perestroika* worked only to half of its potential: it opened the borders for specialists to go abroad, but it did not facilitate a brain-gain from abroad. Only a very small number of foreign scientists came to work in the USSR Such a 'two-way' flow, typical of scientific establishments in Western countries, leads to a dynamic balance of 'brain flow,' which in turn amplifies scientific activities".[89]

Other former scientists and engineers were compelled to enter the business sphere or join newly established cooperatives. Avanesov argued that: "people left not because they were tired of working here, but because they needed a salary. Especially young people needed to feed their families. For example, my postgraduate student left to trade in timber. But the backbone of the Institute remained. I managed to keep the department, which I headed, thanks to the design and production of stellar sensors, which were ordered and paid for by the space industry. Thanks to the fact that we already had a lot of experience, we managed to find our own niche and survive".[90]

According to Avanesov, the Institute was also preserved, in part, due to the presence of 250–300 researchers with academic degrees (PhD and doctors) working at IKI in the late 1980s. On one hand, their high level of academic background and adherence to specific values (as Avanesov put it, "excitement and interest characterise a very limited circle. Only the top layer boils"), and on the other hand, their specialised scientific expertise limited the range of options available to them. In this context, NPO Lavochkin, the main manufacturer of scientific spacecraft (Phobos, Astron, Granat, etc.), experienced more pronounced difficulties, primarily facing the departure of its workforce as a primary consequence of *perestroika*.

> The personnel at NPO Lavochkin were highly skilled professionals. However, with the onset of *perestroika*, individuals began leaving due to inadequate salaries. This departure initially affected workers, leading to a labour shortage. Many opted to join cooperatives, where the compensation was more lucrative than working at the enterprise. Given the unique nature of their work, only a few people possessed the required skills. As staff left,

[89] Interview with L. Gurvits, December 20, 2024.
[90] Interview with G. Avanesov, April 15, 2023.

> gaps emerged, and in the realm of space technologies there was a crucial need for knowledge transfer from the previous generation. The onset of the cooperative movement in the 1980s and the subsequent exodus of personnel exacerbated this issue, as there were fewer individuals available to pass on their expertise. Additionally, the location of the NPO Lavochkin on the outskirts of Moscow (Khimki), beyond the ring road, compounded the problem, as the metro had not yet reached that area, and transportation was limited to buses. This geographical constraint deterred individuals from Moscow, and recruiting from the small town of Khimki alone proved impractical. As a result, NPO Lavochkin experienced a severe staffing shortage, a challenge common to many enterprises in the space sector.[91]

The very transition to market relations within the space sector was perceived differently by its direct participants. For instance, according to Konstantin Feoktistov, a decisive shift to market relations was the necessary way out of the crisis. He argued that the "situation, where an enterprise within the defence industrial complex received state funding irrespective of the quality of its products, was untenable". Feoktistov believed that the space industry should undergo further privatisation. In his view: "it was essential to separate the customer from the executor, freeing the latter from the administrative constraints imposed by a powerful agency". He suggested that the "military should independently order and purchase rockets, and firms, while benefiting from such orders, would have to compete for them". Feoktistov emphasised that privatisation should extend to industrial enterprises in the space industry and the military and space complex at large.

However, he expressed concerns that, through various manipulations, enterprises might become the private property of their existing managers, among whom, he believed, there were: "quite a few dishonest people". Some months before the Soviet Union's collapse Feoktistov hoped that, with proper management and a transition to a market economy, cosmonautics had the potential to prevent the collapse of the USSR. He believed that "the country was in a challenging situation and that cosmonautics could have played a role in overcoming it if it had been rid of unnecessary showmanship, excessive militarisation, and put on a more business-oriented foundation".[92]

[91] Ibid.

[92] Feoktistov. (1991).

Nonetheless, not all participants in the Soviet space program were positive about the changes taking place within the space sector. Some of them expressed serious doubts about the advisability of introducing market laws into such a complex and directly dependent on government policy field. As Avanesov recalled:

> The shift to market relations disrupted many established practices. The reliance on Gosplan[93] to oversee all the interconnected chains was deeply ingrained. With the absence of Gosplan, individuals began resorting to personal connections. The foundation of cooperation had been established during the Gosplan era, and people were unwilling to abandon this system. These enterprises had individuals dedicated to their work, fostering cooperation through personal connections. I had such a horizontal network for several decades. At other enterprises, there were either peers, classmates, or good acquaintances. Corruption schemes were not prevalent, as there was a mutual understanding that helping one another would be reciprocated. This collaborative approach was actively employed. However, when dealing with a new enterprise, uncertainty prevailed. Previous allies provided a known set of advantages and disadvantages, along with a clear understanding of their capabilities. Integrating a new enterprise into cooperation took decades, as it required significant time to develop a shared language and understanding.[94]

The discussion of this topic will continue below with the specific example of the Phobos science mission. At this point, however, it is worth summarising some preliminary results.

Conclusions

The restructuring of the financial and economic system in the USSR had a significant impact on the space sector, which was profoundly affected by the liberalisation of the economy and its transition to a market-based system. Developments such as the decentralisation of management, the emergence of new research institutes and industrial centres with a gradual shift towards self-financing, efforts to enter the international space market,

[93] The State Planning Committee, commonly known as Gosplan, served as the agency responsible for centralised economic planning in the Soviet Union. Established in 1921 and persisting until the dissolution of the Soviet Union in 1991, Gosplan was primarily tasked with the formulation and execution of a series of five-year plans that governed the economic activities of the USSR.

[94] Interview with G. Avanesov, April 29, 2023.

and the conversion and commercialisation of space production brought substantial changes to the sector. Previously regarded as a stronghold of national security—and therefore among the least susceptible to reforms and external influence—the space industry underwent a transformation as a result of these shifts.

Glasnost brought to the space sector both positive developments (open discussion of urgent problems and the search for their constructive solutions) and excessive criticism from politicians, public opinion, and individuals involved in the space sector. Lacking a clear understanding of the benefits that space research contributed to technological and economic development, public opinion rendered a harsh judgment on cosmonautics, thereby undermining its legitimacy.

Cosmonautics emerged as a domain wherein the countless contradictions of *perestroika* reforms were prominently mirrored. The declassification of the sector led to the expansion of scientific and commercial engagements while simultaneously causing a brain-drain and the erosion of established industrial linkages. Initially envisioned as a means to diversify production and promote closer integration between military and civilian economies, conversion failed to achieve its primary objective and instead weakened existing capabilities. *Glasnost* laid bare the unseemly facets of cosmonautics, stripping away its somewhat mythical veneer. The inclination to adhere to the principles of a market economy by the late 1980s proved premature, as the internal mechanisms of the space industry were yet to fully embrace market dynamics. Consequently, a situation emerged wherein the nascent market laws were not *yet* operational, while the traditional command-administrative system, characteristic of the previous era, was *already* losing efficacy. Nevertheless, the principal achievement of *perestroika* in the space realm lay in its preparation for the transition to new conditions that were already taking shape in the 1990s. This preparatory phase unfolded both in terms of public perceptions—through open press discussions regarding the space field's status and future—and in practical terms, including the experimentation with new relationships, practices, and dynamics.

PART II

ESA-Soviet Collaboration in Space

1.1 Introduction

In addition to the expressions previously discussed—such as *uskoreniye*, *perestroika*, *glasnost*, and the *new thinking policy*—which entered not only Soviet but also global political vocabulary, the international community was introduced to Gorbachev's new concept of the "common European home". This slogan encapsulated several principles that Gorbachev aimed to implement in his foreign policy. Chief among these was the acknowledgment of the cultural and historical interconnectedness of European nations, alongside the feasibility of not only peaceful coexistence between diverse political regimes but also the necessity for close cooperation, shared perspectives, and mutual interests.[1]

As part of this approach, the USSR demonstrated a heightened interest in the evolving processes of European integration and the dynamics of Europe's economic and political development. This "turn towards Europe" was further reflected in the establishment of the Institute of Europe within the Academy of Sciences in Moscow. The Soviet political leadership exhibited a reduced scepticism towards the emerging European Community and an increased willingness to engage with pan-European structures.[2]

[1] See more in Gorbachev M. (1987). *Perestroika i novoye myshleniye dlya nashey strany i dlya vsego mira*. Politizdat, 199–219.

[2] See more in Maresceau, M. (1992). The European Community, Eastern Europe and the USSR. In Redmond, J. (eds) *The External Relations of the European Community*. Palgrave Macmillan; Rey, M. P. (2004). Europe is our Common Home': A study of Gorbachev's

These new political trends had a significant impact on the field of space exploration. There was extensive cooperation with CNES, as well as some contacts and collaboration with Swedish, British, Italian, and Finnish space research organisations. Taking into account the already existing literature and historical sources, the focus of our study will be on ESA-Soviet relations. While initial ESA-Soviet contacts can be traced back to the 1970s, the ascent of Gorbachev to power and the introduction of a more open foreign policy provided these relations with renewed momentum. During *perestroika*, Soviet institutions intensified their focus on ESA activities and adopted a more proactive approach to fostering closer ties, particularly in the domain of scientific research. Despite the complexities and lengthy processes associated with establishing connections and implementing projects, efforts to build strong and enduring relationships eventually began to bear fruit. Over time, as the economic situation in the Soviet Union deteriorated but the international context improved, commercial projects within the framework of cooperation between ESA and Soviet space organisations also started to take shape.

Gorbachev's "turn towards Europe" in foreign policy coincided with a pivotal phase in the internal development of both the European and Soviet space programs. The accumulated experience in scientific research, coupled with the highly constrained capabilities of the space transportation system—particularly as Ariane launches had only recently begun and the reliance on American launch services remained unresolved—compelled the ESA leadership to explore new approaches for advancing their scientific agenda.

Simultaneously, Soviet space science, after decades of active development under conditions of limited contact with Western advanced scientific centres, found itself in dire need of "fresh air", innovative ideas, and, most critically, access to modern technologies. Expanding scientific cooperation became a strategic tool for the Soviet establishment, offering a potential pathway out of the economic crisis that, by the end of *perestroika*, had affected all sectors, including space science. Consequently, the latter half

diplomatic concept. *Cold War History*, *4*(2), 33–65.; Rey, M. P. (2012). Gorbachev's New Thinking and Europe, 1985–1989. In *Europe and the End of the Cold War*. Routledge; English, R. D., Svyatets, E. (2014). Soviet elites and European integration: from Stalin to Gorbachev. *Revue Européenne d'histoire*, 21 (2), 219–233.

of the 1980s marked the convergence of several significant trends. However, this alignment did not preclude the emergence of numerous challenges and obstacles that the political and scientific leadership had to address.

CHAPTER 3

ESA-USSR Relations Before Perestroika

This chapter examines the cooperation between ESA and Soviet space organisations during the period leading up to *perestroika*. Formally established in 1971, this relationship gave rise to a series of scientific exchanges that paved the way for more active collaboration in the years that followed. After reconstructing the history of joint projects implemented between 1971 and 1985, we will analyse the factors that contributed to the rapprochement between the USSR and ESA, the characteristics of its development, and the key influences that shaped this collaboration. Special attention will be given to the international Halley's Comet projects, which were initiated during the pre-perestroika era and whose implementation coincided with the early years of Gorbachev's reforms.

Formalisation of Contacts

On February 12, 1971, the first agreement between the Academy of Sciences of the Soviet Union and the European Space Research Organisation (ESRO), the predecessor of ESA, came into effect. This agreement provided a framework for further discussions and cooperation and served to oversee some activities of scientific collaboration that had previously remained rare and sporadic. The Agreement essentially covered three aspects of collaboration: the exchange of scientific and technical

O. Dubrovina, *The Soviet Space Program and International Collaboration during Perestroika, 1985–1991*, Palgrave Studies in the History of Science and Technology,
https://doi.org/10.1007/978-3-032-03157-0_3

information published without restrictions on use; the coordination or joint participation in the execution of programs of mutual interest—an agreement had to be concluded in each specific case—and finally, the organisation of exchange visits of specialists aiming towards mutual familiarisation with scientific work in progress.[1]

However, the initial years following the conclusion of the Agreement were not characterised by intensive cooperation. As the ESA Head of International Affairs Branch, Jean Arets put it: "Even if the cooperation of the first period was cordial, it was devoid of any concrete effect". He argued that the Soviets expressed "a very reserved attitude with regard to possible cooperation".[2] Twelve years later he admitted that the blame for the sluggish cooperation also fell on the European side which rejected "ambitious projects of cooperation with the USSR".[3]

In internal correspondence, ESA staff acknowledged that the Soviets had expressed interest in cooperating with ESA but had not received a positive response. This is further evidenced by the efforts of the Institute of Biomedical Problems to extend invitations to the European side for collaboration in conducting experiments on biological satellites (see Chap. 7). As a result of a series of unsuccessful attempts to establish cooperation in the field of scientific experiments in the first years after the signing of the Agreement, the Soviets had accumulated "a certain amount of mistrust concerning the seriousness of our [ESA–O.D.] intentions".[4] However, the distrust was mutual. European counterparts harboured doubts regarding the extent of their Soviet colleagues' knowledge about the internal affairs of ESA, suspecting the involvement of Soviet intelligence services. They were surprised by the depth of knowledge demonstrated by the Soviets—an intricacy of information that they could not plausibly have obtained from the free press.

Mikhail Mogilevsky, a leading researcher at the Space Plasma Physics Department of IKI, attributes the slow development of the scientific ties between ESRO and the USSR to two specific circumstances: the scientific unattractiveness of ESRO (as ESA officials themselves would later attest) and the primary focus of Soviet space researchers on CNES and NASA

[1] *Relations with U.S.S.R.*, January 1987. In ESA-28130, HAEU.

[2] Arets, J. *Relations avec l'URSS*, 2 décembre 1975. In ESA-4815, HAEU.

[3] Arets, J. *Prospects of cooperation with U.S.S.R.*, 24 Mars 1987. In ESA-28130, HAEU.

[4] Trendelenburg, E.A. *Report D/Sci in the Moscow mission*, 22 March 1976. In ESA-4815, HAEU.

during this period. According to Mogilevsky, "Firstly, ESRO was a 'paper' structure and had no formalised scientific and technical part that could be used for joint projects. Secondly, at that time Europe had the third largest space structure - CNES. The agreement between the USSR and France had been concluded in 1966 and by the 1970s serious joint work was underway". At the same time, Mogilevsky emphasises the attractiveness of the French partner due to the advantageous position it occupied: "CNES had the opportunity to compare the two approaches to the organisation of the space industry - the USSR and the USA. At the initial stage, the Soviet approach was based on attracting bright individuals and minimal formalism, i.e. low level of bureaucracy. In contrast, the USA approach was very formalised and required a large number of different documents detailing everything. (Later on, there was a convergence of these approaches). The French had the opportunity not to repeat the mistakes of the pioneers and to build their own system on the basis of the positive experience of the Soviet Union and the USA". This led, according to Mogilevsky, to the fact that in the mid- to late 1970s France already "had a well-equipped centre for pre-flight tests (in Toulouse), a delivery vehicle (the Ariane rocket), the necessary services (navigation, telecommunications, data processing, etc.), as well as a serious space industry. There was also cooperation between CNES and French institutes operating under the national scientific agency CNRS. The Soviet-French cooperation was formalised organisationally (regular meetings), which made it possible to carry out high-tech and original space projects (Arcad-1, 1971; SRET 1972; Arcad-2, 1973, Ipocamp 1974; Araks, 1975; Venera-9 and 10, 1975, Arcad-3, 1981; Vega, 1986; Interball, 1995–1996, etc.)".[5] Additionally, the first non-Intercosmos cosmonaut aboard Soyuz was French (Jean-Loup Chrétien, Soyuz T-6, 1982).

With the formation of ESA in 1975, cooperation was strengthened when the leadership of the European agency decided to make attempts to "hold out a hand in the direction of Moscow" since "there appear to be signs that the Russians would not be averse to renewing their contact with us".[6] Arets acknowledged that the primary reasons for ESA's interest in seeking closer ties with the Soviet space sector were the diversification of the launch vehicle market and the desire "not to be left behind during the

[5] Correspondence with M. Mogilevsky, April 24, 2024.

[6] Gibson, R. *Relations with Soviet Union*, June 12, 1975. In ESA-4815, HAEU.

development of the post-shuttle program".[7] This was probably one of the best times to commence a new joint effort, considering that the recently completed Soyuz-Apollo project demonstrated the readiness and ability of the Soviets for scientific and political cooperation in space. Although simultaneously, this "handshake in space" initiated a new phase of gradual escalation of disputes between the USA and the Soviet Union, coinciding with Richard Nixon's resignation from the Presidency in 1974.

Despite the expressed interest in cooperation, over the next few years no real steps were taken towards rapprochement by both sides. The deterioration of Soviet-American relations inevitably had an impact. The leadership of the European Space Agency received confirmation of the cooling of Soviet-American relations in space, with initial signals emanating from their Soviet counterparts. Specifically, Vladlen Vereshchetin, Deputy Director of Intercosmos, hinted to his European counterpart at the gradual decline in cooperation with the Americans: "Americans are becoming less and less open".[8] Similar sentiments were also expressed by American sources, accusing the Soviets of pursuing "only a cosmetic activity" while NASA "had suffered considerable criticism because of the negligible scientific objectives included in the Apollo-Soyuz mission".[9]

Towards the end of the 1970s, there was a resurgence in Soviet-European cooperation, primarily driven by ESA's interest in utilising Soviet rockets to deploy European satellites into orbit and to engage in various scientific research endeavours. Conversely, according to ESA officials, the Soviets were motivated by the imperative to enhance their expertise in cutting-edge technology. That was, for example, the explanation of the Soviet readiness for cooperation in the Very Long Baseline Interferometry (VLBI) field given by E. Peytremann, the Head of Future Scientific Programs Department at ESA and promotor of European-Soviet collaboration in the VLBI area.[10] Arets gave another two explanations. The first concerned the increased role of ESA in space research and consequently, as he argued, this appealed greatly. The second concerned the cooperation with the United States that "was not very fruitful (for

[7] Arets, J. *Relations avec l'URSS*, 2 décembre 1975. In ESA-4815, HAEU.

[8] Bourély, M. *Congrés de la FAI*, 26 octobre 1976. In ESA-4815, HAEU.

[9] Mellors, W.J. *NASA/U.S.S.R. Relations*, 2 November 1976. In ESA-4815, HAEU.

[10] Peytremann, E. *Possible cooperation with USSR on VLBI via satellite*, September 12, 1979. In ESA-4815, HAEU.

Soviets—O.D.) and the snags in the policy of *détente* do not allow the USSR to hope for development".[11]

First Practical Steps

Relations between ESA and the Soviet space institutions were guaranteed primarily through communication with the embassy of the Soviet Union in Paris. Then, during a visit to ESA HQ in Paris in February 1979 by M. Elchin, the first secretary of the Soviet Embassy, the areas of possible scientific cooperation were identified. The Soviet Union was attracted to European developments in the field of solar energy, while ESA sought to obtain the results of Soviet experiments in the field of space medicine and life sciences.[12] Contacts also took place during international conferences directly between ESA and representatives from IKI and Intercosmos, two Soviet entities responsible for international collaboration in scientific exploration of space. The cosmonauts also did not shy away from cooperation with the Europeans: in June 1979 two Soviet cosmonauts, Aleksandr Ivanchenkov and Vladimir Kovalyonok, visited ESA.

The name Intercosmos referred both to the Soviet institute for international cooperation in space and to the program under which this cooperation was conducted. On May 30, 1966, the Council for International Cooperation in the Exploration and Peaceful Uses of Outer Space, commonly known as the Intercosmos Council, was established by the USSR Council of Ministers under the auspices of the Soviet Academy of Sciences. One of its main functions was the coordination of international cooperation in space exploration. Its mandate was also to coordinate joint efforts in space research and applications among various ministries, departments, scientific institutions, and industrial organisations. These included the USSR Academy of Sciences, the Ministry of Public Health, the Ministry of Communications, and the State Committee for Hydrometeorology and Environmental Control. The tasks of the Council also included participation in drawing up existing and prospective plans for cooperation in the study and development of space with foreign countries, familiarising the countries participating in the cooperation with the capabilities of Soviet space rocket technology, assisting in establishing business contacts and

[11] Arets, J. *Commentaires sur le rapport de la visite en URSS du DG*, 10 octobre 1979. In ESA-4815, HAEU.

[12] Arets, J. *Visite de Elchine, M.* In ESA-4815, HAEU.

links between scientific and industrial organisations of the Soviet Union and other countries.[13] From its inception, Intercosmos was led by Academician Boris Petrov. In 1981, Academician Vladimir Kotelnikov succeeded him as Chairman of the Council. It is important to distinguish the Intercosmos Council, established in 1966, from the Intercosmos Program, which was launched 1 year later.

On April 13, 1967, in Moscow, representatives from nine socialist countries (GDR, Bulgaria, Cuba, Czechoslovakia, Hungary, Mongolia, Poland, Romania, and the USSR) convened to adopt a joint program of cooperation in outer space exploration for peaceful purposes, named Intercosmos. The objective of this international program was the development of spacecraft equipped with scientific instruments, contributed to by the participating nations. The Soviet Union provided its spacecraft and launch vehicles for this endeavour at no cost. All program participants were granted access to the results of the conducted experiments. Research spanned various fields including space physics, astronomy, meteorology, communication technology, biology, and medicine. Each participating country also established a national coordinating body, tasked with implementing the program. In the Soviet Union, this entity was the Intercosmos Council within the USSR Academy of Sciences. The inaugural satellite developed under this international collaboration was Intercosmos-1, launched on October 14, 1969. In total, 25 Intercosmos satellites and 11 Vertical sounding rockets were built and launched until 1991.[14] Intercosmos activities were not limited to the socialist bloc but also extended to capitalist countries, with joint projects coordinated under the leadership of the Intercosmos Council.

As previously mentioned, IKI also played an active role in establishing and fostering relationships with foreign partners. Lev Zelenyi recalled the Institute's role in the development of international cooperation activities: "International activities were coordinated by the Intercosmos Council under the Academy of Sciences. IKI was not formally involved in this, although it was the Institute that did most of the work. All international missions of IKI were part of the Intercosmos program. Intercosmos

[13] Istoricheskaya spravka o Sovete RAN po kosmosu. Retrieved February 6, 2025, from http://sovet.cosmos.ru/sites/default/files/history/2_14.pdf.

[14] *Iz Istorii Mezhdunarodnogo Sotrudnichestva V Sfere Osvoyeniya Kosmosa*. Russian State Archive of Scientific and Technical Documentation. Retrieved February 6, 2025, from rgantd.ru/news/pamyatnye-daty/iz-istorii-mezhdunarodnogo-sotrudnichestva-v-sfere-osvoeniya-kosmosa/

received funds for project implementation from the MOM, and from Intercosmos these funds came to us. It was prestigious to work under the authority of the Academy of Sciences, so nobody refused from this scheme".[15] Indeed, Sagdeev played a highly active role in initiating and sustaining international scientific collaborations. It is not incidental that Roger Bonnet, Director (1983–2000) of ESA's Science Program, refers to him and all subsequent IKI directors as "space makers".[16] It was Sagdeev himself who took the initiative to strengthen the ties between IKI and the European Space Agency.

In 1978, unexpectedly for the ESA staff, Roald Sagdeev, then Director of IKI, invited Ernst Trendelenburg, thenDirector of ESA's Science Program, to the Moscow Institute as an honoured guest, without explaining the specific reasons for the visit. A positive opinion was expressed in ESA regarding the future trip, as it reflected the general trend in the Agency towards rapprochement with the Soviet Union and the establishment of closer ties compared to the last "somnolent"[17] years. Subsequently, in the same year, 1978, Roy Gibson, the inaugural Director General of ESA, was approached by Boris Petrov, Head of Intercosmos, Vladlen Vereshchetin and Roald Sagdeev, who demonstrated interest in the nature of European-American cooperation and entertained the "wild idea" that in the future there could potentially be a rendezvous between an ESA unmanned space station and a Soviet spacecraft. Gibson concluded: "there were sufficient indices of interest and approval for our activities to make me think that the Russians are considering some kind of re-opening of our negotiations".[18] According to ESA officials, the willingness of the Soviets to engage in cooperation and the endorsement of closer ties with ESA at the highest levels of authority were evidently discernible.

In the late 1970s, both parties discussed the possibilities of agreements in such fields as biomedicine, space transportation systems, manned space programs, remote sensing, and cooperation on VLBI via satellite. The first ESA-IKI collaboration project regarded European scientific experiments on the Intercosmos satellites. In fact, the collaboration between ESA and

[15] Interview with L. Zelenyi, July 27, 2023.

[16] Bonnet, R. M. (2016) 50 years of vision, science, and international friendship, in L. Zelenyi (Ed.), *Space Research Institute in Times of Change. Glimpses of the Past and Visions of the Future*. IKI RAN, 45.

[17] Kaltenecker, H. *URSS: Invitation de D/SCI par Dr. Sagdejev*, 10 mars 1978. In ESA-6896, HAEU.

[18] Gibson, R. June 9, 1978. In ESA-6896, HAEU.

Institute of Terrestrial Magnetism, Ionosphere and Radiowave Propagation (IZMIRAN) on the development of electric field sensors and associated electronics for the Czechoslovak Magik subsatellite (part of Intercosmos-18) started in 1976, and was followed in November 1977 with two ESA engineers who delivered in Moscow "the satellite hardware produced by ESA for incorporation into Magik satellite"[19] that was launched in 1978.

In 1978, negotiations began on the possible participation of ESA in the Soviet Prognoz program as IKI was preparing the launch of a Prognoz satellite for interplanetary plasma studies, scheduled for 1985. The Prognoz spacecraft would carry a joint ESA-USSR payload consisting of experiments provided by ESA member countries.[20] But the joint project within the framework of the Prognoz M (Interball-1) program was implemented only in 1995, a whole ten years after the last launch of the Prognoz-10 satellite.[21] However, the Interball satellite did carry the payload of experiment RON (Spacecraft Potential Control Experiment) conducted by an international scientific team with researchers from ESA.[22]

The definite interest towards further collaboration with the Soviet Union was expressed by European life scientists who needed "more frequent flight opportunities on manned spacecraft, more frequent missions, easier access to missions, less complicated procedures and lower costs for developing and launching the scientific payload".[23] ESA representatives maintained a direct line of communication with Academician Oleg Gazenko, who served as Director of the IBMP from 1969 to 1988. According to Roy Gibson, Gazenko was "tremendously enthusiastic [...] and anxious to have a collaboration".[24] However, this initial contact, which commenced with the aim of pursuing biological experiments, failed to sustain momentum. Prior to the Space Shuttle Challenger disaster in 1986, which caused the cessation of all scientific experiments on board USA Shuttles, cooperation with American counterparts held significantly greater appeal for Europeans. A noteworthy remark of an ESA official

[19] Letter of Gibson, R. to Petrov, B. November 7, 1977. In ESA-6896, HAEU.

[20] Scenario for a Possible Cooperation. November 22, 1978. In ESA-4815, HAEU.

[21] Harvey, B., Zakutnyaya, O. (2011). *Russian space probes: scientific discoveries and future missions*. Springer, 416.

[22] Riedler W. et al. (1994). Spacecraft Potential Control Experiment RON. In *Interball mission and payload*, RKA, IKI, CNES, 373.

[23] Oser, H. *Report on Mission to U.S.S.R.* 4–8 June 1979. In ESA-4815, HAEU.

[24] Gibson, R. *Report of visit to U.S.S.R.* 28 September – 6 October 1979. In ESA-4815, HAEU.

made in 1986 at an IRAC[25] meeting was: "The attractiveness of cooperation [with the USA—O.D.] was obvious before the stringent limitation of flight opportunities in the USA".[26]

In 1983, Sagdeev again issued an invitation clearly aimed at establishing Soviet-ESA ties. As Roger Bonnet recalled: "in April 1983, one month before I took my duties at ESA, Roald Sagdeev organised a magnificent farewell visit to Samarkand, in honour of Ernst Trendelenburg, an unforgettable event to which both Trendelenburg and Roald had the elegant attention of inviting me. That visit cemented even more strongly the ESA-IKI friendship and triggered an irreversible impulse to the development and the strengthening of future IKI-ESA cooperation, embracing all domains of research".[27]

Halley Armada

The major scientific alliance, notable for its international impact and image, was the observation of Halley's Comet in 1986. This effort involved the "Halley Armada", which consisted of the Soviet Vega-1 and -2 spacecraft, ESA probe Giotto, and two Japanese Suisei space probes. As Lev Zelenyi asserted: "Vega represented the epitome of international cooperation".[28] "The most remarkable aspect of this particular international cooperation was the implementation of the Pathfinder Concept by which data from the Soviet spacecraft Vega, received with the help of NASA's Deep Space Network antennas, was used to optimise the trajectory of Giotto".[29] Coordination between Soviet and European scientists for this mission commenced in the early 1980s. Between September 13 and 15, 1981, representatives from four entities (ESA, NASA, NASDA, and Intercosmos) convened for the first time in Padua (Italy) (Fig. 3.1). They resolved to engage in close cooperation, primarily through information exchange, particularly on flyby days.[30] As Roger Bonnet recalls, on November 7, 1986, celebrating the success of the international joint

[25] ESA International Relations Advisory Committee.

[26] *Relations with U.S.S.R. IRAC* 1/10/86. In ESA-28130, HAEU.

[27] Bonnet. (2016), 45.

[28] Interview with L. Zelenyi, July 27, 2023.

[29] *Relations between the European Space Agency (ESA) and the U.S.S.R. in the field of space cooperation*, June 1987. In ESA-28130, HAEU.

[30] Maffei, P. (1984). *La Cometa di Halley*. Mondadori, 351.

Fig. 3.1 Soviet Vega spacecraft. The National Air and Space Museum Steven F. Udvar-Hazy Center. Photo by the author

efforts, Pope John Paul II called all those involved in that unique venture the "Peace makers" of the twentieth century.[31]

The Vega project was the most successful, spectacular, and extensive in a series of research missions carried out by Soviet scientists in collaboration with foreign counterparts. The design of this project began in the era of Brezhnev's stagnation, and its implementation coincided with the beginning of the *perestroika* period. Given its successful completion, it could be argued that it had absorbed the best characteristics of both political periods. Scientific collaborations continued despite the escalation of the arms race and the deterioration of international relations, marked by events such as the Soviet invasion of Afghanistan, the Euromissile crisis, advancements in the SDI, and the Soviet Union's withdrawal from the Geneva negotiation process, among others. Remarkably, the international project to study Halley's Comet remained ongoing throughout the 1980s. Subsequently, the new political landscape ushered in by Gorbachev's reforms enabled the elevation of international scientific cooperation to new, unprecedented heights.

The Soviet Vega project also exemplified multi-linked cooperation. Several international teams contributed to the project. Notably, nine countries were involved in the design, construction, and providing science payload and its operations of the Soviet probes Vega-1 and -2. Within the framework of the Vega project, Jacques Blamont—a French astrophysicist and the founding scientific and technical director of CNES —played a key role in fostering Soviet-European cooperation in space in general and in ensuring the success of Vega in particular. Additionally, Halley's Comet projects included contributions from probes launched by ESA and NASDA, as well as participation from American scientists through the Pathfinder project and the Venus balloon component of the Vega mission. This initiative served as a powerful propaganda tool for the Soviet Union, allowing its leadership to deliberately highlight its commitment to cooperative scientific research in space against the backdrop of the Reagan administration's promotion of the military-focused SDI project.

Collaboration on the Vega project was extremely valuable, not only for the foreign participants, who were given the opportunity to place their research instruments on board the Soviet spacecraft, but also for Soviet scientists. As IKI staff recalled: "Soviet scientists were unable to conduct many interesting experiments on their own. They lacked sufficient

[31] Bonnet. (2016), 47.

funding, the necessary infrastructure for the design, development, and production of instruments, as well as adequate staffing. Direct contact with foreign specialists allowed them to stay informed about modern trends in experimental space technology developed in the West".[32]

The missions to Halley's Comet revolutionised both the approach to international collaboration and the attitude of Soviet scientists towards their foreign colleagues in terms of greater openness, lowering the barriers of secrecy and surveillance during the joint work on the project. Such high tolerance was unheard for the historically closed Soviet space program. As Sagdeev stated in an interview with the American magazine *Science* in 1987, "*Glasnost* has to apply to space exploration as well. We were especially proud that we were able to contribute in our field of activity to a general framework of what he (Gorbachev) is doing for our society".[33]

Conclusion

Prior to the onset of *perestroika*, the ESA-Soviet cooperation encompassed just one project involving the scientific satellite *Intercosmos-18*, an attempt to launch ESA *Marecs* satellite with the Soviet Proton rocket (see Chap. 7) alongside a solitary large-scale international endeavour focused on studying Halley's Comet. In the first fifteen years following the conclusion of the Agreement, the Soviet Union expressed interest in joint activities spanning various areas of space science. For its part, the ESA leadership considered initiating comprehensive cooperation with Soviet space institutes. However, tangible results did not materialise until the mid-1980s.

Despite the expressed intentions, European space officials moved very cautiously along the path of rapprochement with the USSR. In the words of Roy Gibson, they were determining "what action needed to be taken gently to push the ball along".[34] ESA officials also had to take into account numerous factors influencing international relations in general and ESA policy in particular. The cautious and deliberate approach in the initial stages of ESA's engagement was primarily driven by concerns about the reaction of the United States. The stance of the United States, particularly NASA and certain American agencies, responsible for overseeing the

[32] Galeev, A., Tamkovich, G. (1999). *Space research Institute of the Russian Academy of Sciences. The 35th anniversary*, IKI, 50.

[33] Waldrop, M. (1987). Soviet space science opens to the West. *Science*, 236 (4807), 1429.

[34] Gibson, R. *U.S.S.R.*, October 11, 1979. In ESA-4815, HAEU.

transfer of advanced technology and sensitive knowledge, played an important role in shaping the Soviet-ESA relations. This aspect will be further elucidated in the subsequent analysis, particularly in the context of the data exchange amongst space science data archives (see Chap. 5). Secondly, the hesitation of Member States also influenced ESA's approach. Indeed, Jean Arets cautioned about the political ramifications of any collaboration with the USSR, emphasising that "the agency must take care not to go against the current of the policy of its Member States".[35] Thirdly, uncertainty regarding the true intentions of the Soviets concerning cooperation, coupled with a reluctance to be seen as indirectly endorsing Soviet propaganda, also contributed to the cautious stance adopted by ESA's leadership.

[35] Arets, J. *Commentaires sur le rapport de la visite en URSS du DG*, Octobre 10, 1979. In ESA-4815, HAEU.

CHAPTER 4

Winds of Perestroika

At the onset of *perestroika*, space research in the USSR was thriving, as evidenced by the numerous domestic projects and joint international missions undertaken, which indicated promising prospects for the future. The primary research activities were conducted under the Intercosmos program, which facilitated both cooperation between the Soviet Union and the countries of the socialist camp, as well as collaboration with a diverse range of international partners. Notably, joint projects were implemented with various Western European countries, the United States, India, and Australia. The relations with ESA were also developed across various scientific fields. A key factor driving the rapprochement with Western partners in the latter half of the 1980s was the temporary withdrawal of the USSR's main competitor from the space race. As Western observers noted, following the 1986 Challenger shuttle disaster, the Soviet Union sought to capitalise on this opportunity and seize the initiative. This chapter examines the scientific space projects undertaken during the *perestroika* period, including those conducted in collaboration with ESA.

O. Dubrovina, *The Soviet Space Program and International Collaboration during Perestroika, 1985–1991*, Palgrave Studies in the History of Science and Technology,
https://doi.org/10.1007/978-3-032-03157-0_4

Space Science Program

According to Soviet reports,[1] the Intercosmos program covered five main directions: space exploration, space biology and medicine, space meteorology, space communications, and rocket and space remote sensing of the Earth. In all these domains, the Soviet leadership aimed to engage foreign counterparts, showcasing all facets of their extensive program of comprehensive space exploration during international meetings.

As far as planetary exploration was concerned, the "Martian" direction received a new breath. The British *Flight International* quoted a Soviet scientist who stated that the Soviet Union intended to land two men on the planet Mars between 2005 and 2010. It was considered as the ultimate goal of the program, which would be preceded by extensive exploration with automatic probes. The Phobos mission (see Chap. 6) foreseen for 1988 was the first step in Martian moon landing. The first robotic mission to Mars was initially planned to launch in 1992. It was expected that that year, the Mars orbiter would conduct extremely detailed surveys of the surface of the planet. The roving vehicle and soil sampler could be launched on a 1996 mission. The rover would carry a very sensitive soil analyser. The return of a soil sample to Earth was scheduled for the early twenty-first century, during the next phase of the Mars program, between 2000 and 2005.[2] According to the Soviet plans presented to the international space community, the manned mission would crown the Soviet Mars exploration program.

Moreover, under discussion in the 1980s was the 1994 launch of spacecraft to small bodies in the Solar System (*Vesta* project) and the Lunar Polar Satellite, where a launch was scheduled between 1992–1994. The investigation of the Sun was expected to be conducted by the *Phobos* project and by the launch of three *Koronas* satellites with a wide participation of countries from the Intercosmos community and India. Their launch was scheduled in 1992–1993.[3] However, the project faced challenges during the downsizing of the space program in the 1990s; nevertheless, like *Interball*, it became one of the few scientific programs successfully implemented. Awarded to NPO Yuzhnoe[4] in Dnepropetrovsk (modern Dnipro,

[1] "Intercosmos" program. Annual report for 1989–1990. In ESA-28130 (2), HAEU.

[2] Furniss, T. (1989). Soviet set Mars timetable. *Flight International*, January 21.

[3] "Intercosmos" program, Annual report for 1989–1990. In ESA-28130 (2), HAEU.

[4] "A significant portion of the Soviet space and rocket sector, administered under the MOM, was located in Ukraine, particularly in Yuzhnoe. Yuzhnoe's facilities were responsible

Ukraine) the first *Koronas* was launched in March 1994 and the second one, *Koronas F* (*Fizika*), was launched in 2001. The third *Koronas* (*Foton*) lifted off on the top of a Tsyklon-3 rocket in January 2009. Historically, the project was officially established by a government decision regarding the future of the space program in 1964. It took 45 years for the three missions to materialise—a timeframe that, as noted by Brian Harvey and Olga Zakutnyaya, Director of Communications at IKI, is likely a record-breaking achievement.[5]

The Soviet-French astrophysical observatory *Granat* was launched in 1989 from Baikonur by a Proton rocket. The project envisaged studies of cosmic sources of X-ray and Gamma-ray emission. The science payload suit was developed in the 1980s by the Soviet Union, France, Denmark, and Bulgaria[6] and included the Soviet ART-P and French Sigma gamma-ray telescopes, which observed a broad range of radiation sources. It was supposed that it would complement observations by the Kvant astrophysics laboratory already docked to Mir. In September 1994, after almost five years of operation in orbit in the directed observation mode, the working fluid for the turn engines came to an end and the observatory was transferred to the scanning mode. Data transmission from the observatory was completed on November 27, 1998, the observatory was destroyed during entry into the atmosphere on May 25, 1999. From a scientific point of view, *Granat* was considered as the most productive astrophysical mission of Soviet and Russian space science.[7]

According to Russian scientist and Head of Laboratory IKI Aleksandr Lutovinov: "the late 1980s in the USSR witnessed significant achievements in high-energy astrophysics. We had two observatories: the *Roentgen* observatory on the *Kvant* module of the orbital station Mir, and an even more successful observatory named *Granat*, which was the result of collaboration with French, Danish, and Bulgarian scientists. The initial findings from the *Granat* observatory generated great enthusiasm in Europe, prompting immediate discussions about the subsequent project,

for producing more Intercontinental Ballistic Missiles (ICBMs) than the United States and the remaining territories of the Soviet Union combined. Additionally, Yuzhnoe played a pivotal role in the production of over 1000 spacecraft". Correspondence with M. Gruntman, March 16, 2024.

[5] Harvey, Zakutnyaya. (2011), 432.

[6] Mezhdunarodnaya Astrofizicheskaya Observatoriya "Granat". Retrieved February 6, 2025, from http://hea.iki.rssi.ru/GRANAT/

[7] Harvey, Zakutnyaya. (2011), 389.

code-named *Granat-2*".[8] Also, in collaboration with French scientists, the launch of a gamma telescope, as part of the *Gamma-1* project, was initially scheduled for 1988. However, the launch was ultimately carried out in 1990 from Baikonur. The unrealised *Granat-2* project, along with the experience gained from international cooperation on the *Gamma-1* mission, contributed to the development of the joint INTEGRAL project, whose history traces back to the Soviet era.

The launch of satellite *Relikt-2*, designated to study relict radiation, was planned for 1991–1993. The Intercosmos annual report sustained: "Scientific instruments of the project were developed with participation of specialists from socialist countries and ESA and its implementation was supposed in 1993".[9] Ultimately, the project was not realised: although *Relikt-2* received formal approval, it did not go ahead for lack of funding.[10]

The field of space plasma investigations were also abundant with various projects. On September 28, 1989, the satellites *Intercosmos-24* and *Magion-2* were placed in orbit from Plesetsk Cosmodrome. They carried instruments in the frame of *Aktivny*, an international space project that carried out investigations of processes in the magnetosphere of the Earth. An orbital plasma-wave laboratory was equipped with a 20 m-diameter antenna that transmitted very-low-frequency radio waves towards the Earth's magnetosphere, to study their interaction with charged particles in the radiation belts.[11] The scientific instrumentation was developed by socialist countries under the auspices of the Intercosmos program.

The *Apex* (Active Plasma Experiments) project included a simulation of geophysical phenomena in the near-terrestrial plasma by injecting charged-particle beams, as well as by studying beam-plasma processes in the near-Earth space. It followed *Aktivny* in the release of plasma and its instruments were similar or modified versions.[12] This satellite (*Intercosmos-25*) was supposed to eject a sub-satellite (*Magion-3*) to study the interaction of electron beams, emitted by the main craft in the magnetosphere. The project involved scientists from socialist countries and represented the sole

[8] Observatoriya Integral: Nedremannoye Oko Na Zemnoy Orbite. Retrieved February 6, 2025, from https://prof-ras.ru/index.php?option=com_k2&view=item&id=933:observatoriya-integral-nedremannoe-oko-na-zemnoj-orbite&Itemid=126.

[9] "Interkosmos" program, Annual report for 1989–1990. In ESA-28130 (2), HAEU.

[10] Harvey, Zakutnyaya. (2011), 137.

[11] Furniss, T. (1987). Countdown to co-operation. *Flight international*, 5 December.

[12] Harvey, Zakutnyaya. (2011), 407.

dedicated Soviet scientific mission of the year. It was launched from Plesetsk aboard a Tsyklon booster (SL-14) in 1991.[13]

In the field of space meteorology, the activities were subdivided into three main areas: development and sophistication of methods of interpretation and use of satellite data, investigation of the top atmosphere, and improvement of technical means and techniques of data acquisition and processing.

In the field of remote sensing there were four international purpose-oriented comprehensive projects: the *Susha* (Dry land) project—remote sensing of the dynamics of geosystems with the objective to develop procedures of systems for landscape analysis. The project was aimed at using the results obtained in the national economies of CMEA[14] countries. The *Geologiya* (Geology) project carried out research in remote sensing of geological structures. The implementation of the international space project *Priroda* (Nature) was also foreseen. A set of scientific instruments would be launched on a module of the Mir orbital station. The aim of the project was to develop the remote sensing technology from visible to microwave range for remote sensing of the ocean, atmosphere, and the Earth's surface.[15] It was planned that the *Priroda* module would join Mir in 1991.[16] The *Kosmometriya* project regarded the development of a set of spectrometric and photographic equipment for studying spectral characteristics of natural formations. It had to be implemented on the *Kvant-2* and *Priroda* modules of the Mir orbital station.

From the roster of Soviet projects spanning various domains of space research, several key observations emerge. Firstly, Soviet space science was in a phase of active development at the onset of *perestroika*. Secondly, it leveraged resources for advancement, not only from four decades of accumulated experience, but also from new collaborative ventures with numerous partners from both the socialist bloc and the West. Thirdly, scientific activity was closely linked to the imperatives of the country's economic

[13] APEX-MAIN. NASA Space Science Data Coordinated Archive. Retrieved February 6, 2025, from https://nssdc.gsfc.nasa.gov/nmc/spacecraft/display.action?id=1991-086A.

[14] Council for Mutual Economic Assistance. An intergovernmental economic organisation established in 1949 by the representatives of Albania, Bulgaria, Czechoslovakia, Hungary, Poland, Romania, and the USSR, CMEA was entrusted with the task of organising socialist economic integration.

[15] Interkosmos to J.F. Danisse, February 26, 1987. In ESA-28130, HAEU.

[16] *Priroda*, the sixth module of the Soviet-Russian orbital complex Mir was finally launched only on April 23, 1996, and docked with Mir on April 26, 1996.

progress, particularly the technological development of national industry, as evidenced by the research program. Applied space research was declared one of the top priorities, alongside fundamental scientific research. Thus, the prevailing issue, as we will uncover later, was focused on the challenge of popularising the endeavours of space scientists in the realm of space exploration for the benefit of the civil economy, rather than a genuine absence of connection between purely theoretical projects and their practical applications.

Furthermore, the discrepancy between the planned launch dates for scientific satellites and modules and the actual project implementation indicates, on one hand, the optimistic outlook of scientists during project development and initiation. State financing commitments, a longstanding practice throughout the Soviet space program's history, fostered hope for successful project completion. However, towards the end of the 1980s it became evident that the state was no longer capable of funding costly space exploration endeavours. On the other hand, it is conceivable that agreements and obligations (made primarily with Western European counterparts) compelled the Soviet side to devise ways, albeit with delays, to finalise ongoing projects. Additionally, financial assistance from foreign partners themselves played a crucial role, as will be discussed further.

Working Together with ESA

With the beginning of *perestroika*, relations between ESA and the USSR underwent a noticeable change towards closer cooperation, to which both internal and external factors contributed. From inside, Gorbachev's "new thinking policy" manifested its effects in such a closed and largely classified area as the space sector. From outside, the change in the general political climate also contributed to the Soviet-ESA rapprochement as European officials made no secret of the fact that cooperation with the USSR was largely determined by Soviet-American relations and reflected their evolution. As Jean Arets argued: "no substantive cooperation will be achieved if there is not a minimum of *détente* in the West-East relationship". Expectations of positive changes were associated primarily with the forthcoming Gorbachev-Reagan summit scheduled for 1986. In connection with this event, the ESA officials noticed some revival of Soviet-American relations in the field of space research. The desire not to yield to the Americans the priority of active cooperation with the Soviet side was also an important factor that pushed them towards rapprochement with the

USSR. As Jean Arets pointed out, "it would certainly not be in Europe's interest to be absent from the space dialogue for many reasons related to our relations with both the USA and Soviet Union".

Moreover, the Soviet side's renewed approaches to cooperation with Western partners also facilitated rapprochement. Jean Arets emphasised the new Soviet attitude towards the bilateral cooperation: "The Russians are becoming more flexible and recognise the constraints of a useful cooperation (access of foreigners to the launch site, guarantee of respect of intellectual property rights, to receive all the data, etc.)".[17] Thus, during *perestroika,* meetings had become regular between representatives of ESA and the Soviet Academy of Sciences (mostly from IKI), and many new joint projects had appeared in various scientific fields of space exploration, as well as manned flights.

Nonetheless dramatic changes in the ESA-Soviet collaboration in space during *perestroika* and the fact that "the relations became somewhat less sensitive than they were in the past", Arets emphasised that "every contact with the Soviet Union continued to keep a political connotation and there was a need to have a well-coordinated approach to these contacts". ESA staff was anxious to find a happy medium between not being a Soviet propaganda tool and not spoiling relations with the United States, and at the same time, using "opportunities for the good implementation of the ESA goals".[18]

Mir Station

The launch of the first core module of the Mir station in 1986 provided new opportunities for the astrophysical projects. In March 1987 *Kvant-1*, the first module to dock with Mir's Core module was launched from Baikonur and successfully joined the station, carrying a *Roentgen* astronomy telescope, which was produced by the Netherlands, UK, FRG, ESA (Sirene-2, a gas scintillation proportional spectrometer), the Soviet Union, and socialist countries. ESA paid the full cost of developing and manufacturing these experiments, while the European scientists pointed out that the cooperation with the Soviets allowed scientists to quickly obtain a large return at a relatively low cost. As Alan Smith, ESA's project scientist for the mission put it: "This was the fastest way to get an X-ray

[17] Relations with the U.S.S.R. IRAC, October 1, 1986. In ESA-28130, HAEU.

[18] Arets, J. *Relations with the U.S.S.R.*, October 6, 1987. In ESA-28130, HAEU.

observation experiment launched. The next ESA X-ray satellite would not be in orbit until the 1990s".[19] The Mir astrophysics package on board the *Kvant-1* module gave scientists a wide-spectrum X-ray survey and long-term surveillance.

In 1989–1990, preparatory work was carried out on the *Kvant-2* module, which was intended to be added to the Mir station with equipment and instrumentation designed to significantly extend the "investigations and experiments conducted in the interest of science and the national economy".[20] *Kvant-2* was launched in November 1989 from Baikonur.

Additionally, space medicine, a parallel field of research, was advancing with the new opportunities provided by the Mir station. Soviet scientists began preparations for the development of Medilab, a space laboratory dedicated to medical research. Medilab was intended to function as part of the orbital station, with its research program focused on human physiology. Soviet scientific authorities anticipated that by combining the scientific and technological capabilities of international space institutions (Intercosmos Council, NASA, ESA, and CNES), a high level of cost-effectiveness could be achieved. In fact, as a journalist from the British *Flight international* stated: "Medilab could house life scientists from other countries, which so far were allowed to supply military pilot or engineer cosmonauts"[21] (see more in Chap. 7).

At the conclusion of the 1980s, materials science and technology in space experienced a significant development in the Soviet Union. During this period, Soviet scientists were actively involved in preparing experiments to be conducted on board *Kristal*. The *Kristal* module was successfully launched on May 31, 1990, and autonomously docked with Mir on June 10, 1990. ESA conducted its materials science experiments aboard the Mir space station during the Euromir 94 and 95 missions, which took place in the post-Soviet era. ESA experiments in the field of materials science were also conducted aboard the *Foton* automatic spacecraft.

[19] New Mir astrophysics module to include European Experiments (1987). *Aviation week & Space Technology*, February 16.

[20] "Interkosmos" program, Annual report for 1989–1990. In ESA-28130 (2), HAEU.

[21] Furniss. (1987).

Materials Science

Emerging in the 1970s, this relatively new field of science assumed a prominent position within the realm of space exploration since the mid-1980s. Furthermore, compared to the previous period, the field of material processing in microgravity represented a new area for cooperation. Factors such as weightlessness and vacuums created a unique environment that offered opportunities for enhancing existing materials and developing new ones. Utilising outer space and the conditions of space flight, it became feasible to produce crystals, metals, composite materials, and biological materials with properties unattainable on Earth. To address these challenges, specialists from TsSKB and the Progress plant in Kuybyshev developed the unmanned spacecraft *Foton*.

The first *Foton* spacecraft was launched in 1985 and remained in orbit for 13 days. The experimental results garnered significant interest, leading to the successful deployment of twelve *Foton* spacecraft between 1985 and 1999, with launches occurring almost annually. These missions lasted between 13 and 18 days, with descent vehicles employing soft landing systems. The instruments aboard the *Foton* spacecraft conducted research programs across various disciplines, including semiconductor crystals and optical materials, biotechnology, cell biology, and the effects of space factors on samples returned to Earth. Starting in 1989 with Foton-5, which carried French commercial microgravity experiments,[22] foreign scientists and specialists began participating in the program.[23] In the following two years, CNES conducted its own experiments aboard *Foton*-6 in 1990, followed by the DFVLR (the German Centre for Air and Space Flight), which carried out experiments on Foton-7 in 1991. As Valery Abrashkin, technical manager for international payloads on *Bion* and *Foton* at TsSKB-Progress in Kuybyshev, recalled, "The French colleagues supplied their Crocodile experiment, and then Germany's DFVLR joined the project in 1991",[24] with the Flugdatenrekorder technology demonstrator provided by Kayser-Threde, the West German company.[25] Regarding ESA

[22] Foton 5. NASA Space Science Data Coordinated Archive. Retrieved February 6, 2025, from https://nssdc.gsfc.nasa.gov/nmc/spacecraft/display.action?id=1989-032A.

[23] Abrashkin, V. (2017). Tak my nachali svoye sotrudnichestvo s zaru- bezhnymi partnerami. In Yu. Izyumova, S. Semenov (Eds.), *Chtoby nasha novaya rabota luchshe prezhney sdelana byla, Sbornik vospominaniy veteranov*. AO «RKTS «Progress», 146–148.

[24] Ibid.

[25] Harvey. (2021), 68.

participation, before the collapse of the Soviet Union, ESA took part in the *Kashtan* experiment aboard *Foton*-7 in October 1991 on a pay-by-flight basis.

According to Antonio Verga, a former ESA engineer and head of the *Foton* program, the initial cooperation agreement in the field of material sciences was established between the ESA and Splav, Technical Centre in Moscow, with the blessing of Glavkosmos, and the technical support of KBOM (General Engineering Design Bureau named after V. P. Barmin in Moscow).[26] As Antonio Verga put it: "Splav, that means alloy (or metal alloy), was a pseudo-commercial name given to a company in Moscow, which at that time dealt in special metal alloys. They collaborated with KBOM, which built *Kashtan*, and, for some unknown reason, Splav guaranteed a commercial relationship with ESA".[27]

Notably, a significant achievement of Glavkosmos' initiative to attract European interest in the commercial utilisation of the Soviet space industry was demonstrated through the ESA's participation in material sciences experiments. ESA had to pay for flying protein samples as part of a crystallisation experiment on the seventh *Foton* mission in 1991. This could be regarded as the first commercial flight of an ESA payload aboard a Soviet spacecraft. In contrast, previous ESA experiments conducted on Soviet spacecraft, such as *Bion*, were carried out on a no-fund-exchange basis.

The experimental facility offered by Glavkosmos to ESA was a protein crystalliser known as *Kashtan* (Russian for "chestnut"). The *Kashtan* was capable of hosting 52 protein samples, 32 of which were reserved for the Institute of Crystallography of the Soviet Academy of Sciences (IK AN SSSR). The remaining 20 were offered to the ESA at $5000 per sample, a price which ESA considered as extremely attractive, especially in comparison to the expenses that had previously been made for protein crystallisation experiments on the Space Shuttle.[28] Included in the price was the shared use of *Kashtan*, access and utilisation of the laboratory of CMB (Centre for Medical Biotechnology of the USSR Ministry of Public Health), in Moscow for filling of the crystallisation chamber, transport of the loaded chamber under controlled environmental conditions to the

[26] The cooperation between ESA and KBOM continued thereafter with joint materials science experiments in the POLIZON furnace, built by KBOM, and flown with some of ESA's experiments until 2007.

[27] Interview with A. Verga, April 14, 2024.

[28] Harvey, B. (2024). Forgotten legacy. *Space Chronicle*, 4, 23.

launch site Plesetsk, integration of the chamber into the spacecraft, the orbital flight, post-landing retrieval, re-transport under controlled environmental conditions, and hand-over of the processed samples to ESA in Moscow.

ESA's investigators for *Kashtan* were recruited from scientists already involved in the APCF project (Advanced Protein Crystallization Facility) with a flight foreseen on Spacelab mission IML-2 (STS-65). IML-2 finally flew in July 1994, three years after *Kashtan*. Lead investigator was Dr. Naomi E. Chayen from the Blackett Laboratory at Imperial College, London (UK).[29]

The ESA part of the *Kashtan* mission campaign was carried out by just two people: Wim Jansen (ESA project manager) and Naomi Chayen (lead scientist). Jansen was selected because of his pioneering work in the USSR during the five-week *Bion-9* mission campaign in 1989. Some experience was certainly required because, until 1994, there was hardly any infrastructure available for ESA personnel working in Moscow. While difficult to understand today, telephone calls from Moscow to the European Space Research and Technology Centre (ESTEC) were only possible through a 24-hour advanced booking from one of the international hotels. Mrs. Chayen and Mr. Jansen were in charge of loading two *Kashtan* chambers (flight and ground) with protein samples at CMB in Moscow. After the hand-over to Splav, one chamber was transported to Plesetsk in a cooling box while the other chamber remained in Moscow in a cold room at CMB.

During the largest part of the orbital flight Chayen and Jansen were back in Western Europe. Sparse information about the progress of the space flight was obtained via telephone from ESTEC to KB Foton (contact person: Valery Abrashkin). Within half a day after the landing of the *Foton-7* capsule in Kazakhstan, *Kashtan* was back in Moscow to be handed-over to Jansen and Chayen, who never caught a glimpse of the spaceport (Plesetsk), the spacecraft, the launcher rocket, the flight control centre, or the landing site. Their task in Moscow was confined to the preparation of sample materials and to receive them back after the flight.[30]

The post-flight evaluation revealed that certain scientific goals had not been fully reached because only half of the 20 space-flown samples had grown crystals sufficiently large for a successful x-ray analysis. The initial

[29] Chayen, N. E. (1995). Microgravity protein crystallization aboard the Photon satellite. *Journal of Crystal Growth*, 153, 175–179.

[30] Demets, R. (1991). *Kashtan on Foton-7*. Internal report.

pessimism was later compensated by a serendipitous finding when crystals were further analysed. To trigger crystal formation of transverse transcriptase, a variety of salt concentrations had been used in *Kashtan*. It turned out that in micro-*g*, salt concentrations that yielded the best crystals were not the same as on Earth. This discovery was sufficiently novel to warrant a scientific publication.[31]

Life Sciences

One of the most promising areas of cooperation was within the field of life sciences. Experiments conducted aboard biological satellites of the Cosmos series (*Bions*) and on board of the Mir orbital station constituted a significant component of a comprehensive program of biomedical research undertaken by the USSR. The objectives of these investigations were multifaceted: to assess the adverse effects of microgravity on the human body, to develop scientifically grounded principles and methods for medical support during space missions of varying durations, and to address fundamental issues in general biology, such as the role of gravity in the growth, development, ageing, and evolution of living systems.[32]

The Soviet Union launched its first *Bion-1* in 1973. NASA and CNES participated with their experiments two years later in 1975 on board *Bion-3*. The leading role in the frame of the biological missions was played by the IBMP of the Soviet Union Ministry of Public Health that collaborated with many other Soviet research centres, scientists from the member states of the Intercosmos program (Bulgaria, Czechoslovakia, GDR, Hungary, Poland, and Romania) and from the USA (Ames Research Centre, NASA), France (CNES, Paul Sabatier University), and ESA. Foreign participants were involved in the preparation of biological samples, development of flight equipment, and the conducting of experiments.[33] Between 1973 and 1985, seven *Bion* satellites were launched. However, during the *perestroika* period, only two launches occurred: *Bion-8* in 1987 and *Bion-9* in 1989. Following the successful completion of these two flights, preparations for the tenth *Bion* satellite began in the late 1980s, with its launch scheduled for 1992.

[31] Chayen, N. E. & Helliwell, J. R. (1999). Space-grown crystals may prove their worth. *Nature*, 398, 20.

[32] "Intercosmos" program, Annual report for 1989–1990. In ESA-28130 (2), HAEU.

[33] Intercosmos to J.F. Danisse, February 26, 1987. In ESA-28130, HAEU.

Before *perestroika* there had been sporadic contacts in this area between ESA and the USSR for several years. The main common interest laid in basic research, e.g. gravitational biology, human physiology, and exobiology.[34] Another important point of common interest was radiation dosimetry and radiobiology, i.e. to understand and quantify radiation hazards in space. The period of *perestroika* marked a new phase in the relationship. It was no coincidence that in 1986, the ESA leadership received a clear indication that the Soviet Union welcomed enhanced cooperation with ESA in the field of biological research in space. The Soviets recognised this as an opportune moment to strengthen collaboration with the ESA, capitalising on the challenges faced by the United States following the tragic Space Shuttle accident that occurred in 1986. ESA-affiliated researchers participated in three biomedical missions: *Bion 8*, *−9*, and *−10* (see Chap. 7). ESA officials emphasised the "extreme usefulness" of this type of cooperation with the USSR in the field of life sciences, citing the Soviet Union's unique facilities and their advanced program in this domain.[35] During the post-Soviet period, ESA gained access to the Mir orbital station to conduct long-duration experiments in space medicine and biology as part of the Euromir 94 and Euromir 95 missions. These missions involved ESA astronauts who served as both operators and test subjects.

Spektr Projects

The 1980s also saw the development of three astrophysics Soviet projects collectively known as Spektr (or Spectrum)—observatories for radio, ultraviolet, and X-ray/gamma ray observations—and originally intended as a series of telescopes to follow Astron,[36] Kvant, and Granat.[37] The IKI management proposed involving ESA scientists in two projects: the RadioAstron mission and Spectrum-RG astrophysical space observatory.

The implementation of the "highly ambitious" international RadioAstron project was foreseen for the beginning of the 1990s. The aim of the project was to conduct scientific radio astronomical observations using a radio telescope mounted on the *Spektr-R* spacecraft, which was built at the NPO Lavochkin. It included the launch of a 10-metre radio

[34] Relations with the U.S.S.R. IRAC, October 1, 1986. In ESA-28130, HAEU.
[35] Ibid.
[36] Soviet ultraviolet observatory launched in 1983 using Proton rocket from Baikonur.
[37] Harvey, Zakutnyaya. (2011), 468.

telescope aboard the *Spektr-R* spacecraft, with which would create, together with Earth-based radio telescopes, a VLBI system. The participation of many Earth-based radio observatories of the USSR, Europe, the U.S., India, and Australia were planned.

Starting in 1985, international meetings dedicated to the project began to take place. During an IKI-ESA meeting held in July 1989 in Moscow, IKI informed the others about a plan to launch the RadioAstron spacecraft by the end of 1993.[38] European colleagues replied that they could discuss the possibility of cooperation only if the launch was postponed for some reason to the 1995/1996 timeframe. In this case, one or more member states could be involved with specific hardware items or with other forms of support. During the meeting, IKI staff found several issues that were of interest to them in the field of possible cooperation. They were interested in the possible use of active cooling provided by ESA scientists. IKI expressed their need of potential ground station support and support to develop precise orbit determination software. Unfortunately, none of the equipment requested by the Soviet experts was available at ESA. The solution was to provide additional ground station support from existing/planned VLBI-station of an ESA member state. What concerns precise orbit determination software, the ESOC,[39] was at that period developing such type of software, thus IKI was given all the information to identify the requirements applicable to RadioAstron.[40]

The Russian-European cooperation on the RadioAstron project ended in 1998 with limited ESA participation in the mission. Essentially, ESA offered its infrastructure to support Russian space research. The tests of antenna petals from Russia's RadioAstron spaceborne radio telescope[41] were conducted in the 3-metre-diameter HBF3 thermal-vacuum chamber at the ESTEC Test Centre.[42] ESA has also provided funding for the

[38] The scientific satellite was launched on July 18, 2011, aboard a Zenit-3F launcher from the Baikonur Cosmodrome.

[39] The European Space Operations Centre (ESOC) serves as the primary mission control centre for the ESA and is situated in Darmstadt, Germany.

[40] Meeting minutes for the ESA-IKI Meeting on Cooperation in Space Radio Astronomy, Moscow, IKI, 11–12 July 1989. In ESA-28051 (3), HAEU.

[41] Uchastniki proyekta RadioAstron. Retrieved February 6, 2025, from http://www.asc.rssi.ru/radioastron/organization/coop_ru.html.

[42] RadioAstron antenna petals tested at ESTEC. Retrieved February 6, 2025, from https://www.esa.int/ESA_Multimedia/Images/1998/01/RadioAstron_antenna_petals_tested_at_ESTEC2.

development of the on-board Rubidium frequency standard, which has been manufactured by the Neuchatel Astronomical Observatory in Switzerland.[43] Due to repeated postponements of the mission on the Russian side, the Swiss side ultimately withdrew from the project.

The launch of the second specialised orbital observatory Spektr-Roentgen-Gamma (Spectrum-RG) designated for astrophysical studies at X- and Gamma-ray domains of the spectrum and roentgen diapasons was initially planned in 1992–1994.[44] It had to be a major step towards understanding the physical properties of different classes of astronomical objects. The UK was supposed to participate in the creation of the telescope for this observatory. The project also included specialists from Australia, Denmark, Italy, Canada, USA, Finland, France, FRG, and Switzerland.[45] During the same period, ESA was working on the XMM (X-ray Multi-Mirror) mission, which was synergistic with the Soviet *Spektr-RG* project. IKI was initially invited by ESA to submit proposals for collaborative effort, but the very first idea, an alternative mirror concept developed jointly by Soviet and Danish experts, was declined by the Europeans since "XMM was base-lined with multi-mirror system being presently developed and that at that stage it was not foreseen to use an alternative system".[46]

IVS Mission

Between 1988 and 1989, ESA, the IKI, and NPO Energya, in collaboration with NASA's JPL, developed a project known as the International VLBI Satellite (IVS). The IVS was conceived as a space-based radio telescope, designed to co-observe with ground-based VLBI arrays.[47] Such a space VLBI system would provide a sharper view of celestial radio

[43] Rochat, P., Schweda, H., Mileti, G. and Busca, G. (*1994)*. Developments of rubidium frequency standards at Neuchatel Observatory. *Proceedings of IEEE 48th Annual Symposium on Frequency Control, Boston*, 716–723; Radioastron User Handbook. (2019). Prepared by the RadioAstron Science and Technical Operations Group Version 2.94.

[44] Ultimately, it was developed into a Russian–German high-energy astrophysics space observatory, which was launched on July 13, 2019. For further details, refer to P. Friedrich, P. Predehl (Eds.). (2024). *eRosita. The Journey of the Companions.*

[45] "Interkosmos" program. Annual report for 1989–1990. In ESA-28130 (2), HAEU.

[46] Protocol of the meeting between representatives of the U.S.S.R. Academy of Sciences and ESA, Coca Beach, November 17, 1988. In ESA-28051 (3), HAEU.

[47] Bonnet, R. M. (1991). Towards the Selection of ESA's Next Medium-Size Scientific Project (M2). *ESA Bulletin*, 66.

sources, a capability fundamentally limited by the diameter of Earth for ground-based VLBI. The spacecraft, equipped with a 30-m parabolic antenna, was intended to be launched by a Soviet super-heavy Energya rocket to a distance exceeding one million kilometres from Earth. Boris Chertok, a rocket designer instrumental in the founding of Soviet cosmonautics and Deputy General Designer of NPO Energya in the 1980s, described the project as both ambitious and an engineering marvel.[48] Gury Marchuk, president of the Academy of Sciences in the USSR, was also highly enthusiastic about the initiative, stating: "We regard the IVS project as one of the most remarkable scientific challenges of the coming century. Our scientists look forward with great enthusiasm to studying fundamental problems of the universe using IVS".[49]

From ESA's perspective, IVS was among the proposals for the medium-sized mission (M2) under the long-term Horizon 2000 plan. The proposal, submitted by 74 astronomers in November 1989, was selected as one of six for an assessment study in January 1990. In April 1991, the study was presented to the scientific community for the selection of a maximum of four proposals for the Phase A study.[50] ESA invited representatives from NPO Energya, as well as delegates from IKI and the Academy of Sciences to attend the presentation in Paris on 10 April.[51]

The Space Science Advisory Committee (SSAC) ultimately determined that proceeding to Phase A was not advisable at the time, particularly due to the unproven design of the large antenna. Furthermore, the SSAC concluded that the outcomes of the projects under development, such as the VLBI Space Observatory Program (VSOP) and RadioAstron, could significantly influence the scientific and engineering priorities for any subsequent mission of the same nature.[52] In 1993, the final selection process resulted in the INTEGRAL project being chosen instead.

Despite the mission's unsuccessful conclusion, Soviet-ESA cooperation during its development reached an unprecedented level. Discussions on the joint project began in the late 1980s, with prominent Soviet scientists such as Nikolay Kardashev, Leonid Gurvits, and Vyacheslav Slysh playing

[48] Chertok, B. (1991). *Rakety i Lyudi. Goryachie budni Kholodnoy voyny.* Mashinostroienye.

[49] Marchuk, G. to Luton, J.-M. March 12, 1991. In ESA-28024, HAEU.

[50] Pilbratt, G. (1991). IVS - A second generation space VLBI mission. *Radio interferometry: Theory, techniques, and applications.* Proceedings of the 131st IAU Colloquium, Socorro, NM, Oct. 8–12, 1990. San Francisco, CA, Astronomical Society of the Pacific, 102–106.

[51] Volonté, S. to Kardashev, N. February 6, 1991. In ESA-28051, HAEU.

[52] Bonnet. (1991).

key roles. In addition to astrophysicists, leading engineers from NPO Energya actively contributed to the mission. Unlike most scientific initiatives, where NPO Lavochkin was the primary industrial partner from the Soviet side, this project was intended to utilise the unique Energya rocket. In this project NPO Energya played a significant role, as it sought to leverage an innovative space vehicle and recover the substantial investments made in its development.

During the meetings, NPO Energya was represented at the highest levels, including Boris Chertok, a prominent Soviet spacecraft designer Yuri Denisov, and one of the leading Soviet rocket designers Yakov Kolyako. Soviet scientists and representatives from NPO Energya were received at ESTEC in Noordwijk, while their ESA counterparts, for the first time, were permitted to visit the design bureau and test/assembly facility of the Buran orbital plane at the major NPO Energya site in Kaliningrad (now Korolev), which remained inaccessible to other foreign visitors.[53]

Chertok attributed the termination of the collaboration to financial constraints on both the Soviet (Russian) and ESA sides, as well as declining interest from the American partners. He stated, "The project required approximately one billion dollars for its realisation, to be equally divided among the USSR, Europe, and the United States. Both Russia and Europe were unable to meet this financial demand. American scientists had secured funding for the highly expensive Hubble Space Telescope project and were no longer concerned with our ambitions".[54] Conversely, Leonid Gurvits argued that the primary reason for the project's failure was the ESA officials' perception of the USSR's inability to fulfil its obligations due to the country's economic and political instability, as well as the fragile state of its rocket and space sector in the early 1990s.[55]

Interball Project

The Interball Soviet project involved research on solar wind. The main objectives of the project were to study the interaction of solar plasma flows with planets and small bodies of the solar system. In the history of solar-terrestrial physics research in the Soviet Union and Russia, the *Interball*

[53] Interview with L. Gurvits, March 17, 2025.

[54] Chertok. (1991).

[55] Interview with L. Gurvits, March 17, 2025.

multi-satellite project was considered as one of the most successful missions to study physical processes in near-Earth space.[56] The project succeeded in creating and realising a system of two pairs of satellites, targeting the auroral zone and the tail of the Earth's magnetosphere. The project entailed the deployment of a system comprising two satellites (the Tail Probe *Prognoz-1* and the Auroral Probe *Prognoz-2*) each equipped with *Magion*-type sub-satellites to be jettisoned.[57] The sub-satellites, named *Magion-4* and *Magion-5* respectively, were developed by the Institute of Atmospheric Physics of the Czech Academy of Sciences with the participation of specialists from the Soviet Union (Russia) and Austria.[58] Ultimately, the *Interball-1* satellite was launched in August 1995 from Plesetsk, followed by the launch of *Interball-2* in 1996.

The project was a cooperative work of specialists from the Soviet Union (Russia and Kyrgyzstan), Bulgaria, Czechoslovakia, Hungary, GDR, Greece, Cuba, Poland, Romania, Slovakia, Austria, UK, Italy, Canada, Finland, France, Sweden, and ESA. The project started in 1982, when a meeting of scientists specialising in plasma processes in the Earth's magnetosphere was held in Bulgaria. At the meeting, Albert Galeev expressed the need to develop a mission-oriented theory. Later, he was appointed head of the mission and Lev Zelenyi was nominated as the General Coordinator of the whole *Interball* project, while Rostislav Kovrazhkin and Mikhail Mogilevsky were project scientists of the Interball Auroral Probe. A planning conference was held in Moscow in January 1987 with the total amount of signatories eventually reaching twenty.[59]

The Soviets also made attempts to involve ESA scientists in work on the *Interball* project. In August 1989, the IKI director, Albert Galeev sent a fax to Roger Bonnet with a concrete proposal of cooperation on the *Interball* satellites. He asked ESA specialists to "consider two things: providing Soviet probes with S-band or X-band transmitters that would be installed on board of the vehicles and helping Soviets scientists with tracking from ESA ground stations".[60] David Dale from ESTEC replied to the request of Bonnet that ESA had "two possibilities to use the spare transponder from *Giotto* and the spare transponder from the *Ulysses* that could

[56] From materials kindly provided by L. Zelenyi.

[57] "Interkosmos" programme. Annual report for 1989–1990. In ESA-28130 (2), HAEU.

[58] From materials kindly provided by L. Zelenyi.

[59] Harvey. (2021), 50.

[60] Galeev A. to Bonnet, R.M. August 22, 1989. In ESA-28051 (3), HAEU.

be available only after *Ulysses* launch in November 1990".[61] In any case, he warned Bonnet about eventual restrictions applied by the Coordinating Committee for Multilateral Export Controls (CoCom) on the export of these units to the USSR. Concerning tracking support, he outlined that he needed more technical information about the orbit and the amount of tracking required in order to study the compatibility of the Soviet *Interball* satellites with the ESA ground stations.[62] However, these initial contacts were not further developed.

Nevertheless, researchers from ESA took part in *Interball*. A. Pedersen and R. Schmidt from the Space Science Department of ESA (ESTEC), together with Russian scientists from IKI and Austrian researchers from Austrian Research Centre Seibersdorf, collocated their Spacecraft Potential Control Experiment (RON) on board of a Russian Auroral probe.[63] Simultaneously, another potential area of collaboration between ESA and IKI was emerging, albeit not under the *Interball* framework, but within the broader scope of the International Solar-Terrestrial Physics (ISTP) Program.

Indeed, the mission itself was a key element in the implementation of a more extensive scientific program realised by ten satellites: the Japanese-American Geotail (NASDA/NASA), the Russian-Ukrainian *Koronas-I* and *Koronas-F*, the NASA Polar and Wind satellites, the ESA-NASA solar observatory SOHO, and four ESA *Cluster* satellites. All these international projects of various origins were part of the ISTP research program, coordinated by the cooperative Inter-Agency Consultative Group (IACG) (Fig. 4.1).

Set up in 1981 in Padua, the IACG was an informal body composed of agency representatives and primarily had the primary task of coordinating all matters related to observations of Halley's Comet from space. The establishment of the IACG for Halley's Comet facilitated a significant collaboration among the United States, the Soviet Union, Japan, and ESA. This collaboration enabled the gathering of a maximum amount of scientific data during the various observations of the comet, notwithstanding political differences among the respective participating countries. Following the conclusion of the mission, all participants collectively

[61] Dale D. to Bonnet, R.M. September 11, 1989. In ESA-28051 (3), HAEU.

[62] Ibid.

[63] Riedler W. et al., (*1998*). Experiment RON for Active Control of Spacecraft Electric Potential. *Cosmic Research*, 36 (1), 49

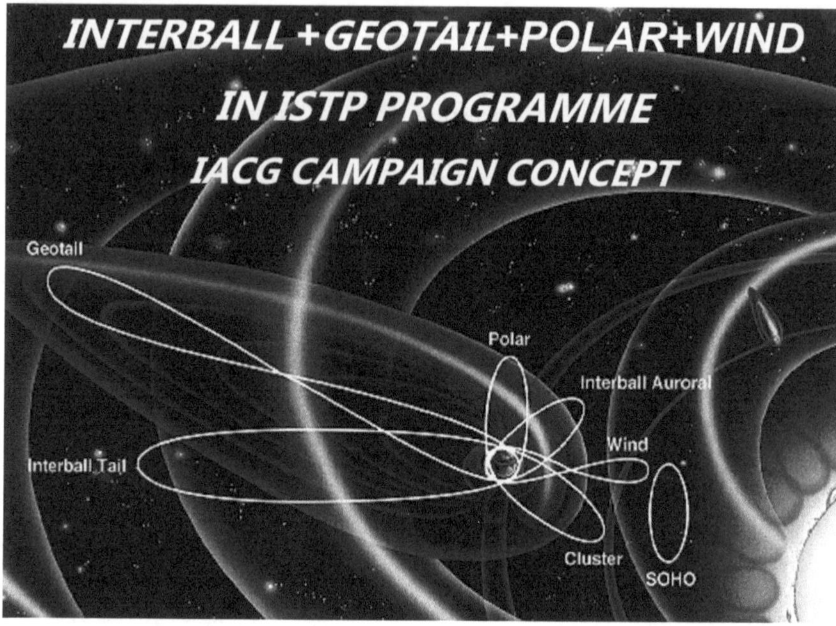

Fig. 4.1 Configuration of international satellites under the ISTP program. Source: L. Zelenyi

decided to broaden the mandate of the Inter-Agency Consultative Group to coordinate future international endeavours across other domains. Thus, in the 1980s, the solar-terrestrial science sector was receiving special attention of various agencies (ESA, NASA, and NASDA) that were planning several missions to study interactions in the solar-terrestrial physics under the effort called the ISTP Program. Soviet scientists also expressed their interest in solar-terrestrial missions and coordinated its efforts with other countries. Therefore, the IACG was considered as a proper body to coordinate future international cooperation.[64] The IACG coordinated more than ten projects among four agencies in the domain of the exploration of the interplanetary medium and in the study of the relationship between the Sun and the Earth.[65]

[64] Members consider extending scope of international group to coordinate future scientific space efforts. *Aviation week & space technology*, March 31, 1986, 119.

[65] Bonnet, Manno. (1994), 94.

Lev Zelenyi recalls the start of the program this way: "Everyone enjoyed the cooperation on Halley's Comet so much and nobody wanted to split up. All the participants of the mission that had already ended decided to choose a new topic. Since there was already *Interball*, a Japanese project, an American project and two ESA projects, the idea was to create a whole satellite armada to study the cosmic manifestation of solar activity, what is called space weather. And it was decided within the framework of IACG to organise joint work of the satellites that complemented each other well. This constellation of satellites became a major breakthrough in the study of near-Earth space, and campaigns were created to facilitate the coordination of satellite operations. For example, there was a campaign for studying the tail, that is, when the satellites were located in the tail, or a campaign for studying the magnetosphere. In total, there were four such campaigns. During this time, the satellites worked synchronously, data was collected from them, and articles were written. In short, everything was excellent".[66]

The Soviet/Russia's contribution to the ISTP was determined by the results of the *Interball* project—the Tail and Auroral probes with Czech sub-satellites. As Mikhail Mogilevsky argued, "the results of measurements on the *Interball* project satellites are used until today, indicating their high quality".[67]

Cluster Project and Regatta Satellite

Another point of contact between the Soviet and ESA programs was the attempt to incorporate the Soviet *Regatta* project into the ESA *Cluster* mission. The Soviet project *Regatta* was involved in plasma studies with a specialised Small Space Laboratory (SSL) (Fig. 4.2). It was developed at the end of the 1980s by a team of IKI specialists. In 1994–1998 the space *Regatta* was supposed to permit high-accuracy detailed measurements of the parameters of plasma and magnetic fields in the magnetoplasma sheet of the Earth (Fig. 4.3). Initially, it was planned that specialists from the USSR, Hungary, GDR, USA, Canada, and ESA would participate in the international *Regatta* project. Later it was decided to use two *Regatta* satellites for participation in *Cluster* programs jointly with ESA.

ESA missions, the *Cluster* and SOHO (the Solar and Heliospheric Observatory) projects were part of the ISTP Program coordinated by the

[66] Interview with L. Zelenyi, July 27, 2023.
[67] Correspondence with M. Mogilevsky, April 23, 2024.

Fig. 4.2 Regatta project logo. Source: G. Avanesov.

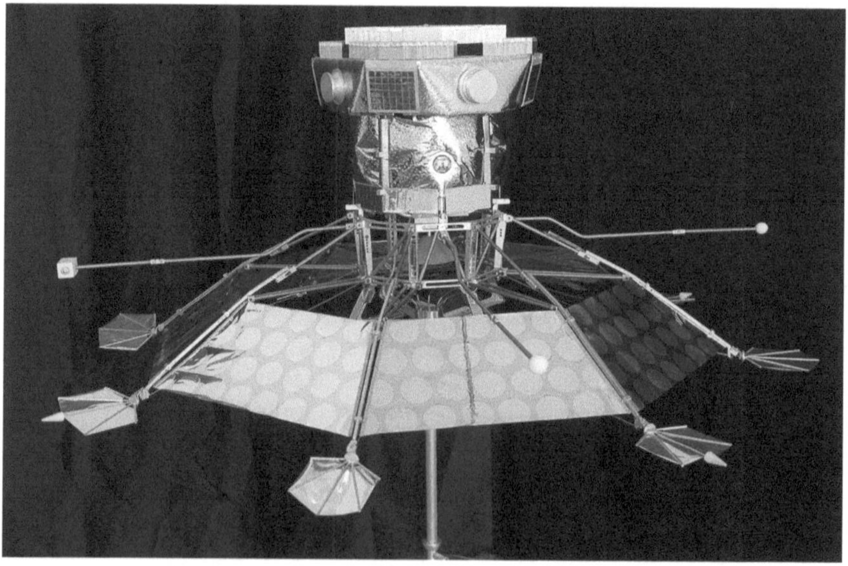

Fig. 4.3 A model of the *Regatta*. Source: G. Avanesov

IACG. SOHO and *Cluster* represented the first cornerstone of the Horizon 2000 program of ESA's scientific development, in cooperation with NASA. *Cluster* comprised four identical satellites, dedicated to investigating the Sun and its impact on the Earth's environment. ESA committed to financing approximately two-thirds of the combined projects, with NASA contributing the remaining funding.[68]

When Soviet scientists expressed their willingness to join the *Cluster* project, ESA leadership got worried essentially about two issues: the advantage of cooperation with Soviet partners for all Member States and the clear definition of the interfaces to solve the problems of technology transfer.[69] However, notwithstanding the doubts expressed by the ESA management, collaborative efforts on the project commenced and persisted until the Soviet side withdrew due to financial constraints.

In the framework of a cooperative agreement and at the end of long discussions, the Soviets were supposed to launch a fifth satellite named *Regatta-Cluster* in the 1995 time frame, so that it could add another point of measurement to the four already provided by *Cluster* and increase the redundancy of the overall mission. *Regatta* was supposed to be developed entirely by IKI, under Genrikh Avanesov's supervision and operated jointly with ESOC. As much as possible, the instrumentation was identical to that of the *Cluster* payload and was developed jointly between ESA and the Soviet Union.[70]

In May 1986, an ESA delegation visited IKI and made presentations on the general outline of the ISTP research program which incorporated both projects. The presentation contained information on the scientific goals of the program, preliminary studies of the technical and ballistic requirements, spacecraft hardware design, and general time schedules for the implementation of the program. Roger Bonnet outlined his perspective on prioritising options, ranging from maximum to minimum implications. He suggested constructing two spacecraft for each party (Soviet and European) or even potentially expanding the *Cluster* mission with additional spacecraft. Bonnet intended to proceed by inviting Intercosmos to participate in the ISTP Program. To go ahead with scientific and technical decisions, three Working Groups were set up (on science matters and

[68] Collins, G. (1990). *Europe in space.* Macmillan, 148.

[69] Relations with the U.S.S.R. IRAC. October 1, 1986. In ESA-28130, HAEU.

[70] Bonnet, Manno. (1994), 144.

instrumentation, on ballistics and operations and on spacecraft design) under the direction of the Joint Steering Committee.[71]

David Dale, Head of ESA's Scientific Projects Department at ESTEC, believed that, "the purpose of this joint activity was to explore in the areas of science, project design and flight operations, to what extent cooperation may be possible and how both projects could benefit".[72] Dale underlined that this joint study activity should be completed in January 1988 following the final Working Group session to take place in Moscow.

The second meeting between IKI and ESA representatives was held in Padua in December 1986. An agreement was reached to explore collaboration on the *Cluster* portion of ISTP and, eventually, to abandon the idea to cooperate on the SOHO one. After the meeting, Roald Sagdeev was determined to participate in the project and considered concrete steps towards cooperation. As he wrote to Reimar Lüst, ESA Director General, "it was really important to begin by now to do this work on the level of technical groups which were responsible for developing of spacecraft for *Cluster* mission".[73]

At an ESA-IKI meeting in January 1987 in Moscow, the Soviet side presented two spacecraft, each with an approximately 75 kg payload capacity and facilities for orbit adjustment. The Soviet mission, although concerned with the same scientific objectives as *Cluster*, had to be independent.[74] As Roger Bonnet pointed out, "this option was also one which almost eliminates technology transfer issues since the spacecraft are built independently and the payloads on the Soviet spacecraft are placed under Soviet responsibility. Should European scientists wish to bid for room on the Soviet spacecraft, it will be on an individual basis, as was the case for the Phobos, Vega and Vesta projects".[75]

Possible orbits would nevertheless be studied to such an extent that the two Soviet spacecraft would be in close proximity to the *Cluster* spacecraft during specific regions of the magnetosphere. Moreover, it was agreed that the two Soviet spacecraft and *Cluster* should contain common

[71] Protocol. Space Research Institute (IKI), May 21–22, 1986. In ESA-28057 (1), HAEU.

[72] Dale, D. to Galeev, A. October 6, 1987. In ESA-28057 (1), HAEU.

[73] Sagdeev, R. to Lüst, R. December 23, 1986. In ESA-28057 (1), HAEU.

[74] Protocol of meeting between ESA and Soviet representatives to discuss the possibilities of a cooperative STSP Program, January 27–29, 1987. In ESA-28057 (1), HAEU.

[75] Bonnet, R.M., Cooperation on ISTP, September 22, 1986. In ESA-28051 (3), HAEU.

experiment elements. This would make it possible to measure the same parameters at the same format.[76]

In June 1987, both sides convened in Hungary to address a series of practical issues, categorised into three working groups. However, a definitive decision regarding collaboration on the *Cluster* project had not yet been reached. The Soviet side anticipated reaching a decision "within the next year".[77] In October, during the annual meeting of representatives of ESA and the USSR Academy of Sciences in Moscow, it was mentioned that in the course of the last year, consultations were held between experts on both sides to consider the mission scenarios and to clarify possible shares of duties and define the technical requirements to the space vehicles, their orbits, and scientific equipment. At this stage, the ESA representatives informed the Soviet side of the possibility to enlarge their cooperation into the International Solar-Terrestrial Physics Program, which involved also Japan and the USA.[78]

In the letter sent after the Moscow meeting in 1987, David Dale contemplated several potential scenarios for the progression of cooperation, anticipating the outcomes of the working groups and the eventual conclusion of the final bilateral meeting. Like Bonnet before, on the base of results already obtained from the discussions, he considered the possibilities for cooperation from minimum scenarios (discussion and coordination of the scientific objectives and some general coordination of ESA/IKI spacecraft flight operations planning) to extensive scenarios (full scientific collaboration, joint operations planning, joint spacecraft performance and schedule alignment, inter-experiment calibration, joint in-orbit assessment and coordination, and full data exchanges in near real time).[79] At the final meeting in March 1988 in Moscow, IKI indicated that their preference laid with the option that foresaw full flight operations and science interaction between the ESA and the IKI during the mission. Sagdeev

[76] Protocol of meeting between ESA and Soviet representatives to discuss the possibilities of a cooperative STSP Program, January 27–29, 1987. In ESA-28057 (1), HAEU.

[77] ESA-IKI STSP Collaboration Steering Committee Protocol, June 1987, in ESA-28057 (1), HAEU.

[78] Protocol of the meeting between representatives of the U.S.S.R. Academy of Sciences and the ESA, Moscow, October 3, 1987. In ESA-28051 (3), HAEU.

[79] Dale, D. to Galeev, A. October 6, 1987. In ESA-28057 (1), HAEU.

added that it would be opportune to expand the "excellent cooperation started during the Giotto mission".[80]

After her visit to Moscow, Karin Barbance from the ESA International Affairs office reported the details of IKI's situation. As she put it, "IKI had initially intended to procure the spacecraft through Glavkosmos. Glavkosmos have not responded with constructive proposals and therefore IKI was actively considering to change the original procurement completely". During the meeting with ESA representatives, Soviet participants were supposed to retain the design authority and to procure the needed sub-systems directly from industry. They planned to procure the propulsion unit only from Glavkosmos. The spacecraft design was intended to provide the IKI with a bus-type modular design to be used in many missions during the 1990s.[81]

In March 1989 IKI informed ESA that funding had been approved for the first two spacecraft of the *Regatta* plasma physics program. The missions chosen for the two spacecraft were SSL-E (Equator) for a launch in 1993 and SSL-A (Cluster) for a launch in 1995/1996. The decision on the three other spacecraft in the *Regatta* program was postponed.[82]

The first evidence of financial difficulties with the Soviet party in fulfilling the obligations assumed under the *Cluster* project appeared in connection with the arrival of a Soviet delegation for a meeting at ESTEC in July 1990. ESA supported the living expenses for three Soviet scientists supposed to attend the ESTEC meeting. As John Credland, *Cluster* project manager, wrote to his IKI counterpart, Deputy director of IKI, Vyacheslav Balebanov: "I can agree to cover the expenses of three of your team here in the Netherlands. I offer you this because of our previous understanding on the subject but must draw your attention to Art.7 clause 2 of the Agreement between the ESA and Government of the USSR concerning cooperation in the field of exploration and use of outer space for purposes signed on April 25, 1990. This agreement states that the respective parties will bear the costs of visits of their own personnel".[83] Following meetings and extensive correspondence in January 1991, "John Credland and his

[80] Barbance, K. ESA/IKI meeting on *Cluster*, Moscow, March 9–10, 1988. In ESA-28057 (1), HAEU.

[81] Ibid.

[82] Protocol on the meeting between representatives of the IKI and ESA in potential cooperation on *Regatta/Cluster*, IKI, Moscow, March 14–15, 1989. In ESA-28057, HAEU.

[83] Credland, J. to Balebanov, V. June 5, 1990. In ESA-28057, HAEU.

team assembled the elements for the Implementation Arrangement for *Cluster/Regatta-A* cooperation".[84]

Despite the extensive preparatory efforts, detailed negotiations, and discussions, the Soviet satellite *Regatta* ultimately did not join the European armada. *Regatta* suffered a fate common to many *perestroika*-era projects: financial difficulties prevented the project from being realised.

As Bonnet stated: "The *Regatta* Program was invented for the benefits of IKI. It was both unconventional and very clever in many aspects".[85] Bob Farquhar, NASA-Goddard engineer and space flight expert, quoted by Bonnet added: "Perhaps the most unusual aspect of *Regatta* was the revolutionary spacecraft design concept base-lined for all of the missions in the Program".[86]

And yet the USSR and then Russia took part in the *Cluster* project. Initially, it was assumed that all four ESA satellites would be launched into orbit by the European rocket launcher Ariane. However, as follows from the correspondence between representatives of ESTEC and IKI, a proposal was received from the Soviet side to provide a Proton rocket for launching the satellites.[87] ESA's *Cluster* satellites were successfully launched aboard a Russian Soyuz-Fregat rocket from Baikonur in August 2000. This launch followed the unsuccessful maiden flight of Ariane 5 from French Guiana on June 4, 1996, during which the payload, consisting of four *Cluster* satellites, was lost.

Conclusion

The aforementioned examples of cooperation between ESA and Soviet space institutions, primarily IKI and IBMP, demonstrate the existence of a rich, diverse, and active scientific space exploration program in the USSR during the early stages of *perestroika*. Both entities applied programs focused on utilising space resources for the national economy and purely scientific projects were under development. However, funding for such programs in the USSR became increasingly insufficient, and their status remained subordinate to military space initiatives. Nonetheless, at the onset of *perestroika*, Soviet scientists were able to make significant

[84] Barbance, K. January 14, 1991. In ESA-28057, HAEU.

[85] Bonnet. (2016), 43.

[86] Ibid.

[87] Credland, J. to Avanesov, G. October 10, 1990. In ESA-28057, HAEU.

contributions to the advancement of scientific space exploration by inviting European colleagues to collaborate on their projects and by actively participating in ESA initiatives.

ESA-USSR cooperation occurred both on specific scientific projects and within the framework of broader international initiatives. For instance, the *Interball* project was part of an international ISTP program, with an instrument, RON, developed by ESA scientists and included as part of the payload on one of two Russian satellites. Similarly, the *Regatta* spacecraft was intended to be a component of the ESA *Cluster* project which, in turn, was also part of the aforementioned international program.

Three key points emerge from the analysed projects. Firstly, the cooperation could unfold in different ways: in some cases, the original projects were fully implemented (e.g., *Bion, Foton*); in other cases, the nature of the cooperation underwent significant changes during project (e.g., *RadioAstron*); and sometimes, for reasons beyond the control of the coordinators, the cooperation was entirely abandoned (e.g., *Regatta*).

Secondly, in comparison to national agencies, ESA's active collaboration with the USSR began much later. This is clearly illustrated by the cooperation on the *Bion* and *Foton* projects. ESA often relied on established paths, drawing heavily on the experience accumulated by national space agencies—primarily CNES—while striving to minimise potential risks.

Thirdly, the scientific community's desire to collaborate encountered various bureaucratic (and other) obstacles that hindered the development of a fast and effective cooperation. In other words, ideas did not always come to fruition due to factors beyond the control of their originators.

The following chapter will examine in more detail the reasons for and characteristics of the cooperation between ESA and Soviet space organisations.

CHAPTER 5

ESA-USSR Collaboration: Key Features and Insights

On one hand, the onset of *perestroika* and the associated policy of openness of the USSR to international cooperation in space opened new opportunities and unexpected advantages for ESA. On the other hand, there were certain doubts and drawbacks regarding collaboration with a new partner long considered "toxic" for both political and security reasons. As an intergovernmental organisation, ESA united several countries, each with differing perspectives on the potential for cooperation with the USSR. The development of a unified policy was considered a complex challenge for the entire management system responsible for decision-making. Clearly, the classified nature of Soviet space science activities, given their close ties to the VPK, did not make ESA leadership's task any easier. However, these "shortcomings" of Soviet space were largely offset by its numerous advantages, which were recognised by the national space agencies. In the emerging triangle of ESA, national agencies, and the USSR, each had to determine its position based on its interests and capabilities. This chapter will examine the challenges ESA faced in its efforts to establish closer relations with the USSR.

O. Dubrovina, *The Soviet Space Program and International Collaboration during Perestroika, 1985–1991*, Palgrave Studies in the History of Science and Technology,
https://doi.org/10.1007/978-3-032-03157-0_5

Challenges in Establishing a Unified Approach

It is true that with the change in the foreign policy climate and the greater openness of the Soviet Union, the position of most ESA member countries also changed. Jean Arets argued that: "We may conclude that the negative attitude of most of our Member States which penalised during fifteen years now belongs to the past. We can therefore consider the prospects of cooperation without any illusion and without purely political obstacles".[1]

Nonetheless, doubts continued to exist within the ESA leadership about a dynamic and close rapprochement with the Soviet Union as the foreign policy circumstances and the position of individual participating countries were still of major significance. Indeed, active Soviet-ESA alignment aroused the cautious attitude of the ESA Department of International Relations, which monitored not only the regulation of international contacts, but also their compliance with the general political line of the European Community and individual member countries of the Agency. In confirmation of this Reimar Lüst wrote in an inter-office memorandum that "ESA's relations with the Soviet Union have been intensified and have become more varied over the past year. This new situation may have political implications until now unknown in the Agency's relations with non-member States".[2] As Karl Bergquist, ESA Administrator for the International Relations Department, stated "Every international cooperation that ESA engages in must be unanimously adopted by its member states. In 1985, this process was exceptionally challenging and prolonged. Nevertheless, from a scientific perspective, there was always a strong drive towards participating in such activities".[3]

Thus, the differing positions of the European Space Agency's member countries regarding cooperation with the Soviet Union posed a significant obstacle to the establishment of full-fledged cooperation and rapid rapprochement between ESA and the Soviet Union. As Arets wrote in December 1987, "our Member States and the USSR are not ready to go full speed into intensive cooperation. We have therefore to accept that our progress will be slow and our steps prudent".[4]

[1] Arets, J. Prospects of cooperation with U.S.S.R., Mars 24, 1987. In ESA-28130, HAEU.
[2] Relations with the U.S.S.R. IRAC, October 1, 1986. In ESA-28023, HAEU.
[3] Interview with K. Bergquist, May 28, 2024.
[4] Arets, J. *Cooperation with U.S.S.R.* December 7, 1987. In ESA-28130, HAEU.

The primary concern revolved around the necessity for a realistic assessment of cooperation with the USSR. During the Science Program Committee colloquium held in Capri in April 1988, dedicated to the National Space Science Programs, the issue of cooperation between ESA and the USSR was deliberated. The summary reported a discussion concerning the features of collaboration with the Soviet Union from the perspectives of individual members. "A point was made by different speakers on the apparent extreme flexibility of opportunities on the Soviet program. This contrasted somewhat with our rather heavier procedure. Nevertheless, everybody agreed that the two systems could hardly be compared. Under this point discussions took place on the validity and advisability for Europeans to join the Soviet program without in-depth considerations of a scientific and financial nature. The potential danger to the Horizon 2000 program[5] was mentioned, if large bilateral initiatives were entered into without proper assessment".[6] Thus, the correlation and mutual combination of research programs also posed challenges. As Karl Bergquist recalled later,

> Collaborating with Russian scientists has always posed challenges in defining scientific programs. Our approach is fundamentally bottom-up, where scientists identify key questions about the universe, and we construct programs around these inquiries. As a result, we already have established programs in place.
>
> During our bilateral discussions with the Russian Academy of Sciences, integrating their missions proved challenging. They had their own established missions and agendas, making it difficult to find common ground or alignment between our respective scientific priorities.[7]

In the ranks of the ESA leadership, there was no consensus on the usefulness and benefits of cooperation with the USSR in the field of space research. Thus, Roger Bonnet believed that "Soviet rocket launchers were

[5] The Horizon 2000 initiative marked the inception of the ESA Science Program, representing a pioneering endeavour in the realm of space science and exploration missions. Spanning the years from 1985 to 1995, Horizon 2000 facilitated the realisation of eight missions, distinguished by four pivotal "cornerstone missions"—namely, SOHO and Cluster II, XMM-Newton, Rosetta, and Herschel, alongside three medium-scale missions: Huygens, Integral, and Planck.

[6] Summary points of the SPC Colloquium on the National Space Science programs, Capri, April 1988. In ESA-28051 (3)

[7] Interview with K. Bergquist, May 28, 2024.

failure-oriented, while we in Europe are success-oriented. We must be careful not to lose our identity. This is a dangerous phenomenon". He argued that the Soviets were offering possibilities "to fly experiments in space which looked very good on paper". But he felt they had "neither the technology nor the funds to cope with them and so were relying on the participation of the West".[8] It is worth noting that Bonnet's harsh judgment about the Soviet space infrastructure followed the relatively unsuccessful flight of both Phobos spacecraft, which carried European equipment. Another explanation may be the competition posed by the Soviet space industry and science to the emerging ESA space program. European space science projects frequently intersected with Soviet space missions, fostering an atmosphere of either competition or cooperation. It should be added, however, that despite such mild criticism of the Soviet space infrastructure in the free press, Bonnet was one of the most active supporters of Soviet (Russian)-European rapprochement, which was particularly evident during the period of the INTEGRAL project.

Serious doubts were also raised regarding the working style of Soviet scientific space institutes and industry. An important negative factor was the "cumbersome procedures and management difficulties".[9] Difficult access to the space infrastructure, coupled with the lack of transparency in scientific research and production processes, significantly raised doubts among Europeans and bolstered opponents of convergence. The early episode involving the placement of European equipment on board the Mir orbital station highlighted several challenges that European scientists encountered in cooperation with the USSR. As previously mentioned, on March 31, 1987, an astrophysics module named *Kvant-1*, which included X-ray experiments manufactured by ESA and agencies in West Germany, the Netherlands, and the UK, was launched to the Mir station as the first significant addition to the *Core Module*. One month before the launch, the USA magazine *Aviation week & Space Technology* published an interview with Alan Smith who complained about the Soviet restrictions that did not allow "Western scientists to know the exact launch date".[10] Moreover, the issue of data access was not resolved in a clear way. As Smith put it, "Technically we all had rights of access to each other's data, but we did not

[8] Warning over Soviet link-up. *The Independent*, October 4, 1988.

[9] Relations with the U.S.S.R. IRAC, October 1, 1988. In ESA-28130, HAEU.

[10] New Mir astrophysics module to include European Experiments. *Aviation week & Space Technology*. February 16, 1987.

know how it would work out in practice. We had no guarantee of access to their [Soviet scientists'—O.D.] raw data, but we thought we could get some if we wanted".[11] According to the newspaper information the Soviets did not allow Western scientists to observe the launch, though they were allowed to visit the Soviet Union for the experiment switch-on, 4–6 weeks after launch and Western scientists remained in the Soviet Union to monitor the package throughout its lifetime.

The opening of the Soviet space and rocket sector to the international community occurred gradually, beginning with invitations to participate in joint projects and eventually extending to the admission of foreign scientists to previously restricted research centres and production facilities. Thus, only in 1989 French specialists were permitted access to NPO Lavochkin, the spacecraft production plant. Their assignment was to install the *Sigma* telescope on board the *Granat* spacecraft directly within the plant's premises. As recalled by one participant of the joint project, "For the first time, a French team was allowed to work on the premises of this prestigious company, which until then had been shrouded in secrecy".[12]

ESA Vs. National Agencies

ESA staff closely monitored the development of relations between individual national agencies and Soviet space organisations. The Agency's leadership was apprehensive about the gradual convergence of Soviet space institutes and production centres with the national agencies of individual ESA member states, although the degree of participation in collaboration with the Soviet Union varied among the nations. By 1988, an analysis of joint projects separately conducted between national agencies with the USSR revealed the following ranking: France, Finland, Germany, Sweden, and Denmark.[13] The reinforcement and broadening of Soviet-ESA relations during the latter half of the 1980s coincided with heightened collaboration between the Soviet Union and some Western European countries that had either finalised intergovernmental agreements (France, Great Britain), or were on the brink of finalising them (Italy, the

[11] Ibid.

[12] Debouzy, G. (2015). Les programmes de coopération pour l'exploration pacifique de l'espace. In A. Ammar-Israël (Ed.), *50 ans de coopération spaciale France-Urss/Russia*. Tessier&Ashpool, 92.

[13] Summary points of the SPC Colloquium on the National Space Science programs, Capri, April 1988. In ESA-28051 (3), HAEU.

Netherlands, Canada, FRG).[14] In addition to inter-agency agreements, the Soviet Union entered into cooperation agreements with individual research centres. For example, in 1987 the Institut d'Aéronomie Spaciale concluded an agreement with the USSR Academy of Sciences in different areas of cooperation: astrophysics, material science, earth observation, and exchange of personnel. In March 1989, a French delegation, including French PTT representatives, negotiated the possible installation of a Soviet telecom station in Toulouse for the relay of scientific data via Intersputnik. With some regret, Karin Barbance mentioned the missed opportunity for ESA to get ahead of the French. She underlined that this opportunity had been discussed previously also with European officials (D. Dale and ESOC) and that "ESA had not reacted because it considered the possibility premature and too 'daring' vis-à-vis the member States".[15]

A notable aspect of the ESA-national agencies-USSR triangle was the combination of competition and cooperation between the first two parts in relation to Soviet space institutions. On the one hand, the emergence of newly available opportunities to leverage the Soviet space infrastructure engendered an atmosphere reminiscent of a race to attain a privileged position. On the other hand, numerous instances of consultations and exchanges of experiences arose concerning the ascendance of a new powerful player in scientific collaboration and the international space services market. Particularly noteworthy in this regard was the French space agency CNES, which could boast of extensive experience in space cooperation with the USSR over many years.

However, there was no shortage of contrasting instances characterised by mutual accusations and reproaches. As Roger Bonnet put it in 1986, CNES "wanted to have and keep the monopoly of cooperation with the Soviet Union and they were not ready to share it with the Space Science Department (SSD)".[16] The French side also lodged a series of complaints with ESA representatives regarding their prevailing stance towards Soviet space institutions. In particular, CNES officials accused the SSD of delivering scientific equipment to IKI, "while the French have to go through severe controls of their exports to the Soviet Union which are imposed by

[14] Relations with the U.S.S.R., Summary of a meeting with delegates, June 30, 1988. In ESA-28130, HAEU.

[15] Barbance, K. U.S.S.R.-IKI, Mars 23, 1989. In ESA-28051 (3), HAEU.

[16] Bonnet, R. M. to Barbance, K. Warning: SSD-IKI Cooperation, September 22, 1986. In ESA-28051 (3), HAEU.

the military authorities". Moreover, according to CNES staff, "French delegation had always had with SSD, viz. unfair competition for other scientific laboratories and not enough staff of French nationality employed in SSD".[17]

The participation of national agencies in Soviet parallel projects also caused dissatisfaction with the leadership of the European agency. Thus, according to the British *The Independent* that quoted the words of British scientists, "ESA has been upset" by the decision made by the UK to participate in the Soviet Spektr-RG space astronomy mission, due for launch in 1993 because "it diverted resources from the Agency's X-ray astronomy project, XXM, scheduled for launch in 1998".[18]

The positions of ESA and the national agencies diverged in their approaches to cooperation with the USSR, particularly with regards to the transition from gratuitous scientific cooperation to commercial partnerships with the Soviet space sector. Financial matters between ESA and the Soviets persisted unresolved for a significant duration. Primarily, ESA management needed to ascertain precisely which entity was held responsibile for this domain and with whom negotiations regarding the continuation of gratuitous cooperation should occur: Glavkosmos or Intercosmos. While Intercosmos was a structural component of the USSR Academy of Sciences, Glavkosmos was established as the international interface of the MOM. In summarising his efforts to clarify this issue, Arets argued: "The relationship with Glavkosmos could be the easiest one and would be based exclusively on commercial considerations. Intercosmos is more interested in cooperative aspects".[19] Karin Barbance noticed that high-level representatives of the Academy of Sciences and of Intercosmos indicated that they were in favour of the continuation of cooperation on a no-exchange-of-funds basis in areas of fundamental science provided both parties had mutual interest in such cooperation.[20] The ESA's goal was to find a compromise between the reimbursable basis and the cooperative one. The challenge was that the most critical areas of cooperation for ESA—life sciences and material sciences—were also the most "vulnerable", as these fields faced significant challenges without access to the Soviet space infrastructure necessary for conducting experiments.

[17] Ibid.

[18] Warning over Soviet link-up. *The Independent*, October 4, 1988.

[19] Arets, J. *Cooperation with U.S.S.R.* December 7, 1987. In ESA-28130, HAEU.

[20] Barbance, K., Arets, J. *Relations with the U.S.S.R.* In ESA-28130, HAEU.

By then, France and Germany had already entered in commercial cooperation with Glavkosmos, which limited the field for future manoeuvres for ESA by setting a precedent. The Soviets requested that the concluded agreements of cooperation with the French and German governments were reviewed with a view to introduce a commercial element.[21] At the outset, the French space agency rejected this approach as it was unwilling to conflate purely scientific cooperation with commercial dealings. French space leadership maintained a preference for scientific collaboration, grounded in the principle of no-exchange-of-funds. Nonetheless, they entertained the notion that, in certain instances, there might be scope for a commercial-type agreement with Glavkosmos. For instance, this could entail the launch of a commercially or industrially oriented payload necessitating retrieval by a Soviet vehicle. Therefore, CNES concluded such a contract with Glavkosmos for the launch of a material sciences payload on a Soviet retrievable capsule.[22]

Indeed, as already mentioned, France was first on board the *Foton* program, beginning with *Foton-5* launched on April 26, 1989. It was followed by an agreement between CNES and the Soviet import-export company Licensintorg[23] to fly the Crocodile experiment, launched aboard *Foton-6* in 1990.[24] The French agency adeptly defended itself against criticism, addressing concerns regarding both political and commercial inconvenience. It did so by affirming that the scientific instruments transported into space on Soviet rockets were securely sealed, mitigating the risk of potential tampering by Soviet intelligence services. This should have

[21] *Relations with the U.S.S.R.* Summary of a meeting with delegates. June 30, 1988. In ESA-28130, HAEU.

[22] Ibid.

[23] Licensintorg was a foreign economic association established in 1962 to carry out international technological exchange on a commercial basis. Until 1991, the Licensintorg was a state-authorised customer for attracting foreign technologies to national production. In 1985, by the Decree of the Council of Ministers of the USSR, the Licensintorg was instructed to represent interests in the field of international cooperation in the exploration of outer space. The first commercial experiments aboard satellites and space stations, British, West German, and Japanese astronauts were sent into space under the contracts of the Licensintorg. Retrieved March 4, 2024 from https://licenz.ru/o-kompanii/.

Jeffrey Manber explained Licensintorg's role in the commercialisation of Soviet space infrastructure as follows: "Licensintorg for decades had done business with America. It must mean that Licensintorg had the necessary bank accounts and experienced contract negotiators". Manber. (2009), 24.

[24] Harvey. (2021), 66.

reassured Western allies within CoCom, who were apprehensive about the potential leakage of sensitive knowledge and cutting-edge technologies. Furthermore, CNES' decision not to utilise a domestic product such as Ariane but instead opting for a Soviet competitor was rationalised by the modest size of the payload and the necessity to employ capsule return technology pioneered in the USSR.

The long-standing Franco-Soviet project on manned cosmonautics was also transformed into commercial co-operation. On November 26, 1988, a spacecraft carrying an international Soviet-French crew was launched for the second time from Baikonur. This second Soviet-French space flight was executed in accordance with an agreement between the governments of the USSR and the French Republic. Concurrently, during a visit to Moscow in November 1988 by the president François Mitterrand, a new agreement was reached.[25] This agreement outlined plans for four French flights commencing in 1992, with launches scheduled at biannual intervals. For the inaugural two-week flight, CNES remitted $11.6 million.[26] Thus, for the first time since the start of the Franco-Soviet collaboration in 1966 CNES "had to accept the principle of a financial participation: approximately 12 million dollars for the Antares mission", a joint manned flight scheduled for 1992 on the Mir orbital station.[27]

Meanwhile, France was losing its longstanding privileged status in space collaboration with the USSR. Space agreements in that period were being negotiated from Bonn as well. To understand how this shift occurred, we must delve into events rooted from 1970, when an Agreement was reached between Bonn-based Deutsche Forschungsgemeinschaft (The German Research Foundation) and the USSR Academy of Sciences; encompassing all realms of scientific collaboration and serving as a comprehensive framework for most scientific cooperation efforts. On July 22, 1986, an Agreement on scientific and technological cooperation, including a reference to space endeavours, was officially sanctioned at the governmental level. Scientific space experiments were to be conducted by researchers from the Max Planck Institutes and the Soviet Academy of Sciences, focusing on radio astronomy and experiments aboard Phobos.[28]

[25] Nikitin, S. (1989). *SSSR: mezhdunarodnoye sotrudnichestvo v kosmose*. Znaniye, 55.

[26] Baturin, Yu. (Ed.), (2005). *Mirovaya pilotiruemaya kosmonavtika. Istoria. Tekhnika. Lyudi*. RTSoft,481.

[27] Moreau, Ph. (1990). L'ESA va signer un accord-cadre avec Moscou. *Les Echos*, April 6.

[28] Relations with the U.S.S.R. Summary of a meeting with delegates. June 30, 1988. In ESA-28130, HAEU.

Two years later in the Autumn of 1988, an FRG-Soviet space Agreement was signed in Moscow. The document was signed between Gury Marchuk, President of the USSR Academy of Sciences, and Heinz Riesenhuber, the Minister of Scientific Research and Technology of the FRG. This Agreement outlined scientific and technical cooperation in the realm of research and peaceful utilisation of outer space. Additionally, the agreement stipulated collaboration in both manned and unmanned space flights, aimed at the exploration and utilisation of outer space.[29] Cooperation would be expected in several areas: from solar-terrestrial physics to space biology and space medicine. The agreement also predicted the flight of a cosmonaut from West Germany.[30] According to ESA staff information, the Soviet side attempted to implement the first flight on a commercial basis, but the German counterpart successfully rejected this proposal. It was then agreed that the flight would take place in the framework of scientific-technical cooperation and would therefore be implemented on a no-exchange-of-funds basis. Furthermore, the commercial undertakings were not comprised in the Agreement.[31]

Relations between the USSR and the FRG in the field of space exploration and development also progressed at a lower level through private business. However, initially, the FRG government disapproved of the private initiatives taken by German firms that signed contracts with Glavkosmos, seeking to avoid any violations of CoCom rules and to adhere to the policy of "privileged utilisation of Ariane".[32] Notwithstanding this official stance, which mirrored the prevailing dynamics of international relations during the Cold War era, emerging trends were steadily gaining traction. Thus, while the United States called for generally stricter control over exports to Moscow, FRG Foreign Minister Hans-Dietrich Genscher argued for a liberalisation of the CoCom regulations. Acting as a spokesman for the West German government, he claimed that export controls were a product of the Cold War and needed to be adapted to the new spirit of cooperation: "These restrictions need to be rethought and reduced to what is really necessary...It is, after all, in the European and the

[29] Nikitin. (1989), 56.

[30] German astronauts will fly on Mir in 1992 (Klaus-Dietrich Flade) and 1997 (Reinhold Ewald).

[31] Barbance, K. *USSS; German-Soviet space agreement.* In ESA-28130, HAEU.

[32] *Relations with the U.S.S.R.* Summary of a meeting with delegates. June 30, 1988. In ESA-28130, HAEU.

Western interest to overcome the economic and technological division in Europe as a whole…".[33]

Simultaneously, industrial engagements were progressing. Kayser-Threde finalised an agreement for the launch of three spacecraft, weighing between 100 to 150 kg each by Soviet carriers, commencing from 1989. These spacecraft were intended to conduct a variety of biological and materials science experiments under zero-G conditions. Initially, Kayser-Threde had been launching its devices using American Shuttles. However, after a more than two-year hiatus in American Shuttle launches following the Challenger catastrophe, West German companies found themselves compelled to explore the Soviet alternative for "space freight". Furthermore, Glavkosmos guaranteed the complete safety of launched objects, and the cost of Soviet freight was lower than any other option available[34] (see more in Chap. 10).

On October 11, 1988, Nikolay Ryzhkov and Franz Vranitzky, the Federal Chancellor of the Republic of Austria, signed an official intergovernmental Agreement in Moscow between the USSR and Austria for a joint Soviet-Austrian manned space flight. Subsequently, on November 24, a follow-up commercial agreement on cosmonaut training and flight performance was signed in Vienna.[35] This agreement outlined the preparation of an Austrian citizen for space travel with the objective of conducting a series of scientific experiments in a weightless environment.[36]

Italy began to establish official collaboration in space with the USSR with a noticeable delay compared to other countries, such as the USA, France, and Germany. Only in 1987 did Soviet-Italian negotiations start on scientific space cooperation. On October 14, 1988, in Moscow, Eduard Shevardnadze, the Minister of Foreign Affairs of the USSR, and his Italian counterpart Giulio Andreotti, signed an Agreement between the Soviet government and the government of the Italian Republic concerning cooperation in the exploration and utilisation of outer space for peaceful purposes. The agreement envisaged cooperation across various domains, including the physics of the solar system, planetology, astrophysics, cosmology, space materials science, space biomedicine, geodesy, and Earth

[33] Cited in Ketterer, C. (1994). *German Liberalism and Foreign Policy: The FDP's Ostpolitik Under Hans-Dietrich Genscher, 1974–1990*. Submitted for the Ph.D. Degree, London School of Economics, 187. Retrieved February 7, 2025, from https://etheses.lse.ac.uk/1375/.

[34] Sevastianov, Pryakhin. (1989).

[35] Baturin. (Ed.), (2005), 684.

[36] Nikitin. (1989), 55.

observation. As part of the concrete provisions of this Agreement, the Italian side was offered participation of an Italian cosmonaut in one of the Soviet space missions.[37]

The UK was also in no hurry to formalise its relations with the Soviet Union in space programs. Their intergovernmental Agreement on civil space cooperation was signed in 1987. In early 1988, the USSR embassy in London requested its review to take account of the Soviet commercialisation efforts but the UK government resisted the approach indicating that it was not opposed to private companies having contacts with Glavkosmos.[38]

Returning to the financial settlement of relations between ESA and the Soviet authorities, it is noteworthy that the leadership of the Agency held a fundamental opposition to transitioning towards market-oriented relations with the Soviet Union in the field of scientific collaboration. On one hand, efforts were made to identify areas of cooperation that would be mutually beneficial on a non-monetary basis. On the other hand, endeavours were made to explore opportunities for reimbursing the cost of services to the Soviet side, ideally through non-monetary "in kind" means. As a result, ESA managed to negotiate a cooperation agreement on Phobos with Intercosmos instead of Glavkosmos, recognising it as a mutually advantageous arrangement (see Chap. 6). Furthermore, ESA was authorised to "leave behind some limited ground support equipment necessary for carrying out its experiments provided that this is in conformity with the CoCom rules".[39] Additionally, European experiments conducted on Biocosmos (*Bion*) satellites in collaboration with Intercosmos/IBMP were also viewed as a mutually beneficial area.

Still, the ESA leadership did not preclude cooperation on a financial basis. This stance was influenced by several factors. Firstly, in the past, Europeans had already incurred costs for the services of American carriers. Secondly, due to the Soviet insistence, opportunities for purely scientific cooperation were becoming increasingly limited. Thirdly, precedents had already been set by European national agencies that opened the way for commercial relations with the Soviet rocket and space sector. As Arets pointed out: "the commercial approach might be a welcome means for

[37] Ibid., 57.

[38] *Relations with the U.S.S.R.* Summary of a meeting with delegates, June 30, 1988. In ESA-28130, HAEU.

[39] Arets, J. *U.S.S.R.* November 8, 1988. In ESA-28130, HAEU.

flying materials science experiments as 'black boxes' without sharing the final results. ESA has not hesitated in the past to pay NASA for flight opportunities. Member States have flown experiments on Chinese and Soviet space vehicles against payment".[40]

It is evident that, despite the efforts of its staff, ESA lagged behind the national space agencies of its member countries in terms of the intensity of cooperation by the end of the Cold War period. David Southwood, Director of Science and Robotic Exploration at ESA (2001–2011), offers the following explanation for this phenomenon:

> Cooperation with the USSR in the late 1980s was largely driven by shifts in Western states' foreign policies, which transferred control to individual national governments. My presence in Moscow was a result of Margaret Thatcher's personal rapport with Mr. Gorbachev—there were no other motives. She pursued collaboration in space, and I was always eager to participate.
>
> This collaboration was not overseen by ESA but rather by national governments. West Germany engaged because its government aimed for openness towards Russia. Germans, known for their generally amicable relations with Russia, and the French naturally joined in cooperation efforts. The British and Americans also participated in this scientific endeavour.
>
> Nevertheless, our governments supported this cooperation for political reasons, seeking to forge connections and recognising that bolstering scientific and technological ties after the Soviet Union's collapse would enhance global stability. So, ESA was standing a little bit on sidelines.[41]

Data Exchange: A Case Study in Dynamic Collaboration

Despite the above-mentioned series of difficulties that the ESA management had to overcome in cooperation with Soviet partners, a significant leap forward in establishing both scientific and commercial links was made during the 1980s. It is suggested that the evolution of the ESA approach be examined through the example of such a sensitive area of cooperation as a data exchange. This case also offers a clearer understanding of the environment in which European and Soviet scientists were required to operate. Indeed, the risk of a possible transfer of cutting-edge technology

[40] Ibid.

[41] Interview with D. Southwood, May 21, 2024.

was one of the key issues in Soviet-ESA cooperation. The member countries of ESA, being bound by the CoCom regulations, could not avoid being careful and took into account this significant aspect of cooperation in the space field. They had to respect the special constraints imposed by the rules of CoCom and look for satisfactory solutions in every single case of collaboration.

Thus, the first attempt to establish an exchange of scientific data between an ESA institute and a Soviet organisation was unsuccessful. In 1982, the Soviet VNIIPAS[42] sent its request to the ESA IRS (Information Retrieval System) regarding the exchange of computer-based information. The Soviet side offered access to their space data banks in return. The ESA International Relations Advisory Committee (IRAC) advised the Executive against further pursuit of this matter.[43]

In 1985, the issue resurfaced when the office of the USA Secretary of Defence took interest in a statement from the ESA's 1982 annual report, which addressed discussions taking place at that time between the USSR Academy of Sciences and ESA regarding the aforementioned exchange of data. Ian W. Pryke, an ESA representative at the Agency's Washington office, later clarified that "the inquiry appeared to have originated from the Office of the Undersecretary of Defence for Policy".[44] In addition to concerns over the potential exchange of data between the European and Soviet sides, the Americans also expressed apprehension about the possible imminent signing of a "major agreement" between ESA and the Soviet Union, the existence of which was unknown to NASA.

Pryke suggested that USA partners were wary of the possibility of technology transfer related to orbital stations: "It is possible that it could have a bearing on the space station technology transfer issue!"[45] Jean Arets hastened to assure his American colleagues that between ESA and the Soviet Academy of Sciences there was only "a cooperative agreement providing for the exchange of already published documentation" signed in 1971. He also specified that the discussions about the exchange of

[42] The All-Union Scientific Research Institute for Applied Automated Systems (VNIIPAS) was a Soviet research institute that maintained computer networking, including international digital connections.

[43] *Relations with U.S.S.R.* IRAC. October 1, 1986. In ESA-28130, HAEU.

[44] Pryke, I. W. *ESA/U.S.S.R. relations.* April 18, 1985. In ESA-28130, HAEU.

[45] Pryke, I. W. *ESA/U.S.S.R. relations.* April 17, 1985. In ESA-28130, HAEU.

computer-based information "have not been pursued and it was not expected they would re-start in the near future".[46]

However, Arets was somewhat hasty in his promise. As he argued one year later "the only area of cooperation in which it was possible to comply with the requirements for non-transfer of secret technologies were scientific projects". At the same time, even in the field of scientific cooperation, it was necessary for ESA staff to adhere to clear rules for working with Soviet colleagues: simplify the interface, guarantee the presence of ESA staff when equipment had to be integrated in the USSR, and verify the sensitivity of the technology to be used before starting cooperation in this area. Even if, as Arets pointed out: "the problem of transfer of technology was an important issue which was taken very seriously by European officials, it nonetheless had not constituted an unsurmountable obstacle for cooperation".[47]

This stance is prominently exemplified in the episode involving the establishment of a direct fax line of communication between the Soviet IKI and ESOC.[48] The establishment of this link was initiated to support the implementation of Project Pathfinder in the early 1980s, within the framework of the Halley's Comet project. In 1987, the Soviet authorities broached the question of its potential utilisation post-project culmination, as well as the necessity for its maintenance and potential upgrades.

The already mentioned Pathfinder project represented a collective effort, involving the Soviet Intercosmos, NASA, and ESA for the benefit of the Halley's Comet mission. There was the stringent requirement for the European *Giotto* probe to fly past comet Halley at a distance of ~500 km on the sunward side. The uncertainties in the comet's trajectory were such that this requirement could not be fulfilled with the necessary accuracy using only ground-based astronomical observations; early studies revealed that a considerable improvement could be attained by incorporating space-borne observations. As the two Soviet *Vega* spacecraft were to reach the comet a few days before European *Giotto*, optical observations performed by the Vega cameras could be used for this purpose. These, combined with a very accurate determination of the Vega trajectories

[46] Arets, J. *ESA/U.S.S.R. relations*. April 17, 1985. In ESA-28130, HAEU.

[47] Relations with U.S.S.R. IRAC. October 1, 1986. In ESA-28130, HAEU.

[48] The European Space Operations Centre located in Darmstadt, Germany. It is responsible for tracking and controlling European spacecraft, developing and managing ground systems.

using VLBI, permitted a final fly-by distance of 600 km to be achieved with an uncertainty of only ±40 km.[49] Thus, the link was primarily used for the transmission of observation measurements and ephemeris data of Halley's Comet. As Kurt Heftman, Director at ESOC stated, "The Soviets rendered us an excellent service on the Pathfinder mission".[50]

The computer terminal and communications equipment were supplied by ESOC to IKI. However, after the very successful cooperation during the Halley's Comet mission, there was a certain reluctance on both sides to remove the link. Moreover, since the termination of the Pathfinder, the link had practically been idle. Indeed, in 1987, this equipment was already to a large extent technically obsolete, and it was impossible to re-use it in case it would return to ESOC. Thus, the ESA management decided to maintain the link for the time being, because the Soviets demonstrated an interest to maintain it. In evidence, they presented several proposals for future cooperation, centred around a continued use of the computer link. ESA scientists turned their attention to the proposal for an observation campaign of Mars' moon, Phobos. However, Heftman defended this project arguing that it was ESA's scientific interests to provide a "possible support in order to ensure a sufficient orbit determination accuracy of Phobos for the purpose of an optimal navigation of the Russian space probe".[51]

Kurt Heftman continued with a series of proofs of European interest in the Soviet Phobos project. Thus, he argued that with this support, ESOC would contribute to an improvement in the level of confidence for an optimal approach of Phobos by the Soviet space probe. From a scientific point of view, this collaboration in the Soviet project would extend ESA expertise in orbital mechanics related to interplanetary space missions. However, as Heftman put it, it was not only the question of pure scientific cooperation. He considered the European engagement in the Soviet project in terms of "fair recognition of the support which the Soviets have rendered to us for the navigation of *Giotto*". Moreover, he underlined that the total ESA effort in financial expenditure "was certainly less than the Soviets have expended when rendering us a service on the Pathfinder

[49] Münch, R., Sagdeev, R. & Jordan, J. (1986). Pathfinder: accuracy improvement of comet Halley trajectory for Giotto navigation. *Nature* 321, 318–320

[50] Heftman, K. Computer link ESOC-IKI (Moscow), January 16, 1987. In ESA-28130 (2), HAEU.

[51] Ibid.

Mission".[52] Thus, it was decided not only to maintain, but also to extend the ESOC-IKI link for the support of future cooperation projects between ESA and IKI. These were the Soviet's Phobos mission with ESA's active participation in spacecraft navigation manoeuvres (ESOC) and onboard scientific experiments (ESTEC) as well as the future Solar-Terrestrial Scientific Program (STSP).[53]

ESOC undertook the necessary steps through German authorities to obtain the necessary export licences. For ESA, this area of cooperation presented several serious problems, as it concerned a highly sensitive aspect of collaboration, i.e. exchange of data therefore the risk of technological information leakage. As Arets claimed, the very nature of ESA as an International Organisation "prevented the normal inter-relationship between national authorities. This applies, for example, to the procedure for [...] the improvement of a direct line between ESOC and IKI. While, for instance CNES cooperation with the USSR is permanently guided by national security consideration and supervised by competent authorities".[54] Arets warned that "ESA had to be careful enough to avoid being accused of doing something a national entity could not do for security reasons even if we think that we do the right thing. Appearance in this field can be more important than reality".[55]

A decision to maintain the ESOC–IKI connection was made in March 1987, and during the same year, ESA officials supposed to start transferring data on it.[56] It was decided not only to maintain the existing communication link between ESOC and IKI but to ensure improved services for the support of the forthcoming agreed ESA-IKI cooperation projects by modernising the outdated ESA facilities located at IKI. This entailed, in particular, the integration of a telephone system with teleconference capability and Electronic Mail stations at IKI and Glavkosmos. In addition to establishing a direct link between the two institutions, it became necessary to explore the means whereby high-speed electronic data exchange for all unprocessed data of IKI and ESA spacecraft being part of STSP could be established. IKI proposed to make use of Intersputnik, a Soviet Telecommunication satellite, with a special dedicated transponder for the

[52] Ibid.

[53] *ESA-IKI Cooperation*. Proposal for Communication Link. In ESA-28194, HAEU.

[54] Arets, J. *Cooperation with U.S.S.R*. December 7, 1987. In ESA-28130, HAEU.

[55] Ibid.

[56] Bonnet, R. M. *Cooperation with the Soviet Union*. March 23, 1987. In ESA-28130 (2), HAEU.

space activities of the Academy of Science. This proposal was made in Moscow in July 1987. The ESA representatives considered this proposal as a very attractive and far-reaching cooperative program, particularly in respect of ESA's role in future global space science projects. The experimental phase was foreseen for the period of 1988–1989 while the operational phase had to be developed in 1993–1998. Both phases were considered as important steps towards a bilateral cooperation between ESA and IKI and towards the creation of a Global Space Science Network involving ESA in a leading role.[57]

Consequently, the equipment for the Communication system was delivered from ESA to IKI. It was highlighted that all equipment was remaining ESA property and could not therefore be altered, removed, or replaced without ESA's prior approval. IKI had to ensure continuous power to the main communication system rack, to keep the material in working condition by allocating qualified IKI staff, to guarantee the preventive and corrective maintenance and to clean all equipment regularly (particularly facsimile machine, photocopier, magnetic tape drive, and floppy disk drives). Necessary consumables were supplied by ESA in sufficient quantity to cover approximately twelve months of normal operation. It was understood that IKI would ensure future supplies at their own expense in time to allow an uninterrupted service. Some further extensions were considered by IKI and ESA to the already installed communication system. Voice and data extension were foreseen to link ESA and Glavkosmos, the Flight Control Centre in Kaliningrad, CRIP[58] Budapest and a direct computer-to-computer file transfer link. Installed facilities were serving as general-purpose communication devices for cooperative projects based on agreed requirements. Additional requirements could be eventually met by increasing the existing circuit capacity, but this technical change was subject to the analysis of requirements, circuit quality, CoCom licence, and funding of necessary investments.[59]

ESA specialists performed training for IKI staff for fax, phone, and PROFS services. However, the latter had some difficulties to be used as IKI missed dedicated print facilities that seriously hampered the way

[57] *ESA-IKI Cooperation*. Proposal for a Communication Experiment, 1987. In ESA-28194, HAEU.

[58] Central Research Institute for Physics.

[59] Protocol. ESA-IKI Interagency communication system hand-over, Moscow March 9–10, 1988. In ESA-28194, HAEU.

PROFS could be used. Therefore, it was strongly recommended to allocate a printer at IKI which could act as a remote PROFS hardcopy station.[60]

The quality of the established connection, coupled with its scarcity of alternatives within the contemporary Soviet context, predetermined its remarkable success. Nils Oeberg, head of the Stockholm office of the recently founded "International Foundation for the Survival and Development of Humanity" (IFSDH) with offices in Moscow, Washington and Stockholm, turned to Sagdeev for help. Roald Sagdeev who became one of its Board Members addressed Kurt Heftman, in order to ask him for temporary permission to use one of the IKI-ESOC dedicated telephone lines for the communication needs of the Foundation. As Sagdeev himself admitted, "it's crucial for the successful establishment of this organisation" and he saw "few other viable alternatives at the present time. The ones we have tried were too unreliable".[61] In his letter, Nils Oeberg outlined that "the Foundation was a newly established organisation which was international, independent, non-governmental and non-profit". He stressed that the possible use of the IKI-ESOC link "would temporary solve our problems and certainly help us getting started".[62]

In his internal report, Kurt Heftman commented on the technical feasibility of such a project, stating that: "the current note at IKI permits up to six IBM PC compatible workstations to access Western communications networks via the ESOC based Network Control Centre". He recommended that the request be accommodated, underlining that "all extra costs be paid by the Foundation, that priority was assigned to Phobos traffic, and that this service be provided on a best-efforts basis".[63]

However, the issue of the concurrent use of the communication system did not end there. Gradually, a series of projects surfaced, entailing its utilisation for the advantage of both IKI and non-ESA third parties. On each occasion, IKI was obliged to seek authorisation for the link's usage, encountering several objections from European personnel. Notably, during a telephone discussion between Karin Barbance and Klaus Blank, ESA Satellite Controller, on June 8, 1988, several critical issues were brought to the fore.

[60] Memorandum, ESANET Installation at IKI Moscow, Mars 31, 1988. In ESA-28194, HAEU.

[61] Sagdeev, R. to Heftman, K. May 29, 1988. In ESA-28194, HAEU.

[62] Oeberg, N. to Heftman, K. 23 May 1988. In ESA-28194, HAEU.

[63] Heftman, K. *Use of ESOC-IKI line.* June 7, 1988. In ESA-28194, HAEU.

Firstly, they emphasised the fact that due to Sagdeev's active promotion of the link, ESOC received more and more requests for use of the link from the outside. Apart from the already mentioned IFSDH they recalled Prof. Reigber from Munich, the University of Nice, and the University of Antwerp. They expressed doubts about the advisability of providing communication established by ESOC at IKI to external organisations, believing that such additional requests would increase the use of the link and heighten the risk of technical issues. It became dangerous as "the Soviets were not capable of repairing the equipment". Secondly, the demand for ESA technical representatives in Moscow was becoming increasingly urgent. This necessitated a rise in expenditure for the maintenance of the communication system, mostly borne by ESA. Thirdly, additional costs were anticipated for extending the ESOC-IKI line to the Flight Control Centre and the G. Babakin Research and Testing Centre, as suggested by the Soviet side within the framework of the Phobos cooperation. Thus, as Barbance and Blank suggested, a decision of principle was required with regard to further external requests to use the link. They paid special attention to the need to assess the advantages to ESA in terms of political or operation benefits and to the eventual reaction of member states to the fact that "ESA provides equipment to the Soviet Union which might be difficult to obtain from the West otherwise or cost too much in terms of foreign exchange".[64]

This case illustrates the evolution of ESA's policy of developing closer ties with Soviet organisations in the field of communications and information exchange. It provides valuable insights into the changing dynamics of the relationship between the European agency and IKI, a scenario that only a few years earlier seemed improbable. Despite the persistence of previous barriers (CoCom restrictions), cooperation reached a new higher level, which implied the establishment of links in quite sensitive areas. This new factual state of affairs was inevitably reflected in the formalisation of ESA-USSR relations. Approximately fifteen years later, the 1971 treaty was replaced by a new agreement that was more precisely aligned with the realities of the changed circumstances.

[64] Barbance, K. *U.S.S.R. ESOC – IKI link.* June 8, 1988. In ESA-28194, HAEU.

New Cooperation Agreement

The transition to a new relationship necessitated a revision of the outdated framework agreement governing the development of Soviet-ESA scientific contacts. The foundation for any cooperation endeavour between ESA and the Soviet Union was previously established by an Agreement constituted through an exchange of letters between the Soviet Academy of Sciences and ESRO, which came into force on March 12, 1971. This Agreement facilitated the exchange of scientific and technical information without restriction as to its use, as well as the organisation of exchange visits among specialists. However, the execution of joint projects necessitated the conclusion of a special agreement in each specific case. In 1989, the Soviet side expressed a willingness to sign a new agreement with ESA.[65]

The process of convergence of space programs, initiated during *perestroika*, culminated on April 25, 1990, with the signing of a framework Agreement between the government of the USSR and ESA, concerning cooperation in the field of exploration and the use of outer space for peaceful purposes. The Agreement was concluded for a duration of 10 years. Unlike previous agreement, which focused on the intentions of cooperation between the Agency and the Academy of Sciences of the USSR, this treaty comprised a series of specific provisions outlining the form and content of further cooperation.

Thus, the cooperation within the framework of the new Agreement could take the form of exchanging and visiting of scientists and other specialists; exchange by each party of scientific and technological information and literature published without restrictions as to its use on scientific programs and projects; cooperation on joint projects for design development, launching and operation of payloads, and in the conduct of joint experiments in orbit; joint symposia and conferences and other cooperative activities agreed by the parties.

The Article 2 covered a large number of cooperation fields: solar system exploration, space astronomy and astrophysics, solar-terrestrial physics, Earth observation including space meteorology and geodesy, and space biology and medicine, fundamental research in the field of microgravity.[66] For the implementation of cooperation under this Agreement, the parties

[65] Arets, J. U.S.S.R. July 20, 1989. In ESA-28130, HAEU.

[66] Agreement between the government of the USSR and the ESA concerning cooperation in the field of the exploration and use of outer space for peaceful purposes. In ESA-28132, HAEU.

had to form Joint Working Groups covering the listed fields. Joint Groups were supposed to adopt recommendations leading to joint projects. To further establish a closer cooperation with ESA, Albert Galeev proposed in the fall of 1990 that ESA open a representative office of the agency in his Institute (IKI), but the initiative was not received with immediate enthusiasm. During that period, circumstances were not favourable for ESA to establish a representative office in Moscow. However, this situation changed in 1995 when ESA successfully inaugurated an office in Russia.

The above-mentioned difference in approaches to the financial side of cooperation was reflected in the framework Agreement. Article 7 foresaw "as a rule, no exchange of funds unless the Parties agree otherwise". However, that same Article left room for exceptions, i.e., for possible projects to be realised on other terms if both parties agreed: "The cooperation under this Agreement may also include projects which are implemented on a reimbursable basis".[67] The Agreement reached at the dissolution of the Soviet Union served to both solidify the existing practice of exchange and close cooperation while also introducing a novel mechanism of interaction between the two parties. Furthermore, it laid the groundwork for a new form of cooperation on a commercial basis, by opening avenues and providing a legislative framework for such endeavours. Simultaneously, this Agreement encapsulated the evolving state of the late-Soviet space sector. French journalist Ph. Moreau sought to articulate the essence of the new document in *Les Echos*: "These new conditions suggest that Moscow is now seeking to negotiate as effectively as possible what was, until recently, solely a matter of prestige".[68]

For its part, in light of the irreversibility of events and the necessity of proactive preparation for rapidly evolving circumstances, ESA management was compelled to develop suitable strategies and attitudes towards emerging partners.

Before the Soviet Union Collapsed

Within the framework of the new Agreement, in early December 1990, a meeting was convened in Moscow between representatives of ESA, the Soviet Academy of Sciences, and MOM to discuss the concrete

[67] Ibid.

[68] Moreau, Ph. (1990). L'ESA va signer un accord-cadre avec Moscou. *Les Echos*, 6 avril.

implementation of cooperation between the Soviet Union and ESA. The initial four Working Groups were overseen by the Soviet Academy of Sciences, covering areas such as space science, space biology and medicine, fundamental research in microgravity, and Earth observation. Meanwhile, the fifth Working Group, focusing on inhabited transport systems, fell under the purview of the Ministry of General Machine Building.[69] The negotiations unveiled two primary trends in the trajectory of Soviet-ESA cooperation in space: first, a keen enthusiasm on the part of ESA in light of the fresh opportunities presented by the Soviet space sector, and secondly, the dire state of Soviet cosmonautics amidst the backdrop of the political and economic turmoil gripping the USSR.

Indeed, these negotiations were facilitated by the cooperative stance of both negotiating parties. ESA representatives demonstrated significant interest in leveraging the extensive experience of Soviet scientists and engineers in the field of space research and development. Conversely, Soviet scientists faced such considerable financial challenges that, in the words of Karin Barbance: "they were ready to pounce on any contract (consultations, technical support, etc.) that would involve payment in foreign currencies".[70] Indeed, according to the testimony provided by the European side, the Soviet Union was unable to cover even the expenses related to international business trips for its own staff. *Lettre de l'Europe de l'Est* quoted the commentary of a French expert: "The Soviets have only one word left in their mouths: sell. They proposed to put satellites in orbit, or to launch probes to explore Mars".[71]

The ongoing deterioration of the political and economic situation in the Soviet Union inevitably raised concerns among the leadership of the European Space Agency regarding the prospects of a new, promising phase of collaboration. As Jean Arets wrote: "space technology was one of the rare fields in which the Soviet Union could offer something to Western Europe in return for its immense needs. The United States, Europe and Japan considered large-scale cooperation with the USSR obviously if the USSR does not go into chaos". At the same time, Arets assumed that the "space lobby" in European countries would oppose the implementation of large projects in the context of future cooperation. However, he pinned

[69] Barbance, K. *Urss – Rapport de mission*, Décembre 10, 1990. In ESA-28130 (2), HAEU.
[70] Ibid.
[71] L'URSS vend l'Espace. *Lettre de l'Europe de l'Est*, 2 avril 1990.

his hopes on politicians who he believed could "look at things differently".[72] This notion encapsulated a conflict of interest among diverse stakeholders within the European space sector during that era. On one hand, European industrialists and space technology manufacturers were keen on advancing their products and technologies in both European and international markets, while also harbouring concerns about potential market disruptions caused by Soviet dumping practices. On the other hand, politicians, scientists, and ESA staff recognised the political and scientific significance of collaborating with former Soviet republics, valuing the exchange of scientific and technical expertise.

As a temporary solution to this predicament, ESA officials proposed collaborating with the crumbling Soviet Union on smaller-scale projects. Indeed, given the economic and political instability engulfing the USSR, the available options were severely limited. "A sizable and grandiose project would be deemed unacceptable to Member States and could entail significant repercussions if the instability in the Soviet Union results in its failure. Therefore, the enhancement of cooperation should be expedited while expertise remains available within institutes and industries in the Soviet Union, ensuring it does not diminish or vanish due to financial constraints. Moreover, it is imperative to act swiftly to secure advantageous positions before they are occupied by other Western partners".[73]

Another pressing issue for the ESA management was the impending fragmentation of the Soviet Union into multiple independent states, within which the industrial and scientific enterprises of the erstwhile unified rocket and space complex would persist. Indeed, the emergence of new states from the ruins of the Soviet Union was anticipated by the ESA staff before the Soviet Union collapsed. In September 1991, just a few months prior to the dissolution of the USSR, ESA officials foresaw its imminent disintegration and endeavoured to propose several initiatives for continued cooperation within its space infrastructures. However, they encountered numerous difficulties. Primarily, no one knew at that stage what the definitive constitutional organisation of the former Soviet Union would entail. In the framework Agreement, concluded by the Agency with the Soviet government, it was the USSR Academy of Sciences and the Ministry of General Machine Building (represented by Glavkosmos) that were designated as the entities responsible for cooperation with ESA. Yet,

[72] Arets, J. *Union Soviétique et Europe de l'Est.* Septembre 19, 1991. In ESA-28130, HAEU.
[73] Ibid.

at that time, ESA harboured concerns regarding the legitimacy of these counterparts.

Additional concern was to identify the country or countries that could potentially become ESA's new partners in the foreseeable future. Alongside Russia, Ukraine emerged as a prospective partner, owing to the presence of multiple centres for the design and production of rocket technology within its territory. Moreover, Kazakhstan, with its launch site at Baikonur, was also considered. It was also imperative to comprehend the functioning of the civil space sector: would it remain unified as a single entity or would telecommunications and Earth observation be allocated to the former republics, while space science and manned spaceflight programs would continue under Russia's purview. The question of cooperation with newly established private organisations remained unresolved as well. The transition to a market economy might enable firms in the space industry to extricate themselves from the oversight of their parent ministry and engage directly with Western partners.

To bring at least some clarity to the ongoing events, it was proposed to seek assistance from Soviet lawyers. They would explain to their European colleagues the content of the new legislation in the field of astronautics, the newly established and transitional governing bodies, and, ultimately, the potential consequences of a radical transformation for ESA.[74] Nevertheless, the question persists regarding the extent to which Soviet space law experts themselves could comprehend the situation and could provide competent advice to their European counterparts. ESA will inevitably confront these challenges in a new era for the former Soviet republics as they begin establishing independent relationships with foreign partners, including in the field of space activities.

Conclusion

Summarising the final stage of Soviet-ESA cooperation in space towards the end of the Soviet era reveals several notable features. At the onset of *perestroika*, the Soviet Union exhibited exceptional proficiency in the rapid deployment of satellites and human crews into space, achieving remarkable success in scientific space endeavours. However, during this period, signs of the Soviet Union falling behind in advanced technologies compared to the achievements of Western countries had already begun to

[74] Barbance, K. *URSS*. September 19, 1991. In ESA-28130, HAEU.

emerge. As Fabio Favata, ESA Head of the Science Planning and Community Coordination Office (2008–2024), suggested, "the long isolation impeded the Soviet technological progress. Being excellent in some areas (theoretical physics, mathematics) the Soviets lacked expertise and access to cutting-edge computing hardware and more broadly to electronics. While Soviet scientists had an enormous edge in solving problems analytically, this lack of access to hardware was a key disadvantage. The Soviet system also lacked the ability to attract brains, moreover, they gradually lost scientific personnel who emigrated in search of more favourable working conditions. In the ESA astrophysics division where I was working at the time in the early 1990s, we were suddenly flooded with CVs of brilliant Russian colleagues looking for jobs outside of the country".[75]

Alongside the geopolitical imperatives reflected in Gorbachev's concept of a "common European home", this largely explains the Soviet scientific intellectual elite's desire to establish close collaborative ties with ESA. Consequently, from the second half of the 1980s, the USSR actively engaged Western nations in its national programs, particularly through projects under the international Intercosmos initiative. But the realisation of the urgent need for international knowledge exchange and cooperation and its practical implementation came too late. As the *perestroika* period unfolded, the space industry and science began to grapple with the profound ramifications of the political and economic crises within the country, ultimately teetering on the brink of collapse by the conclusion of the Soviet era.

Amidst this backdrop, European space science and the industry underwent a rapid evolution, bridging the substantial gap that had separated it from the American and Soviet space endeavours during the early 1980s. The decade witnessed substantial technological advancements across various fields and disciplines in Europe, positioning its applied science programs as unparalleled on the global stage. The collaboration between the fledgling European Space Agency and the Soviet Union had in its last years therefore naturally evolved into a symbiotic partnership, underpinned by an exchange of services. ESA procured access to low-orbit Soviet platforms in exchange for technological assistance and supportive infrastructure. Remarkably, this period also witnessed the emergence of commercial exchanges between Western nations and the Soviet Union,

[75] Interview with Fabio Favata, April 26, 2024.

marking a departure towards previously unimaginable avenues of cooperation.[76]

Despite initial challenges, stemming from years of entrenched mistrust and the imposition of established political frameworks, the *perestroika* period fostered the establishment of a mutually beneficial cooperation. This collaboration laid the groundwork for further relations that flourished throughout the 1990s.

[76] Zhdanovich. (2012), 327.

PART III

Case Study: Soviet International Space Endeavours in the Late 1980s

Introduction

The analysis of two specific scientific projects—one in interplanetary exploration and the other in space biomedicine—serves to illustrate the structural changes in both domestic and foreign policy, as well as the space reforms undertaken during *perestroika*. The single-use *Phobos* and the reusable *Bion* spacecraft encapsulated a series of transformations characteristic of the late 1980s. Examining these projects in detail, considering all influencing factors where possible, and analysing the interactions among the various stakeholders involved will provide deeper insights into the complexities and contradictions of that era.

Phobos and *Bion* are connected not only chronologically but also by the involvement of ESA scientists. A comprehensive study of their history enables a detailed examination of the international dimension of these missions, as well as the evolution of international cooperation in space. In particular, it sheds light on the collaborative framework between Soviet scientific institutions engaged in space research (IKI and IBMP) and ESA during the final years of the Soviet Union.

Amidst the swell of enthusiasm that followed the successful culmination of the international *Vega* mission to Halley's Comet, Soviet scientists resolved to recommence efforts on the *Phobos* project aimed at exploring the Martian moon. Vasily Moroz, a prominent astronomer and participant in the *Phobos* mission from IKI, noted that groundwork for this endeavour had commenced prior to the start of preparations for the *Vega* mission. However, progress had been impeded due to the impracticability of two concurrently advancing projects of such intricacy and expense.[1] The prioritisation of the *Vega* mission was motivated by the imminent appearance of Halley's Comet. The proposition to dispatch a spacecraft to Phobos was initially unveiled in November 1984. At that juncture, the projected timeline aimed for a launch in 1986, yet this schedule experienced a delay, ultimately shifting the launch to July 1988.[2] Yet, the two *Phobos* spacecraft were compelled to prematurely cease their operations before the culmination of the expedition.

The *Phobos* project symbolised, on one hand, the apex of the Soviet scientific space program, and on the other, the lamentable conclusion of the Soviet era in space exploration. The scientific part of the project was supervised by Soviet scientists from IKI, while its practical implementation, including the design and construction of both spacecraft, was supervised by engineers from the NPO Lavochkin. Leading soviet scientists played a pivotal role in coordinating international cooperation and procuring funding for the mission, with foreign scientists assuming responsibility for financing their respective experiments. The project's unsuccessful outcome brought to light a plethora of unresolved issues that had been festering within the space sector for years. These impending conflicts not only sparked public discourse within the Soviet press, but also transcended the borders of the USSR, rendering foreign mission participants unwitting witnesses to the cross-criticism among Soviet project members.

In the field of biomedical research, the primary driving force behind the development of life sciences at the dawn of the space age was the need to proactively evaluate the potential effects of space conditions—including launch and re-entry—on the human organism. The idea and the concept of launching a human into space took tangible technical form in the Soviet Union during World War II, a period when a rocket capable of launching a payload into Earth's orbit had not yet been realised. Consequently, the need arose to conduct experiments on living beings, with dogs playing a central role, as they had been a traditional subject of laboratory research since the era of Ivan Pavlov.

Due to the extensive experience of Soviet space biomedical scientists and the advanced space infrastructure of the time, cooperation with the

[1] Moroz, V. (2014). Na pyl'nykh tropinkakh dalekikh planet. In T. Mulyarchik (Ed.), *Vasily Ivanovich Moroz. Pobedy i porazheniya. Rasskazy druzey, kolleg, uchenikov i yego samogo.* IKI RAN, 123.

[2] Huntress, Marov. (2011), 369.

USSR attracted the interest of biomedical researchers not only from the United States and France's national program but also from ESA. However, while American and French scientists began launching their experiments aboard biosatellites as early as 1975, ESA researchers had to wait until 1987 to conduct their experiments on a Soviet capsule.

CHAPTER 6

Phobos Mission

This space mission, conducted during the final stage of the USSR's existence, garnered significant attention within the global space community. This was due, firstly, to its ambitious scientific objectives, and secondly, to the extensive involvement of foreign scientists, including those from ESA. A detailed examination of the history of this mission sheds light on the challenges and difficulties faced by Soviet cosmonautics during the closing phase of *perestroika*, particularly in the realm of international cooperation. The failure of the *Phobos* mission exposed the systemic shortcomings of the Soviet command structure and undermined the confidence of foreign partners in the Soviet space sector.

Why Phobos?

What did truly explain the fascination of Soviet science with a seemingly inconsequential cosmic entity to an individual unacquainted with scientific intricacies? As Vasily Moroz recalled, it began with a proposal from V. Davydov, his former graduate student and collaborator, to exploit Phobos, the natural satellite of Mars, as a platform for the remote sensing of Mars. However, Phobos itself was also interesting for scientists as the

O. Dubrovina, *The Soviet Space Program and International Collaboration during Perestroika, 1985–1991*, Palgrave Studies in the History of Science and Technology,
https://doi.org/10.1007/978-3-032-03157-0_6

most accessible sample of an asteroid-like body.[1] Roald Sagdeev, scientific director of the project, confirmed Moroz's hypothesis: "After the success of the Vega, Giotto and Suisei missions to Halley's Comet , it was natural to 'try' an asteroid or a little moon. From this point of view, *Phobos* was an ideal choice, as in its spectral characteristics, it resembles a real asteroid, but is much easier to approach, using the Martian gravitational field to manoeuvre".[2] In fact, the mission also aimed to test the hypothesis that, like Mars' second satellite, Deimos, both moons were originally asteroids that had been captured by the Red Planet.[3] Phobos, however, garnered interest among Soviet scientists not solely for its own characteristics, but rather as a strategic stepping stone towards more ambitious missions to Mars.

The design engineers of the *Phobos* spacecraft had their own vision of the expedition. Vyacheslav Kovtunenko, responsible for building of the *Phobos* spacecraft, aspired to combine scientific and practical components in response to a question about the purpose of the expedition. Thus, as the main goal, he specified the study of the solar system. "Two fundamental problems are clearly defined here, which are of great practical importance. We need to understand firstly how the solar system arose and evolved to understand the history of the Earth itself, and secondly, the causes of the origin of life on Earth and the possibility of its distribution in the solar system".[4] According to Kovtunenko, the results obtained would lead to the explanation of many natural phenomena on Earth, to be able to predict them and finally learn how to control them. In a not-so-distant future he foresaw the opportunity to begin to develop the resources of the solar system. It was the exploration of Mars, a planet that in many respects resembled the Earth, that should, in his opinion, help in solving these problems.

In the 1980s, the expedition to Phobos caused a wave of scientific and would-be scientific enthusiasm. The general community also had far-reaching plans for the exploration of outer space and its gradual settlement by mankind, where asteroids, such as Phobos, played a significant role. In popular science literature it was assumed that in the coming decades, intensive building would unfold in near-Earth space. For the construction of

[1] Moroz. (2014), 123.

[2] Sagdeev, R., Zakharov, A. (1989). Brief history of the Phobos mission. *Nature*, 341.

[3] Nelepo, E. (1986). Ot Vegi k Fobosu. *Znaniye – sila*, 12.

[4] Markov, Yu. (1989). *Kurs, na Mars*. Mashinostroeniye.

powerful solar power plants, long-term or permanent scientific stations, multi-purpose space platforms, large radio astronomical antennas, and a huge number of various materials would be required. It was difficult and expensive to deliver such materials to space building sites from Earth. Besides, it would take too much time to achieve this. As a solution, Soviet scientists proposed utilising materials of extraterrestrial origin and products derived from the processing of asteroid natural resources. Consequently, in scientific circles, the idea of surveying and studying Phobos was becoming increasingly relevant.[5] Another concept involved utilising Phobos as an intermediary space station, prompting scientific journals to publish articles on the design and development of a transportation and energy infrastructure connecting Earth, Phobos, and Mars.[6]

Of particular interest is the fact that, during that period, the press did not address the question of whether using Mars' moon as an intermediary station or as a potential source of building materials was a wise strategy. Nor did it explore the idea of constructing in space. Similarly, authors of scientific articles refrained from offering detailed responses, wary that such inquiries might inadvertently cast doubt on the fundamental significance of space exploration. The general public would pay attention to this topic at a later date. But at that time Soviet society, or rather, the Soviet scientific and would-be scientific community, was euphoric from the success of the mission of Halley's Comet and was in a hurry to move in the Martian direction.

Leaving aside the purely scientific and science-fiction motives of the expedition to Phobos, let us turn to practical and political perspectives. In fact, there is a whole story behind the choice of Phobos as the main goal of the expedition dating back to the 1970s, when Sagdeev, Director of IKI, made a decision—"a real turning point" in Soviet space science, as he himself called it—to concentrate the efforts of Soviet science on the exploration of Venus, leaving Mars in the competence of the Americans. Such a choice reflected the fact that a more rational system of planning overcame short-term political consideration.[7]

However, with the success of the *Venera* and Vega missions and the near exhaustion of what could be achieved by Venus, it was time to

[5] Kolesnikov, Yu. (1987). Primknem k asteroidu. *Znaniye – sila*, 10.
[6] Polyakov, G. (1988). Obuzdanie dikikh sputnikov, *Tekhnika – molodezhi*, 4.
[7] Harvey, Zakutnyaya. (2011), 287.

reconsider a return to Mars. IKI was keen to select goals that would not directly challenge the US, so the small but unexplored Martian moon of Phobos was chosen as a target.[8] Looking back, Jacques Blamont, who for a long time collaborated with Soviet scientists on various scientific space projects, called the abandon of Venus for Mars a "major mistake of the Academy" that led to the failure of the Soviet planetary program, "first with *Phobos*, then with *Mars-96*".[9] But the disappointment and bitterness among the Soviet and international scientific community came later. With the advantage of previously successful experience and still counting on the support of the State, Soviet scientists were in a strong position to take the risk of expanding their research and finally extending it to Mars which, until then, had not been accommodating to Soviet science. The perpetual sense of competition was further fuelled.

Thus, Roald Kremnev, director of the G. Babakin Research and Testing Centre,[10] validated the importance of Mars exploration, citing the scientific programs of Western colleagues as an example. "The work programs of the USA National Aeronautics and Space Administration (NASA) and the European Space Agency (ESA) are scheduled to launch to Mars in the 1990s. The most representative community of space explorers - the International Astronautical Federation highlights the preparation of unmanned and manned expeditions to Mars as one of the main directions".[11] In his view, the rationale for the expedition was more political than scientific: to align with the international trend and, if feasible, maintain a competitive edge.

Indeed, at that time, the Soviet scientific community made a bold bid to participate in a joint effort to explore Mars. The ESA official magazine *ESA Bulletin* presented the *Phobos* mission as a starting point of the major Soviet project of Mars exploration. As the *ESA Bulletin* articulated "the next two Soviet spacecraft would be launched towards Mars in 1994, to

[8] Ibid.

[9] Blamont, J. (2014). Friendly remembrances of Vassili Ivanovich Moroz. In T. Mulyarchik (Ed.), 203.

[10] The spacecraft's development was undertaken by the G. Babakin Research and Testing Centre, established in the mid-1980s, leveraging the expertise of specialists from diverse departments. In 1985, Roald Kremnev assumed the role of Chief Designer and Director. The centre operated as a division of the NPO Lavochkin, under the leadership of Vyacheslav Kovtunenko, who served as Director-General and Chief Constructor from 1986 onwards.

[11] Markov. (1989).

make the most comprehensive study of Mars ever undertaken".[12] The large international cooperation was foreseen in the 1990s which saw joint activity from the USSR, the USA, and some European countries such as France. For European observers, numerous possibilities existed for cooperation between the Soviet Mars mission and the American Mars *Observer*, scheduled for launch in 1992. It was also presumed that this cooperation could be the precursor of much more extensive US-Soviet cooperation, which would have the goal of returning several kilogrammes of Martian soil to Earth-based laboratories around the year 2000. Following this planned trajectory, the Soviets intended to develop a manned mission to Mars in the early twenty-first century. Such a mission was also under study by NASA.[13] For this reason, the Soviet *Phobos* mission was a part of a huge (inter)national program that presented the Soviet Union as a protagonist of a new wave of "Mars attack".

The USSR viewed the opportunity to join the Martian race, not only for prestigious purposes but also to take advantage of international cooperation in the field of Martian research. Hence, in the mid-1980s, Gorbachev's call to avoid unreasonable duplication and to follow the principle of collective space exploration was, to some extent, realised in the case of the *Phobos* voyage. From a political and technological point of view, the flight of the *Phobos* spacecraft would be the last attempt by the USSR to make a breakthrough to Mars and take a leading position in the scientific exploration of space.

Therefore, numerous expectations were vested in the *Phobos* mission. Firstly, the scientific community anticipated fresh data to validate existing hypotheses. Secondly, the successful revival of the Soviet Martian program was poised to enhance the prestige of Soviet science and technology. Thirdly, the mission was seen as a practical demonstration of Gorbachev's aspiration to unify efforts in space exploration under the leadership of the Soviet Union.

Mission Description

The existence of two Martian moons had been predicted by Johannes Kepler in 1610, without any firm scientific justification. The scientific discovery of Phobos and Deimos occurred two and half centuries later in

[12] ESA Bulletin (1989), 58.
[13] Ibid.

1877, by the American astronomer Asaph Hall. During the twentieth century, the two bodies were intensively studied by ground means. However, due to the small size of the Martian satellites, no observations from Earth had been able to estimate either their mass or even their size. Only measurements taken from spacecraft would make it possible to provide a definite judgement.

The main objective of the Soviet project was to study Phobos, considering that the American satellite, *Viking-1*, had already flown past both Mars satellites in February 1977. It passed at a distance of around a hundred kilometres from the surface of Phobos. By 1988, Phobos was the first small body in the solar system to be reached and explored by a Soviet spacecraft. In addition to Phobos, the objectives of the new international mission included the exploration of Mars, the Sun, and interplanetary space in general. The timing of the launch was not arbitrary. Every 2 years, the Earth aligns with both Mars and the Sun along a straight line, and 1988 was identified as the optimal year for the spacecraft to enter Mars' orbit aligned with the plane of Phobos' orbit. Preceding its approach to Phobos, the spacecraft would spend several months orbiting Mars, during which time the program would focus on studying its surface, atmosphere, and magnetosphere.

The spacecraft was planned to fly over the surface of the satellite of Mars at an altitude of tens of metres, but this could be done only when the satellite and the spacecraft were rotating around Mars synchronously.[14] The strategy was to conduct both passive and active remote sensing using lasers and particle guns. For the first time in the history of planetary experiments, it was proposed to study the mass and isotope composition of the soil of a celestial body using a laser and ion fluxes. Furthermore, during the flight, the descent vehicle was supposed to separate in order to study the body directly on its surface, by contact methods. One of two small landers for Phobos was built by IKI and the other by the Vernadsky Institute of Geochemistry and Analytical Chemistry.[15] The mission was expected to last about 15 months.

Since the *Venera* spacecraft design were considered obsolete (to quote Vasily Moroz: "The good old-fashioned 'Venera' has already exhausted itself"[16]), the *Phobos* spacecraft was planned as the archetype of a new

[14] Balebanov, V., Zakharov, A., Linkin, V. (1986). Proiekt "Phobos". *Nauka i zhizn'*, 10.
[15] Huntress, Marov. (2011), 370.
[16] Moroz. (2014), 124.

generation of Soviet scientific spacecraft with new subsystems, design and even components and the extensive use of computer technology capable of solving complex problems arising in flight.[17] Unlike the *Venera* project, the new vehicle was built to a completely different design with extremely complex technologies.[18] This spacecraft was expected to serve as the baseline for the next 20 years of Soviet planetary exploration.[19]

According to Leonid Ksanfomaliti, "the rapid development of computer technology and software very quickly led to their widespread use in spacecraft control and in communication systems already in the 1970s. However, the use of information technology brought with it specific inherent features".[20] One of these challenges, as he articulated, stemmed from failures during operations due to imperfect programs. He contended that the complexity of the program directly correlated with the likelihood of errors occurring during its development. Ksanfomaliti asserted that the situation became even more intricate when dealing with interplanetary missions. Unlike Earth satellites, which could be launched more flexibly, interplanetary missions adhered strictly to predefined years, months, and days for launch opportunities. This strict adherence to specific launch dates frequently resulted in demanding work schedules and critical time constraints.

Throughout the project implementation, technical challenges emerged repeatedly, primarily as engineers had to grapple with the new technologies associated with the innovative design of the *Phobos-1* and *Phobos-2* spacecraft. This included advanced onboard and ground systems representing the new generation of space technology. Ground tests of both spacecraft, from March to July 1988, were carried out with the participation of numerous foreign specialists.[21] The launch was finally scheduled for July 1988. However, as the deadline approached, it became evident that the flight entailed high risks. Nevertheless, the *Phobos-1* and *Phobos-2* spacecraft were launched as planned, on July 7 and 12, 1988, respectively, utilising Proton rockets from the Baikonur. Subsequently, on July 16 and 21, the first trajectory corrections were executed. However, on September 1, *Phobos-1* failed to establish radio contact during the scheduled session

[17] Zakharov, A., Rogovskiy, G. (1987). Proiekt "Fobos" – Novaya expeditsiya k Marsu. *Zemlya i Vselennaya*, 4.

[18] Ksanfomaliti. (2010), 155.

[19] Huntress, Marov. (2011), 369.

[20] Ksanfomaliti. (2010), 154.

[21] Ibid., 155.

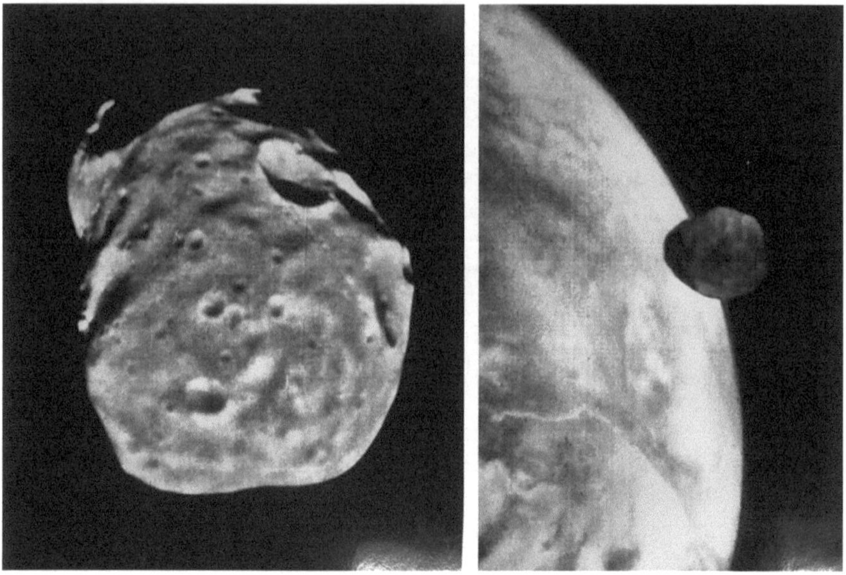

Fig. 6.1 Pictures taken by *Phobos* cameras. Source: NASA Archives, 15301

with Earth and did not respond to multiple radio requests.[22] The *Phobos-2* spacecraft safely reached Mars at the end of January 1989 (Fig. 6.1), first entering an elliptical orbit, then switching to a circular one, gradually approaching Phobos, and at the end of March became its artificial satellite. After 2 months of successful remote sensing of Mars and Phobos, the most ambitious part of the expedition remained: the flight at a minimum distance and the landing of landers. It was scheduled for the end of March—or by the beginning of April. But by March 27 communication with the second *Phobos* was subsequently lost.[23]

Postflight Analysis

The loss of the sophisticated spacecraft was a profound disappointment for Soviet space authorities and a significant setback for the international community of planetary scientists eagerly awaiting results. Following the

[22] Sagdeev, Zakharov. (1989).
[23] Moroz. (2014), 124.

disaster, criticism from scientists began to surface in the free press, sparking open discussions in both Soviet and Western scientific publications, newspapers, and press conferences, often within an international forum. These discussions broached topics that had long been shrouded in secrecy within closed institutions. Furthermore, top leaders of IKI, such as Roald Sagdeev and Albert Galeev, along with their staff, actively engaged in public analysis to uncover the reasons behind the mission's failure. The primary conflict ensued between scientists and engineers.

The Academician Mikhail Marov recalled that "it was a shock and set off an acrimonious debate of blame between Soviet scientists and engineers in the full light of international attention. The team of international scientists was summoned to Moscow in May 1989 for a post-mortem in which the Deputy Director of NPO Lavochkin delivered a messy excuse unrelated to the spacecraft that fooled no one and angered everyone. He could not resort to the old Soviet habit of concealment, not admitting a fault with system".[24] In his memoirs, Moroz more specifically commented on the causes of the malfunction in the spacecraft. As he remembers, "the first reaction of the NPO Lavochkin was odious: a meteorite hit. Nobody believed this version. Much later it was explained to me that in the onboard computer of the *Phobos* spacecraft there were, as it turned out, unreliable radio components subject to ageing. Someone in our electronics industry received an award for a form of streamlining that made certain types of capacitors cheaper to produce, and the 'improved' products were subsequently incorporated".[25]

Sagdeev did not limit himself to searching for reasons of a purely technical nature but called on public opinion and the Soviet authorities to consider the need for fundamental reforms in the system of interaction between the scientific and industrial sectors in the space field. In his opinion, the reasons for that failure related to the functioning of the entire scientific and production system in the Soviet Union in general. As one of the main reasons, Sagdeev named the way of operation within the space industry, which did not provide open competition and relied just on the largest and most reputable spacecraft manufacturers.

As he argued, "the failure of the *Phobos* spacecraft before completion of the mission invites a discussion on the interaction of scientists with industry (customers with contractors) to ensure reliability and survival of a

[24] Huntress, Marov. (2011), 369-370.

[25] Moroz. (2014), 125.

mission. Glavkosmos (the prime contractor) has promised to change its approach to future missions. The scientific and engineering team must discuss technical problems in the early stages of project preparation".[26] Sagdeev endeavoured to contextualise the entire space field within the broader political landscape of the period, emphasising the imperative of practically implementing the principles of Gorbachev's reforms. According to him, the transparency introduced in the relationship between scientists and engineers would help create a real *perestroika*.

While Sagdeev advocated for transparency in the interactions between scientists and engineers involved in spacecraft construction for scientific research, Galeev emphasised the necessity of altering the format of cooperation with foreign partners. In an interview conducted following the failed *Phobos* mission, he further associated the integration of *perestroika* ideals into the space domain with the reduction of excessive secrecy that had previously hindered Soviet scientists from collaborating with their foreign counterparts. Despite the extensive involvement of foreign scientists in the project, their access to essential information concerning the spacecraft's condition was restricted: "*Glasnost* could only contribute to the elaboration of a correct collegial decision. In a narrow team, without the involvement of foreign colleagues, we discussed the current situation. We had to change the mode of operation of the spacecraft in order to avoid difficult situations as much as possible and reduce the load on systems that failed. For foreign partners, the transition to such a sparing regime caused natural bewilderment, but they were never told anything".[27]

Transforming Internal Relations

Returning to the interplay between scientists and industrialists within the USSR, it is worth quoting a firsthand account from Leonid Ksanfomaliti, a key participant in the project representing IKI. He offers valuable insight into the dynamics of interaction between design engineers and scientists. An intriguing detail that emerged was that projects were typically initiated "by the spacecraft constructors themselves, who often possessed a preliminary technical blueprint of the proposed project. Spacecraft for all deep

[26] Sagdeev, Zakharov. (1989).

[27] Lapina, M. (1990). Fobos: povod dlya razdumiy. Interview s Albertom Galeevym. *Nauka v SSSR*, 2. Retrieved February 26, 2025, from https://epizodsspace.airbase.ru/bibl/nauka-v-ussr/1990/fobos.html.

space missions were developed by the renowned NPO Lavochkin". Only after the technical capabilities of the proposed new mission were fully understood did institutions within the Academy of Sciences, notably IKI, along with other institutes and scientific organisations specialising in space-related research, become engaged in the discussion of the mission. "In order to tackle the most pressing scientific challenges, a selection of instruments for scientific research was assembled".[28]

In the 1970s, NPO Lavochkin's specialised focus on producing scientific spacecraft facilitated the assembly of a team comprising technically skilled specialists, arguably the sole group within the Soviet Union capable of constructing spacecraft for scientific research. However, according to some critical experts and space field observers, this situation had a downside: the manufacturer's monopoly. They pointed to the example of the American system of customer-executor interaction, where NASA would call a competition for each project, inviting proposals for the development of probes and satellites from multiple specialised firms simultaneously. In contrast, in the USSR, the selection of a manufacturer adhered to the well-known Soviet principle of "choice of one". Konstantin Gringauz, Head of the Laboratory for Interplanetary and Near-Earth Plasma Research at IKI, further emphasised that "throughout the world, scientific institutions acted as customers, while industrial enterprises functioned as suppliers—wherein the institutes provided the funding and specified their desired requirements. However, in our country, the institutions of Academy of Sciences lacked control—they received the spacecraft without financial contribution. Consequently, industrial enterprises determined what was produced".[29]

Ksanfomaliti argued that the decision-makers, in this case the Military-Industrial Commission overseeing all VPK, once delegated these responsibilities to NPO Lavochkin, effectively leaving scientists with no alternative options. He observed that "other enterprises within the same space specialisation were not overly enthusiastic about engaging in research endeavours within deep space; instead, they displayed notable contentment with activities confined to 'near space'".[30] This observation implies a

[28] Ksanfomaliti. (2010), 155.

[29] Covault, C. (1989). Soviet Space program strife threatens Mars Mission plans. *Aviation week & Space technology*, May 22.

[30] Ksanfomaliti. (2010), 155.

pronounced preference among space sector enterprises to prioritise activities aligned with military space objectives over scientific pursuits.

Albert Galeev described more specifically the difficulties the Academy of Sciences faced as a customer for its scientific missions. "When it comes to producing spacecraft for planetary research, G. Babakin Research and Testing Centre has a complete monopoly". He suggested that the money allocated by the Council of Ministers for a specific project should be given directly to the Academy of Sciences who would then order spacecraft for industry. "Then we will be able to set our own conditions, look for competing firms". He argued that if the money remained in the Ministry of General Machine Building (which oversaw all design bureaus), then the Ministry would again be following faulty logic. This contention arose from the fact that because the large proven design bureaus already had experience, test benches, and qualified employees they did not require substantial financial investments. The Ministry therefore economised by placing orders for space technologies with the same manufacturers instead of investment with new firms that required more funding. With this approach, no competition was expected and Galeev maintained that the Ministry was not particularly interested, either, since its main task was to fulfil the state order and not to efficiently implement the project. This left the direct customer—the scientists—at the mercy of the Ministry for the execution of the project.[31]

Such anti-market relations between the customer and the contractor ultimately led to the fact that, as Vice-Director of IKI General Gennady Tamkovich stated, the Academy of sciences was forced to accept the offer from the contractor, the Babakin Research and Testing Centre. Given that the contractor was self-financed, it did not perceive any reliance on the target program. However, in certain instances, it assumed the role of dictating to the scientific community what actions it could and should undertake throughout the project implementation phase.[32] This perspective is particularly intriguing as Tamkovich, a military figure, assumed the role of deputy director at IKI, representing the interests of military circles within the institute and overseeing mission coordination.[33]

Sagdeev went as far as to address the issue of a "dictatorship" of industrials in light of the *Phobos* failure. He posited that individuals involved in

[31] Lapina, M. (1990). Fobos: povod dlya razdumiy. *Nauka v SSSR*, 2.

[32] Tamkovich, G. (1989). Zavershena li programma "Phobos"? *Zemlya i Vselennaya*, 5.

[33] Interview with M. Gruntman, March 19, 2024.

the space field keenly experienced the ramifications of the prevailing flawed economic system. Soviet scientists had overseen the launch of 16 space probes to Venus, projects for which spacecraft manufacturers left no room for alternatives: "It was either Venus (or at least Mars), or nothing".[34]

It is possible that the Phobos project scientists, frustrated by the loss of years of hard work and the missed chance to collect valuable data, may have intentionally placed greater blame on the spacecraft design engineers for the mission's failure. Their statements strongly suggest there was a serious imbalance of power in how decisions were made during the project.

Nonetheless, alternative perspectives to those outlined above emerged. Soviet experts justified the monopoly on space manufacture by pointing to the elevated degree of specialisation and inherent complexity in space technology. Overall, as asserted by some engineers and even scientists, the command-and-control system, alongside the associated industrial rigor and discipline, positioned the Soviet rocket and space industry as one of the most globally advanced. In contrast, the transition to market-oriented relations, the erosion of state oversight over the industry, and the dissolution of established production connections precipitated a substantial downturn in various industry metrics.

Even in their involvement in the *Phobos* mission, the engineers refrained from accepting liability, motivated both by self-preservation and by a commitment to learn lessons even from failed missions—a principle inherent in the field of astronautics. It is unsurprising that the engineer involved in the construction of the automated *Phobos* spacecraft held a considerably more optimistic perspective regarding the technological performance of the hardware compared to the scientists. Yuri Markov, engineer at NPO Lavochkin, believed that what was done under the *Phobos* program "for our level of development of science, technology, and economic management was a success". Among the accomplishments achieved, Markov listed the creation of "a new interesting space machine, the fact that it accurately entered the flight path, went all the way to Mars, entered the orbit of its artificial satellite, performed the most difficult manoeuvres and substantially did a good job".[35]

The already mentioned Roald Kremnev also defended the engineers' position. He stated that "space technology designers must comply with a set of restrictions relating to funds, the weight and size of spacecraft, etc.

[34] Sagdeev, R. (1989). Ne stanovit'sya v pozu mladshego partnera. *Priroda*, 1.
[35] Markov. (1989).

Of course, Project *Phobos* has taught us a lesson, and we are determined to rectify the defects we have identified, notably those relating to power supply, operational errors, and the possibility of probes functioning in automatic mode. Actually, I think such spacecraft are capable of efficient performance".[36]

Even scientists had to concede that the *Phobos* mission yielded some positive outcomes. As Vasily Moroz put it: "In just two months of operation, the *Phobos* mission achieved far more significant results than all previous Soviet expeditions to Mars combined. The significance lies not only in the quantity of data gathered but also in their quality. The mission successfully conducted a series of novel experiments that had not been undertaken in previous American missions".[37]

According to Lev Zelenyi, notwithstanding the project's outcomes which were deemed less than wholly satisfactory, space researchers garnered valuable new insights into various phenomena. These include the composition of interplanetary and near-planetary plasma, the physical properties of solar wind particles, the magnetosphere of Mars and its interaction with the solar wind, as well as the geometrical and physical parameters of Phobos, along with the kinematics of its motion. Additionally, the acquisition of numerous photographic images of Phobos, captured from various angles and distances, notably exceeding a minimum of 180 km, has generated considerable interest.

Another pivotal outcome, Zelenyi argued, lies in the "affirmation of the remarkable manoeuvring capabilities exhibited by the novel 'space robot'. Never before have Soviet spacecraft engaged in interplanetary expeditions executed such intricate flight manoeuvres—comprising multiple adjustments to their orbital parameters while within the gravitational influence of a planetary body—aimed at approaching its natural satellite, which yields minimal gravitational influence".[38]

Further information regarding these scientific achievements was provided by ESA scientist R.J. Grard who stated that: "the 800 kg scientific payload carried by *Phobos-1* and *-2* was allocated to 23 different experiment packages, 17 of which were flown on both spacecraft. Despite the premature loss of the mission, many of the *Phobos* instruments (16 out

[36] APN. (1989). Phobos Loss – Spacecraft Designers Blamed. *Spaceflight*, 31, July. In NASA HQ Archives.

[37] Moroz. (2014), 124.

[38] Zelenyi, L. *Boleye chetyrekh desyatiletiy partnerstva na zemle i v kosmose.*

of 23), including ESA's Plasma-Wave System, returned new and very important measurements".[39] Initial results were published in the prestigious English-language science journal, *Nature*, in 1989 and a dedicated international conference was held in Paris in October 1989.[40]

International Collaboration

As previously mentioned, the efficacy of cooperative international exploration was exemplified by the Halley's Comet project, with the vast majority of teams involved in the Vega mission also participating in the *Phobos* mission. Framing the Mars exploration program as an inaugural endeavour to match the successes of Venus exploration, the Soviet Union aimed to replicate the significant scientific and political achievements derived from the internationalisation of the Vega mission. This strategy was reiterated in the *Phobos* mission, with heightened involvement from the United States and Western European countries. Notably, the participation of the United States, through the provision of an instrument, several scientific co-investigators, and tracking support, constituted a breakthrough, amidst the backdrop of easing Cold War tensions. Consequently, each spacecraft carried a record number of instruments for comprehensive scientific exploration of Mars and Phobos: twenty-four experiments provided by the USSR, ESA, and thirteen other countries, including the United States. To quote Marov: "the *Phobos* mission constituted an aggressive, innovative and very impressive scientific attack on both Mars and Phobos, especially when compared to the modest orbiter that the USA was then planning to send to Mars in 1990".[41]

The intention to involve ESA scientists into the *Phobos* mission had first been announced in November 1984, at the annual meeting between ESA and the Soviet Academy of Sciences with the final decision concerning the collaboration taken during the annual ESA-USSR Academy of sciences meeting in Leningrad, May 1985. The conditions of collaboration were signed during the same meeting: funds between cooperating parties were not exchanged.[42] Roger Bonnet preferred to maintain a cautious distance

[39] Grard, R.J.L. (1989). First Results from ESA's Plasma-Wave system aboard Phobos. *ESA Bulletin*, 59.

[40] Harvey, Zakutnyaya. (2011), 290.

[41] Huntress, Marov. (2011), 370.

[42] Protocol of meeting. Leningrad, May 15-17, 1985. In ESA-28023, HAEU.

and underscored the restricted extent of cooperation with the USSR regarding the *Phobos* mission. He pointed out that ESA "did not participate in all the activities of the European states cooperating with the Soviets on *Phobos*. The only involvement of the Agency in the *Phobos* mission was through SSD acting as an independent laboratory and working on plasma wave experiments and experiments on solar oscillations. That SSD experiments were conducted in cooperation with other European institutes".[43] He further clarified that there existed no formal arrangements between ESA and Intercosmos, with cooperation limited to interactions between the SSD and the Academy of Sciences of the USSR.

Political considerations likely underpinned the cautious response of ESA's leadership to the Soviet proposal for cooperation. Indeed, the political bias of the project became apparent during an episode recounted by Bonnet, which transpired during one of the meetings of the IACG held between 1983 and 1986. "Roald Sagdeev who was heading the Soviet delegation, once gave a mastery description of a laser beam device being developed in IKI for the forthcoming *Phobos* mission, and remarked that its power was a thousand times less than that foreseen within some projects of the Strategic Defence Initiative. That elicited some sour smiles from our American friends!"[44]

David Southwood also recalls the political effect of the Soviet scientists' idea to use a laser to study Phobos:

> The laser idea was brilliant. It was a wonderful counter to SDI! The USSR would use lasers in space for science, while the Americans were putting lasers in space for warfare. It was a very clever idea. I am sure Sagdeev advocated for this strategy strongly. I remember the announcement being made at a COSPAR meeting in Toulouse in 1986. For all the scientists who listened to him, it was amazing. Not only were they planning to land on Phobos, which was already a difficult task, but they were also going to use lasers to investigate what Phobos was made of.[45]

Nevertheless, notwithstanding the caution necessitated by political circumstances, ESA participated in several experiments conducted aboard the *Phobos* spacecraft. A press release issued the day before the launch of

[43] Bonnet R.M. to Pryke I. (ESA Washington office), March 26, 1986. In ESA-28051 (3), HAEU.

[44] Bonnet, Manno. (1994), 93-94.

[45] Interview with D. Southwood, May 21, 2024.

the first spacecraft briefly described ESA's participation in the international project. "ESA is involved in several experiments through its Space Science Department. ESA Scientist Dr. R.G. Marsden is leading the effort to fly the so-called Low-Energy Telescope (LET) for solar-particle observations. This experiment is performed in collaboration with scientists from the Max-Planck-Institute for Astronomy in Lindau (Germany), the Central Research Institute for Physics in Budapest and IKI of the USSR Academy of Sciences".[46] Moreover, ESA was engaged in a Swiss-led experiment, IPHIR (Inter-Planetary Helioseismology Experiment with Irradiance Observations). ESA's SSD contributed to the design of this experiment, which was built in collaboration with the French Laboratoire de Physique Stellaire et Planétaire in Verrières-le-Buisson. The third ESA experiment, carried out by ESA Scientist R. J. Grard, regarded a plasma and wave investigation that sensed the immediate environment of both Mars and its moon Phobos. It was developed in collaboration with the Laboratoire de Physique et de la Chimie de l'Environnement in Orléans, the TRW Corporation in Los Angeles, the Aviation Institute in Warsaw, and the Soviet IKI. In addition, ESA had a leading role in contributing to the navigational support for the *Phobos* Spacecraft.[47]

The Soviet side placed significance on cooperation with the European Space Agency, a sentiment underscored by the personal invitation extended to the Director of the Agency to attend the spacecraft launch. Thus, Remar Lüst received an official invitation from the vice president of the USSR Academy of Sciences Vladimir Kotelnikov to "take part as a guest in a trip to the Baikonur for the launch of the *Phobos* spacecraft".[48] It is noteworthy that this event was not the inaugural launch from Baikonur, where foreign guests were permitted to attend. On May 18, 1988, foreign journalists were given the opportunity to witness the launch of the Kosmos-1944 navigation satellite.[49]

The more fervently Soviet authorities endeavoured to garner attention from the Western space community for the launch—through initiatives such as inviting leadership and organising celebratory events—the more pronounced the international repercussions of the mission's failure

[46] The European Space Agency takes part in Phobos mission experiment, July 6, 1988. In ESA-28051 (3), HAEU.

[47] Ibid.

[48] Kotelnikov, V. to Lüst, R. May 31, 1988. In ESA-28051 (3), HAEU.

[49] Ducrocq, A. (1988). Ouverture russe. L'Urss recherche une coopération scientifique tous azimuts. *Air&Cosmos*, 18 Juin, 1195.

became. Anatoly Pevzner, Chief Specialist of the Department of Scientific and Technical Support for Spacecraft Autonomy at IKI, testified regarding the media scope of the project: "Especially for foreigners, air-conditioned working rooms were equipped in the assembly and test building of space instruments at site N. 31 [at Baikonur—O.D.], toilets were repaired, furniture was brought in, a communication centre for telephone connections with Moscow was equipped, meeting rooms, rest rooms, a canteen, and a conference hall were prepared. Official representatives of the countries participating in the project, foreign journalists, scientists and specialists were acquainted with the technical and launch complexes of sites N. 31, 2, 95, visited the Museum of Cosmonautics, houses of Academician Sergei Korolev and the first cosmonaut of the planet Yuri Gagarin. During the launch of the interplanetary stations *Phobos-1* and *Phobos-2*, foreign specialists were present on the observation deck. Both launches were accompanied by live television broadcasts, as well as film and photographic documentation".[50]

The episode recounted by David Southwood also aligns harmoniously with the aforementioned context. It was an intriguing event involving the reception of Western scientists who participated in the preparation of the *Phobos* mission. "When I was visiting IKI in 1987, I did one political thing. That year, Tengiz Abuladze's *Repentance* was released. The movie was also shown in England, and I went to watch it with my wife. It was an important element of *perestroika*. When we were visiting the Soviet Union, we were taken to the opera and ballet. I knew this film was shown in Moscow because Shevardnadze insisted on its release. So, I asked if I could see the film. Instead, they brought the film to IKI and showed it in a lecture theatre there with simultaneous translation from Russian to English because we were an international scientific committee for the *Phobos* mission. Thus, the film came to us, and simultaneous translation was provided for us".[51]

The tragic conclusion of the mission contrasted substantially with its thorough preparation, compelling presentation, and high expectations surrounding it. Tamara Breus discussed the significant ramifications of the termination of the project, alongside the subsequent *Mars-96* mission that followed eight years later. She highlighted that: "The cessation of the *Phobos* and *Mars 94/96* projects proved to be detrimental not only to the

[50] Pevzner, A. (2006). Eto bylo nedavno, eto bylo davno…In A. Pevzner (Ed.), 105.

[51] Interview with D. Southwood, May 21, 2024.

scientific community but also to the broader space industry and the potential for collaborative efforts. Confidence in the NPO Lavochkin within the international community experienced a precipitous decline".[52]

The failed mission prompted foreign partners to draw varying conclusions. Thus, Thomas Donahue, a noted planetary scientist and past chairman of the NASA Space Science Board, concluded that Soviet planetary scientists did not receive strong support from the leadership of the Soviet Academy of Sciences. NASA leadership was forced to re-examine the level of cooperation the USA was willing to pursue with the Soviet Union on future Mars missions. American officials expressed their concern about what they believed was a lack of reliability in Soviet space systems and inflexible Soviet management policies that perpetuated engineering deficiencies. The latest mission failures reinforced the view of many NASA officials that US/Soviet Mars planning should be complementary—"but should avoid joint hardware operations that could jeopardise success".[53]

Conclusion

Curious was the remark of Igor Mitrofanov, a Soviet and Russian astrophysicist, and a member of the *Phobos* mission: "The Soviet command-administrative system worked well under three important conditions: the presence of a political support, high professional level of performers and coincidence of interests of all major participants. These conditions were fortunately formed during the preparation of the *Venera-Halley* space project, the highest achievement of the Soviet planetary research program".[54]

Considering the analysis above, it becomes evident that at least two elements within the listed trio underwent transformations during the implementation of the *Phobos* project. Firstly, political support gradually waned as a result of escalating domestic and international challenges. The acquisition of data from the *Phobos* spacecraft became secondary in significance compared to the pressing shortages of essential goods and the imperative to address mounting centrifugal forces of Soviet republics. Faced with these urgent priorities, the Soviet leadership found itself compelled to allocate decreasing attention to scientific imperatives. Mitrofanov expressed

[52] Breus. (2002), 151.

[53] Covault. (1989), 20-21.

[54] Mitrofanov, I. (2014). O roli lichnosti v istorii. In T. Mulyarchik (Ed.), 208.

this aptly: "Although *Phobos* was created only a few years after Vega, another era has already begun in the country. The disintegrating command-administrative system, by inertia, ensured the implementation of space projects, but in the new *perestroika* reality, 'apple trees on Mars'[55] somehow became useless to anyone".[56] Secondly, concerning the final component of a successful endeavour, i.e. the convergence of interests among all major participants, the previously discussed phenomenon of "airing one's grievances in public", as evidenced by public debates between scientists and engineers, further underscores the erosion of this condition.

Regarding international cooperation within the project framework during the mid-1980s, ESA's reluctance to engage in comprehensive cooperation with the Soviet Union stemmed from political considerations. Despite this, at the national level, particularly within countries such as France and the United States, the *Phobos* mission facilitated the establishment and operation of scientific collaborations. Yet, at the level of interaction between the USSR and ESA, the project revealed potential avenues for future cooperation. This trend was similarly evident in another substantial project, *Bion*, which will be addressed subsequently.

[55] This refers to the lyrics of the song '*And on Mars Will Apple Trees Bloom*,' which was specifically written for the Soviet science fiction film *Mechte Navstrechu* (*Towards Meeting a Dream*, 1963).

[56] Mitrofanov, I. (2014), 210-211.

CHAPTER 7

Bion Endeavour

The extensive experience in working initially with various biological samples and then subsequently with dogs, monkeys, and cosmonauts, coupled with the development of scientific hardware and equipment, brought Soviet medical and biological research to the attention of the international scientific community. This interest catalysed close cooperation in the early 1970s, not only with scientists from socialist bloc countries but also with those from the USA and France in the realm of space biomedicine. The participation of ESA scientists materialised at a later stage for several reasons, which will be expanded upon in this chapter, along with the Soviet interest in cooperation with Europeans and the factors that prompted the latter to embark on joint projects.[1]

[1] The chapter is based on the paper presented at 75th International Astronautical Congress (IAC), Milan, October 2024. Dubrovina, O. (2024). Science and politics: Soviet-European cooperation on the Bions. *IAC 2024 congress proceedings*. https://dl.iafastro.directory/event/IAC-2024/paper/81221/.

O. Dubrovina, *The Soviet Space Program and International Collaboration during Perestroika, 1985–1991*, Palgrave Studies in the History of Science and Technology,
https://doi.org/10.1007/978-3-032-03157-0_7

At the Dawn of Space Biomedicine

The question of sending a human being into space arose in the mid-1940s in the USSR, immediately after the end of World War II (excluding the pre-revolutionary theoretical concepts of Konstantin Tsiolkovsky). For the first time in the Soviet Union, Soviet engineers then seriously thought about the possibilities of human space flight using the technical means available at that time. The development of such a program is attributed to the Soviet scientist and engineer Mikhail Tikhonravov. At the Rocket Research Institute (later NII-1), founded by Mikhail Tukhachevsky in 1933, Tikhonravov organised a group of specialists with the task of developing a design for a high-altitude manned rocket vehicle (pressurised cabin with two pilots) based on a single-stage liquid-propellant rocket.[2] By 1945, this work resulted in the project BP-190. The rocket was to lift a top-sealed cabin with crew to a height of 200 km[3] without going into orbit. In March 1946, this concept was sent to Mikhail Khrunichev, minister of the Aviation Industry, and on April 12, 1946, Tikhonravov's project was reviewed by an expert commission of the Ministry of Aviation Industry, but the favourable conclusion was put on the back burner by one of the deputy ministers. Just in June 1946, Khrunichev reported to Stalin on the possibility of a manned space rocket, but Stalin rejected the resolution.

Nevertheless, this turn of events did not stop the flight of Tikhonravov's technical thought. Moving away for a while from the idea of launching a man into space, his research group focused on the possibility of a craft orbiting the Earth. Based on the idea of a "rocket package", having carried out some calculations, in 1948 he came to the conclusion that it was technically possible to put into orbit an artificial satellite around the Earth with the existing technology. In March 1950, Tikhonravov made a public report at a scientific conference in which he touched upon the prospect of building an artificial satellite.[4] Starting in 1948, Tikhonravov's group at NII-4 focused on theoretical issues. The group then shifted its work to align with the capabilities of rockets developed by Chief Designer Sergei

[2] Bazhinov, I. (2002). O rabotakh gruppy M.K. Tikhonravova v NII-4 Ministerstva Oborony SSSR. *Kosmonavtika i raketostroenie*, 1, 159-175.

[3] Feoktistov, K. (2000). *Traektoriya zhizni*, Vagrius, 17-18.

[4] Baturin, Yu. (Ed). (2008). *Sovetskaya kosmicheskaya initsiativa v gosudarstvennykh dokumentakh 1946—1964 gg*. RTSoft, 7.

Korolev[5]After overcoming a series of bureaucratic hurdles, the founding father of Soviet cosmonautics, Korolev, achieved the approval in 1954 by the Council of Ministers of the USSR for the study of scientific and theoretical issues related to space flight.[6] In 1956, Tikhonravov went to work for Sergei Korolev at OKB-1, creating the Department No. 9 working on three promising areas: automatic reconnaissance satellites, automatic vehicles for planetary exploration, and vehicles for manned orbital flights.[7]

Simultaneously to the development of purely technical means for human space flight, tests on living organisms began. The question that clearly confronted scientists and engineers seeking ways to send humans into space was whether a living organism could survive the launch and landing overloads, as well as the conditions of outer space, weightlessness, and cosmic radiation. It was towards the end of the 1940s that research on the possible effects of space on living beings started. Vladimir Yazdovsky, a physician and space medicine pioneer who, together with the president of the Academy of Sciences Mstislav Keldysh, was part of the informal "Board of Directors" set up and directed by Korolev, recalled the first meeting with the Chief Designer in 1948: "Outstanding aircraft designer Andrey Tupolev suggested that Korolev appoint me as head of a new direction (space biology and medicine). At that time, I was running the laboratory of airtight cabins and suits at the Aeromedical Institute of the Ministry of Defence of the USSR [...]. Korolev told me that they had rockets capable of lifting a load of over 500 kg to a height of 100 km, that geophysical research was already underway at this altitude, and believed it was time to start animal experiments to prepare for human missions".[8]

Biological research in the Soviet Union on automatic spacecraft began at the very dawn of scientific exploration of space. In 1949, Yazdovsky started to develop a methodology and a program of medical and biological research to support animal space flights. In Yazdovsky's recollection, he remarked that, "The substantial complexity associated with conducting experiments to investigate the impact of various factors of space flight on both animals and humans, coupled with significant risks and substantial moral and material expenditures, compelled us in 1949 to initiate preparations for experiments involving animals on rocket aircraft flights within

[5] Feoktistov. (2000), 22.
[6] Baturin. (Ed), (2008), 8.
[7] Feoktistov. (2000), 38.
[8] Yazdovsky, V. (1996). *Na tropakh Vselennoy*. Slovo, 9.

near-Earth space. Human flight on a rocket aircraft was deemed impractical at the time".[9]

The experiments with animals on rocket flights began in 1951, i.e. after rockets capable of reaching altitudes of 100 and 200 km had been created.[10] The first successful launches of animals up to 110 km on ballistic rockets R-1 and R-2A were carried out under Yazdovsky's leadership with the constant support of Korolev. A total of 29 sounding rocket launches were carried out between 1951 and 1960, with the duration of weightlessness ranging from 3.7 to 10.0 min at flight altitudes of between 100–473 km. Dogs as well as mice, rats, and rabbits were launched into space as research subjects.[11] These experiments investigated the effect of acceleration and weightlessness on living beings and prepared the first flight of an artificial satellite in-Earth orbit with a dog on board.[12] Separately, it is worth noting that before Laika's famous one, 59 suborbital flights were made with dogs, some of which have been rocket-launched twice, and the dog Dare has been rocket-launched five times. Some of the dogs died on either the first or second flight. The main cause was a malfunction in the parachute system. The work was carried out under conditions of strict secrecy, and the first publications describing the conducted experiments and the obtained results appeared only in 1958 in the collection "Preliminary results of scientific research with the help of the first Soviet artificial rockets". The names of authors of the publication in this collection were actually fictitious.[13]

A milestone was the flight of the first dog into space. On November 3, 1957, i.e., a month after the successful launch of the world's first artificial Earth satellite, the second Sputnik was launched, with the dog Laika on board. In his recollections, René Demets, former Project Scientist for Biology at ESTEC, noted that Sputnik-2 was displayed in 1958 at the EXPO 58 World Fair in Brussels. Hence, "the size, shape, and configuration of Sputnik-2 were by 1958 well known in the Western world. By me

[9] Yazdovsky. (1996), 9.

[10] Ilyin, E. (2007). *Biology in flights of unmanned spacecraft*, Acting Speech. Retrieved February 26, 2025, from http://biosputnik.IBMP.ru/download/pdf/Biologiya%20v%20poletax%20bespilotnyx%20kosmicheskix%20apparatov.pdf.

[11] Historical Background. IBMP Official Website. Retrieved February 26, 2025, from http://biosputnik.IBMP.ru/history.html.

[12] Yazdovsky. V. IBMP Official Website. Retrieved February 26, 2025, from http://www.IBMP.ru/WebPages/WIN1251/History/Yazdovsky.jpg.

[13] Ilyin. (2007).

(as a six year old boy) as well, because my father went to EXPO 58 and had taken photographs".[14] However, detailed publications devoted to biological experiments in rocket flights and to the second Sputnik appeared only in 1962, i.e., after the successful performance of the first manned space flights. Laika's flight was the first time a living creature had spent hours in weightlessness. The conclusion about the possibility of life in weightlessness, made on the basis of Laika's flight, accelerated the construction of the USSR spacecraft Vostok for human space flights.[15]

However, there were still many medical and biological problems that had to be solved before launching a man into space. In November 1960, Korolev wrote in *Pravda* that: "At present, the conditions and means necessary for a Soviet explorer to make a space flight are already in place. But it is necessary to accumulate practical experience in launching spacecraft and realising their safe and reliable landing back on Earth".[16] Therefore, during the next period preceding human space flight between 1960 and 1961, five 4-ton spacecraft/satellites were launched.[17] Biomedical research continued to be carried out in parallel with hardware and equipment development. Thus, in these flights of spacecraft-satellites *(korabl'-sputnik)*, physiological and biological experiments were carried out on dogs, rodents, and several other biological samples, as was widely announced in the Soviet press.[18]

Of particular importance was the one-day flight (27 h) of the dogs Belka and Strelka on a spacecraft, carried out in August 1960. The results obtained in studies on Laika, Belka, Strelka, and other dogs made it possible to come to the fundamentally important conclusion that orbital flight in a spacecraft does not cause any persistent and irreversible disorders of the basic physiological functions. These experiments made it possible to select the most reliable and effective measures necessary to ensure the vital activity of various organisms during space flight, as well as to develop research methods, ways of recording physiological functions, and most importantly, to ensure the safety of the flight and return to Earth. The results of these studies allowed to substantiate the possibility of human space flights. Yazdovsky and Oleg Gazenko, who since 1955, had been

[14] Correspondence with R. Demets, February 3, 2024.
[15] Ilyin. (2007).
[16] Pishchik, V. (1988). Biosputniki vchera, segodnya, zavtra. *Zemlya i Vselennaya*, 2, 34.
[17] Harvey, Zakutnyaya. (2011), 440.
[18] Ilyin. (2007).

among the key visionaries, leaders, and active contributors to scientific research programs on artificial Earth-bound biological satellites, played a direct role in the preparation and execution of Yuri Gagarin's historic first human spaceflight on April 12, 1961.[19]

During the next period, a significant institutional development of biological research in space occurred. In October 1963, the Institute of Space Biology and Medicine was founded in the Soviet Union under the USSR Ministry of Health upon the initiative of Keldysh and Korolev to conduct research and development work about medical and biological support of manned spacecraft and fundamental research in the field of space biology and medicine. In 1965, the Institute was renamed as the Institute of Medical and Biological Problems (IBMP). Since then, the Institute has played a key role in the area of space biology and medicine in the USSR and later in Russia.[20]

The next and latest experiment with two dogs in space was prepared at IBMP and was conducted in 1966 on the *Cosmos-110* spacecraft, using one of the unused Vostok spacecraft. The flight lasted 22 days, a record-breaking duration for living creatures—the dogs Veterok and Ugolyok—in orbit. By this time, the duration of manned space flights in the Soviet Union had not exceeded five days. Studies conducted by specialists of IBMP during the *Cosmos-110* spacecraft flight and during the period of readaptation revealed numerous disorders in many functional systems of dogs and above all, in muscle and bone tissues. Although all these phenomena did not threaten the life of the animals, they were reversible and completely disappeared within a short time after the flight. It became obvious that to further increase the duration of human space flights, to ensure their safety and reliably prevent the adverse effects of weightlessness, required a deep knowledge into the essence of the changes occurring in the human body. It was necessary to understand the mechanisms of weightlessness acting at different levels of organisation of living systems.[21] This gave grounds for the first time to seriously raise the question of the necessity to develop and introduce into manned cosmonautics means and methods of prevention of the unfavourable effect of weightlessness on the organism. At the same time, the developers of space technology demanded

[19] Oleg Gazenko. IBMP Official Website. Retrieved February 26, 2025, from http://www.IBMP.ru/WebPages/WIN1251/History/Gazenko.pdf.

[20] Zhdanovich. (2012), 317.

[21] Pishchik. (1988), 35.

from physicians and biologists' scientific substantiation about the possibility of long-term human space flights, in view of the emerging plans to build orbital stations for long-term human presence in space. This is how the *Bion* program ultimately came about.

Bion Program: Pioneering Biological Research in Space (1970–1997)

In the early 1970s, a comprehensive program of biological experiments in space missions was initiated. These experiments were conducted aboard both manned spacecraft and orbital stations, as well as through unmanned missions. Among the latter, a significant role was played by artificial satellites in Earth orbit, specifically designed to carry out experiments during orbital spaceflights. These experiments involved a wide array of biological specimens, including animals, plants, isolated cells, and tissues. On January 13, 1970, the Central Committee of the CPSU and the Council of Ministers of the USSR issued a decree mandating the conduction of biological research aboard specialised satellites. This research aimed to address the challenges of medical support for manned spaceflights of varying durations and objectives.[22] The *Bion* satellites were alternatively referred to as *Biocosmos*, *Biosputnik*, and *Biosatellite*, although they were often designated as Cosmos satellites with sequential numbering in public announcements. The *Bion* spacecraft were based on the Zenit reconnaissance satellite and were launched beginning in 1973. These missions focused primarily on studying the effects of radiation and weightlessness on living organisms, with indirect implications for human space exploration (Fig. 7.1). Between 1973 and 1997, a total of eleven *Bion* biosatellites were launched (Fig. 7.2).[23]

The subjects of research in biosatellite missions were bacteria, cells and tissues, seeds, lower and higher plants, invertebrates, reptiles, fish, and mammals (rats and monkeys). A total of 37 different biological species were used for research on eleven biosatellite missions, but rats and rhesus monkeys were the main test subjects. For the first time monkeys were launched into space on *Bion-6* (1983) and *-7* (1985). The first monkeys that flew into space on the *Bion* satellite program were pets of the scientific

[22] Ilyin. (2007).

[23] Bion-9. Description. Retrieved February 26, 2025, from https://nssdc.gsfc.nasa.gov/nmc/spacecraft/display.action?id=1989-075A.

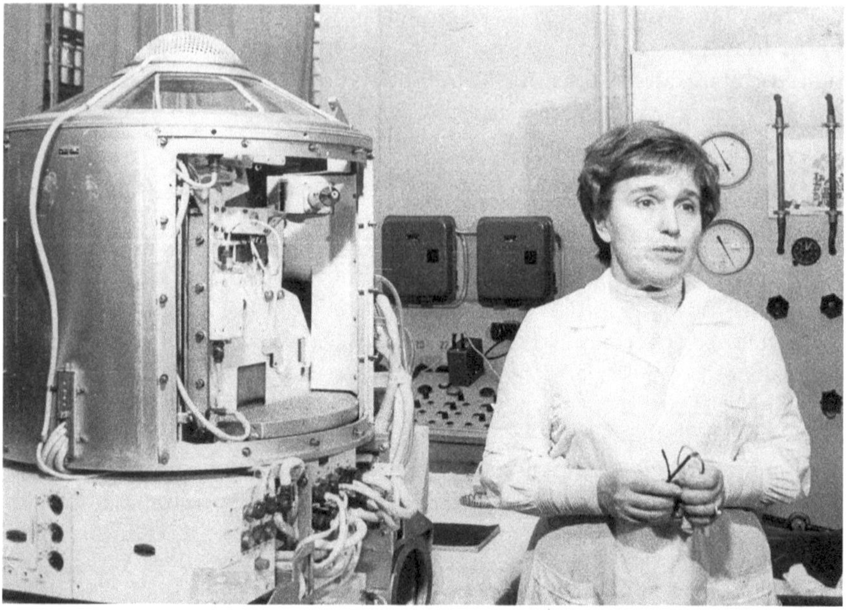

Fig. 7.1 A monkey cage aboard a *Bion* mission. Source: IBMP

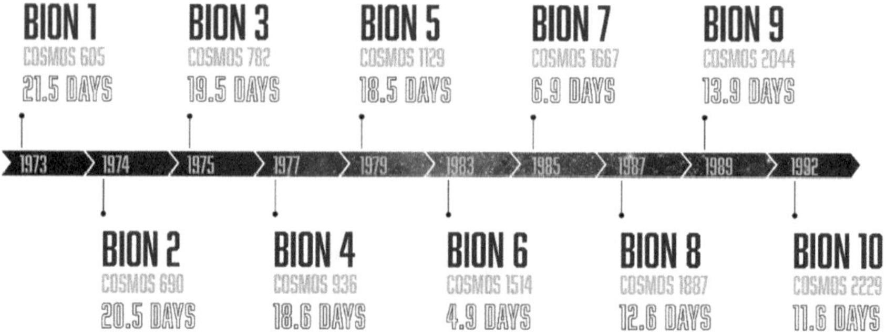

Fig. 7.2 A chronology of *Bion* launches. Data retrieved from http://biosputnik.IBMP.ru/history.html. Graphic design N. Dubrovina

research institute Experimental Pathology and Therapy of the Academy of Sciences of Abkhazia (NIIEPiT). Soviet scientists were proud of their then state-of-the-art developments in this field. To keep biological specimens on board the biosatellite and conduct research in flight, original equipment was developed that had no analogues in the world practice of biological research in space: units for single and herd keeping of rats, capsules for keeping monkeys and conducting physiological studies on them, three types of centrifuges for conducting physiological studies on rats, and three types of centrifuges for creating artificial gravity in space flights, devices for dosed irradiation of animals, thermally controlled containers, greenhouses for growing plants, incubators for quail eggs, various types of biocompatible containers, implantable bio-telemetric body temperature transmitters, dosimeters, and some other equipment.[24]

IBMP was tasked with the development and implementation of scientific research programs and the design of onboard scientific equipment. The *Bion* program involved contributions from numerous domestic institutions for the creation of advanced equipment for biological research.

The following is a detailed description of the satellite, as observed by an ESA representative granted access to its presentation. In 1986, Heinz Oser, senior scientist in Life Sciences at ESA HQ (Paris) from 1977 to 1995, was invited by Soviet colleagues to attend a presentation of a flight backup unit of *Biosatellite-7* in Moscow. As he reported, it was an "impressive large complex: payload unit consisted of an approx. 3 m diameter, pressurised sphere, housing the primate and rodent facilities, onboard 1g reference centrifuges, ample place for self-contained or more sophisticated experiment containers for research in gravitational biology, telemetry and electrical power. Attached to it on one end was the parachute recovery system, and on the other end the service module for power and deorbit engines which was separated during the re-entry". Oser, as he confessed, was not provided with many details but he found out that "this satellite was launched at 800 km altitude, rested for 7–14 days in a microgravity conditions and had a soft-landing capability. The lift-off weight is approx. 2.5 tons! The science return is augmented by a field laboratory flown to the recovery site by helicopters for immediate post-flight sample recovery and preservation". Impressed by what he had seen, Oser assured his superiors that "although somewhat crude in its appearance, this *Biosatellite* is a powerful tool for Life Sciences research in space on a frequent (every 1.5 years) mission basis".[25]

[24] Ilyin. (2007).

[25] Oser, H. *Mission to U.S.S.R.* June 16, 1986. In ESA-28042, HAEU.

As René Demets suggests, "Heinz Oser's report is not without flaws. The real diameter of the pressurised sphere was 2.2 m and the structure attached to it on one end was not the parachute recovery system, but a container packed with batteries to provide electrical power. (The *Bions* were not equipped with solar panels). The parachute recovery system was housed inside the sphere. Moreover, *Bions* never reached altitudes higher than 400 km".[26] Indeed, the modified Soyuz launch vehicle used for Cosmos placed the spacecraft between 200 and 400 km of orbital altitude.[27]

As an engineer who worked for many years on the *Foton* spacecraft, also produced at TsSKB Progress, Antonio Verga commented on the technical characteristics of the *Bion* left behind by Oser. "The term 'lift of weight' is also used improperly; the lift-off weight of a fully loaded Soyuz rocket is about 305 tons. Its low-Earth-orbit injection capability is 7.5 tons (this was the mass of the complete *Bion/Foton* spacecraft at that time). It is also funny that Oser said 'rested in microgravity' while actually the spacecraft was 'free-falling' at a speed of about 20,000 km/h!!!".[28]

The *Bion* program was characterised by three primary attributes. Firstly, its defining feature was its comprehensive approach, integrating the simultaneous application of physiological, morphological, and biochemical research techniques across a wide range of biological subjects. Secondly, the program employed near-synchronous control experiments conducted under ground conditions. These experiments were performed in biosatellite mock-ups nearly concurrently with actual biosatellite flights, replicating all physiologically significant aspects of space flight—with the exception of weightlessness. René Demets noted that launch vibrations, accelerations, cosmic ionising radiation, and re-entry conditions were also not simulated, leaving the physiological impact primarily attributable to exposure to microgravity. Thirdly, following the re-entry and recovery of the descent vehicle, the initial phase of post-flight analysis commenced immediately at the landing site. This methodology enabled researchers to begin examining biological material within 3-5 hours after landing. This prompt analysis was critical for identifying changes induced by weightlessness while minimising the influence of re-entry, landing, and subsequent readjustment to Earth's gravity (Fig. 7.3).

[26] Correspondence with R. Demets, February 3, 2024.

[27] Grindeland, R., Ilyin, E., Holley, D., Skidmore, M. (2005). International collaboration on Russian spacecraft and the case for free flyer biosatellites. *Adv Space Biol Med*, 10, 41-80.

[28] Correspondence with A. Verga, February 11, 2024.

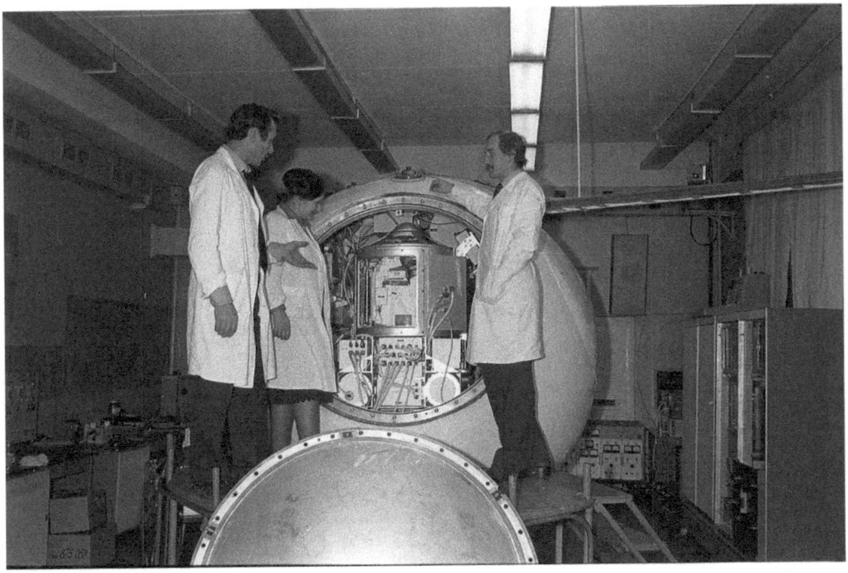

Fig. 7.3 Conduction of control experiments on ground in a mock-up of the satellite. Source: IBMP

Experiments on launching biological samples were carried out under the auspices of three leading Soviet organisations, each of which was responsible for its own area of work. TsSKB Progress provided the spacecraft and launcher. The Moscow IBMP was the lead designer of the experiments. This scientific organisation was responsible for the formation of the research program, its scientific and methodological preparation, the conduct of pre-flight and post-flight tests, and for the final scientific result. Finally, the Special Design Technological Bureau Biofizpribor was responsible for the development and manufacture of scientific equipment. Biofizpribor designed a capsule for experiments on monkeys, blocks for research on small laboratory animals and general biology equipment. All eleven biosatellites were launched into space from the Plesetsk Cosmodrome using a Soyuz-U launch vehicle, which placed them into an elliptical orbit. According to René Demets data "the orbital inclination of *Bion-6*, *-7* and *-9* was 82.3°. All of the other *Bions* were launched into a

Fig. 7.4 Members of the recovery team of *Foton-7* at the landing site. Source: R. Demets

62.8-degree orbit. The higher inclination of *Bion*-6, -7 and -9 was requested by scientists to increase exposure to cosmic radiation".[29]

In the advertising booklet (provided to the European colleagues), it was said that the scientific equipment was loaded with biological samples under laboratory conditions in Moscow or at the satellite launch site, 2 days prior to launch. Some other small devices could be loaded into the satellite, 6–10 hrs prior to launch. 2 or 3 hrs after the landing of the recoverable module (Fig. 7.4), experts could perform necessary primary processing of the biomaterial. For this purpose, a field laboratory was brought to the recovery site; the air temperature within it would not be below 18 °C with lighting and power supply to the laboratory equipment (Fig. 7.5). Separately, the booklet noted that biological experiments on board Soviet Space vehicles could also be performed using foreign-made devices.[30]

[29] Correspondence with R. Demets, February 3, 2024.

[30] Scientific equipment for Biosputnik research (Bion project). In ESA-28043, HAEU.

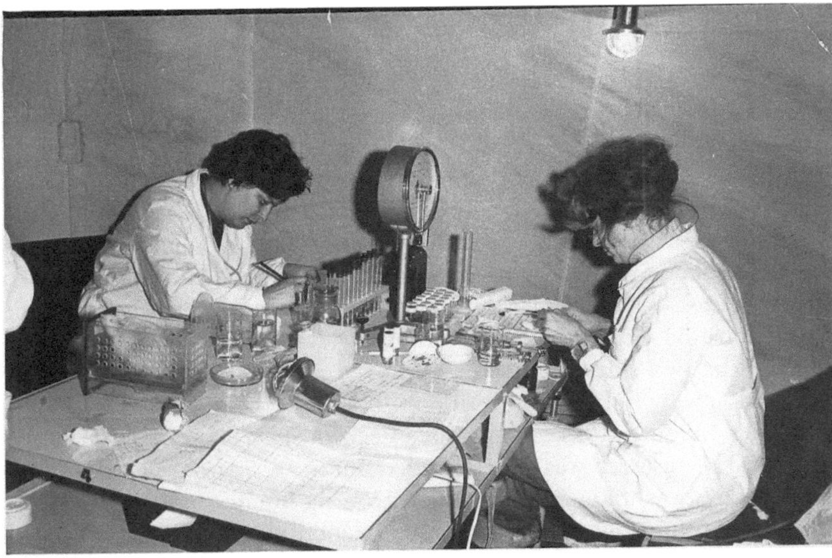

Fig. 7.5 Processing of obtained samples. Source: IBMP

As a result of the *Bion* satellite missions, a substantial volume of scientific and practical data was collected regarding the adaptation of living organisms during the initial phase of space flight, which represents the most challenging stage for crews of manned spacecraft. This information served as the foundation for the development of practical recommendations aimed at reducing the duration of the adaptation period and mitigating its potential harmful effects. Additionally, these recommendations focused on minimising the impact of weightlessness on the muscular and skeletal systems, as well as addressing motor and coordination disorders that may arise during the operational activities of crew members.

The Evolution of *Bion*: Towards International Collaboration

Between 1973 and 1997, experiments under the *Bion* program involved scientists from countries of the socialist bloc, including Bulgaria, Hungary, the GDR, Poland, Romania, and Czechoslovakia, as well as from the USA, France, the Netherlands, Canada, ESA, and China. These international participants joined the studies within the framework of the Intercosmos program.[31] To gain a clearer understanding of Soviet-European cooperation, it will be useful to briefly review the collaboration with IBMP's other major Western partners, the United States and France, for comparative purposes. While this overview does not aim to be exhaustive, we will highlight the key aspects of the joint projects.

The roots of Soviet-American cooperation in the field of space biomedicine date back to 1962, when bilateral agreements were established between NASA and the USSR Academy of Sciences. Space medicine was regarded as an area least likely to generate political or security concerns between the two nations. As a result, a structured exchange of information took place between the two sides during the 1960s, at a time when other forms of cooperation were virtually impossible.[32] Indeed, as the space race unfolded, both sides conducted research on the effects of spaceflight on animals and humans. This research was largely carried out in an environment of limited exchange of ideas. However, there was growing interest in fostering closer collaboration in areas such as space life sciences to enhance and complement the capabilities of both nations. The Soviet Union made use of USA data released from the Apollo program, while the United States drew on Soviet research into the effects of long-duration spaceflight on humans.[33]

Meanwhile, the American biosatellite program concluded in 1969. Following that, the only opportunity for American biological experiments in space was on the Apollo 17 or Skylab 3 missions, which involved mammals that did not require an independent life support system. Once these flights were completed, there was no further access to space for biological research. Although space Shuttles began flying in 1981, they did not carry non-human mammals until 1983, a full decade after the Skylab 3 flight. During this gap, the only space platform capable of supporting mammalian experiments was the *Bion* spacecraft.[34]

An American-Soviet Joint Working Group (JWG) on space biology and medicine was established in 1971 and was formalised under the Nixon-Kosygin Agreement on space cooperation signed in Moscow in May 1972.

[31] Proekt "Biosputnik". IBMP Official Website. Retrieved February 26, 2025, from http://biosputnik.IBMP.ru/about.html.

[32] Harvey, Zakutnyaya (2011), 441.

[33] Doarn, Ch., Nicogossian, A., Grigoriev, A., Tverskaya, G., Orlov, O., Ilyin, E., Souza, K. (2010). A summary of activities of the US/Soviet-Russian joint working group on space biology and medicine. *Acta Astronautica*, 67 (7), 650.

[34] Grindeland, Ilyin (2005), 48.

The primary focus of the JWG in its early stages was the exchange of knowledge on biomedical issues related to human spaceflight. Additionally, there was a need to provide scientific support for the joint docking mission, which came to be known as the Apollo-Soyuz Test Project (ASTP). The JWG agreed to meet annually at alternating locations.[35] However, until the collaboration on the *Bion* missions, cooperation between the two sides had been limited to data sharing, joint conferences, and publications.[36]

At the fifth meeting of the JWG in Tashkent, representatives from the IBMP surprised Ames researchers (with NASA's Ames Research Centre managing USA participation) by offering to include USA experiments aboard one of the Soviet series of Cosmos automatic spacecraft, known as *Bions*. This offer from the Soviet Academy of Sciences came at a pivotal moment for American biomedical researchers, especially considering that NASA had cancelled all biosatellite research in 1969. Following the disappointment of an aborted Skylab program, NASA biologists would otherwise have had to wait until the Shuttle became operational to access space.[37] What made the offer even more appealing was the low cost: the Soviets would cover the entire expense of building, launching, and recovering the satellite, while the USA would only need to cover the costs associated with the experiments.[38]

USA scientists participated with their experiments for the first time in 1975 aboard Cosmos-782 (Bion-3), which carried fourteen experiments from seven countries.[39] In 1977, seven USA biological experiments and medical devices were launched aboard the Soviet Cosmos-936 (*Bion*-4) mission, which also included experiments from France and several countries in the socialist bloc. The primary objective of these missions was to investigate the impact of long-duration spaceflight on living macro-organisms. A subsequent Cosmos mission, Cosmos-1129 (Bion-5) in 1979, carried an additional seventeen USA experiments and devices.[40]

[35] Harvey, Zakutnyaya. (2011), 440-441.

[36] Long Callahan. (2013), 143.

[37] Ibid.

[38] Burgess, C., Dubbs, Ch. (2007). *Animals in space from research rockets to the space shuttle*. Springer, 285.

[39] Bion 3. Retrieved February 20, 2023, from https://nssdc.gsfc.nasa.gov/nmc/spacecraft/display.action?id=1975-110A.

[40] Eisenhower, S. (2004). *Partners in space. US-Russian cooperation after the Cold War*. The Eisenhower Institute, 16.

This series of missions continued despite the suspension of cooperation in other fields during the presidencies of Jimmy Carter and the early years of Ronald Reagan. Evgeny Ilyin, who together with Heinz Oser played a key role in the future IBMP-ESA *Bion* initiative, recalled that in the early 1980s, at the moment of another escalation of the Cold War threats and disputes, all but two scientific projects were shut down, both dealing with human health and safety issues. These were "cooperation on the *Bion*'s to study the effects of spaceflight factors on living organisms and the second project concerned research into heart disease, which was led by Evgeny Chazov[41] through the Soviet Ministry of Health". Ilyin attributed this fact to the inherent nature of collaborative research: "Despite unfriendly relations in politics, the issues of human health and safety continued to be of equal concern to our respective leaders" (Fig. 7.6).[42] There were at least two other crucial factors that favoured the continuation of joint research, particularly in the field of space biomedicine: the absence of risks associated with transferring sensitive technologies, and the financial benefits derived from collaborative efforts.

A total of nine satellites (*Bion*-3 through Bion-11) carried over one hundred U.S.-led investigations into the effects of the space environment on biology and medicine, despite the tense international climate of the 1970s and 1980s. To streamline bureaucratic processes and simplify technological integration, USA and Soviet space program officials adhered to a policy of maintaining independence regarding one another's hardware. American experiments operated autonomously, functioning independently from Soviet spacecraft. They neither drew electrical power from Soviet systems nor utilised their data recording infrastructure—a technical workaround that alleviated mutual concerns about technology transfer for both Soviets and Americans.[43]

As evidence of the low-profile use of advanced technologies in the course of cooperation, Ilyin provided the following testimony: "When conducting our research, neither the Americans nor the Europeans employed high-tech methods of analysis that were then available. They

[41] Yevgeny Ivanovich Chazov (1929 – 2021) was a physician specialising in cardiology, Chief of the Fourth Directorate of the Ministry of Health, academic of the Russian Academy of Sciences and the Russian Academy of Medical Sciences. Minister of Health of the U.S.S.R. in 1987-1990.

[42] Interview with E. Ilyin, April 17, 2023.

[43] Long Callahan. (2013), 144.

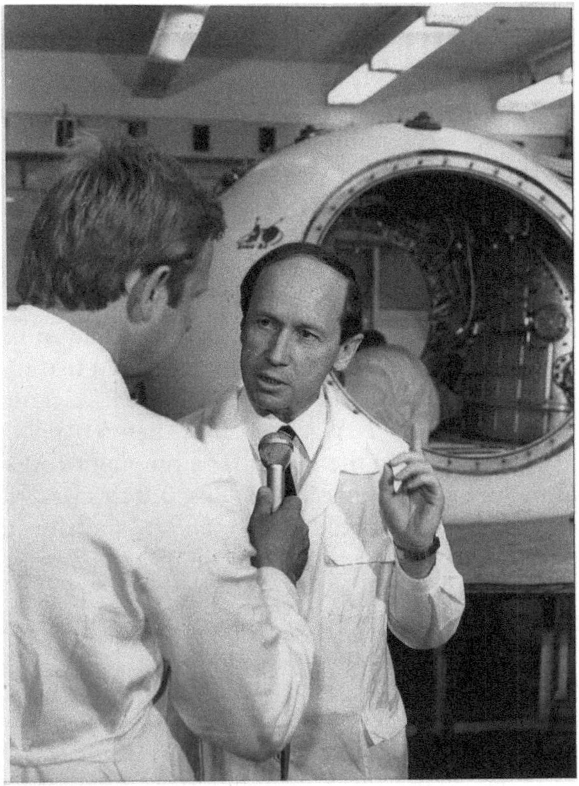

Fig. 7.6 Evgeny Ilyin. Source: IBMP

relied on classical, traditional methods: recording scientific information on magnetic media, small-sized devices, and hard disks".[44]

A typical example of Soviet-American collaboration on a biological satellite is exemplified by the description of the *Bion*-8 flight, which carried twenty-six USA experiments that included monkeys and rats to study the effects of zero gravity and space radiation on animals. It was also planned that several USA researchers would be in Moscow during the two-week flight to participate in the dissection and analysis of tissues from the rats within 24 hrs of the spacecraft's landing.

[44] Interview with E. Ilyin, April 17, 2023.

According to the Soviet-USA agreements, immediately following the biosatellite's landing, the Soviets established a field laboratory in tents beside the descent capsule, where the rats were retrieved, euthanised, and dissected. No USA investigators were permitted at the landing site; however, Soviet scientists separated the rats' tissues for analysis by both Soviet and American researchers. The tissues were then flown to Moscow within approximately 24 hours of landing, at which point both USA and Soviet investigators began examining them for the effects of zero gravity and space radiation. Both USA and Soviet medical data were to be shared, and the Soviets covered all the flight costs.[45]

Regarding economic advantages, experts in the field of space exploration history Colin Burgess and Chris Dubbs sustained that the motivation for NASA to continue the partnership was clear: the scientific payoff was great and the cost was low. NASA spent less than $1 million dollars on each of the Cosmos flights in the 1970s. The pre-flight cost would slowly climb to $3.7 million by the end of the 1980s; but this was still considered bargain pricing. NASA estimated that a similar, unmanned biosatellite program conducted by NASA alone would have cost 10–20 times as much.[46]

With the advent of a new era in Russia's history, scientific projects underwent deep structural changes. Unlike all previous collaborative research projects performed without any exchange of funds, *Bion-11/12* missions were to be implemented under a NASA/Russian Space Agency contract signed by both parties in December 1994. Due to various programmatic and political pressures, including that of People for the Ethical Treatment of Animals (PETA), NASA discontinued the *Bion-11/12* contract in April 1997.[47]

Franco-Soviet cooperation in space biology and medicine began in 1971 with the establishment of the Franco-Soviet Working Group on Biology and Space Medicine. The primary aim of this collaboration was to address major scientific challenges and the development of specialised scientific equipment. The studies were conducted through ground-based

[45] U.S., Soviets to cooperate on September launch of Biosatellite. *Aviation week and space technology*, July 13, 1987.

[46] Burgess, Dubbs, (2007), 292.

[47] Doarn *et al.*, (2010), 656.

experiments, biological satellites of the *Bion* series, and manned missions aboard orbital stations.[48]

Several joint research programs, involving various research centres, had been pursued within the framework of cooperation. The first French experiments conducted under the Franco-Soviet bilateral agreements between CNES and IBMP were driven by a focus on radiation protection and radiobiology. The aim was to assess the consequences of exposure to cosmic radiation on life processes. The Biobloc and Exobloc programs were created in this context. The experiments were carried out aboard *Bion-3*, -4, -5, the Salyut-7 orbital station (1982–1983), *Bion*-6, -7, -8. All these experiments have confirmed the notion of radiation hazards during spaceflight and contributed to the knowledge of the space environment.[49]

Under the aegis of CNES, the Centre d'Enseignement et de Recherches en Médecine Aérospatiale du Service de Santé des Armées (CERMA) collaborated, between 1980 and 1989, in the *Biocosmos* space biology program developed by IBMP. The interest of this participation, shared with the physiologists of IBMP, lay in furthering the validity of the rhesus macaque model with a view to long-term space studies, then envisaged by the international community either on board manned spacecraft or within automatic satellites. The experimental program was carried out during three flights on the *Bion-5, -6*, and *-7* satellites.[50]

The collaboration between the Neuromuscular Physiology Laboratory in Lille and IBMP began in 1983. The Franco-Soviet group launched the collaboration on the study of the mechanisms involved in triggering muscle atrophy in microgravity. The initial foundations for integration into the next biosatellite have been laid. The first *Bion* in which the team participated was *Bion-6*.

As the French participants in the program recalled, "space programs followed in regular succession: *Bion-7, -8* and *-9*. The different flight durations of one and two weeks were very interesting because they enabled us to demonstrate a close dependence between the duration of exposure to microgravity and the degree of development of atrophy. Equally

[48] Grigoriev, A. (2015). La coopération franco-russe dans le domaine de la biologie et de la médicine spatiales. In A. Ammar-Israël (Ed.), 165.

[49] Gasset, G. (2015). Radiobiologie et radioprotection. In A. Ammar-Israël (Ed.), 164-165.

[50] Milhaud, C. (2015). Le Cerma et la biologie spatiale. In A. Ammar-Israël (Ed.), 173.

interesting was the fact that the Soviet colleagues organised symposia the year after each space flight to present and discuss the results obtained".[51] The group of scientists from the Lille laboratory continued to take part in the *Bions* (-10 and -11 in 1992 and 1996), even after the fall of the Soviet Union.

The Franco-Soviet team also worked on research into the skeletal response to microgravity. In the early 1980s, very few results and even fewer physio-pathological hypotheses were available in the literature. French scientists recognised that "it was the development of close collaboration with our Soviet colleagues that gradually enabled us to understand the cellular mechanisms behind demineralisation through animal experiments over a period of ten years. French scientists took part in *Bions-5, -7* and *-9*, during which rats at the end of their rapid growth phase were subjected to periods of increasing duration in microgravity".[52] After the end of the Soviet Union, research continued on board the Spacelab (STS58-LSL2) and on the flights of the *Foton* spacecraft.

Two distinct types of cooperation were practiced aboard the *Bion* satellites. On the one hand, there was discreet and carefully managed Soviet-American collaboration, characterised by meticulous precautions to prevent the transfer of sensitive technology. On the other hand, Franco-Soviet cooperation exhibited a more integrated approach, marked by closer interactions between national scientific communities, joint preparation and execution of experiments, and the shared use of scientific instruments. In this context, it is particularly relevant to analyse Soviet-European cooperation in space biology and medicine. The collaboration with ESA on the Bion program began during the late *perestroika* period, a time when many of the barriers that had previously hindered systematic work on international projects had been overcome, and strategic shifts towards engagement with the Western world were underway. ESA researchers participated in two *Bion* missions during *perestroika*—*Bion-8* in 1987 and *Bion-9* in 1989—as well as in *Bion-10*, launched in 1992. Although the launch of *Bion-10* occurred in the post-Soviet era, its preparations began during the Soviet period.

[51] Mounier, Y., Falempin, M., Holy, X., Stevens, L., Canu, M.H., Desplanches, D., Oganov, V., Koslovskaya, I., Shenkman, B., Nemirovskaya, T. (2014). L'Adaptation neuromusculaire à la micropesanteur. In A. Ammar-Israël (Ed.), 174-175.

[52] Alexandre, C., Oganov, V., Morukov, B., Vico, L. Homéostasie squelettique et micropesanteur. In A. Ammar-Israël (Ed.), 183-184.

Two observations can be made regarding Soviet-ESA relations. Efforts towards gradual rapprochement between the two sides began well before formalisation, dating back to as early as 1979. Moreover, the onset of active cooperation between the USSR and ESA was preceded by years of accumulated knowledge and research experience in the life sciences, which notably positioned the Agency as an attractive partner for IBMP.

Before *Bion*: The Evolution of ESA's Life Sciences Program

The research of European scientists in the field of space biology and medicine began long before the start of cooperation with the USSR. The invitation of the USA side to participate in the Shuttle program provided an opportunity for specialists in this field to take their research directly into space. In fact, the European scientific community entered into a new area for research opportunities on life sciences with the advent of Europe's commitment to contribute the European Space Agency to the NASA Space Transportation System (STS), by means of the Spacelab program. The goals of life sciences research in space had soon been defined by ESA's advising Life Sciences Working Group in the beginning of 1977. The program proposals of the various sub-disciplines concentrated on the advantageous use of the microgravity environment, to study in more depth the gravity relevance of biological systems.[53]

Space biology experiments were part of the Microgravity program that the ESA Council agreed to undertake with the objective of developing advanced Spacelab facilities and ensuring the necessary flight opportunities on future NASA and German missions. Within the Program, established in February 1982, ESA's biology facility Biorack[54] was included. ESA provided the European science community with mission opportunities

[53] Oser, H. (1984). The Life Sciences Programme of the European Space Agency, and opportunities for radiation biology experiments. *Adv Space Res.* 4(10), 265-275. Abstract.

[54] Biorack was an ESA developed multi-user facility dedicated to the investigation of microgravity and radiation effects on small biological samples which could range from unicellular organisms to plant seedlings and insects. Biorack was first flown on board the Spacelab D1 Mission in November 1985 and for the second time in January 1992 on STS-42. Brillouet, C. A. (1994). The Second flight of the ESA Biorack facility on the Spacelab IML-1 Mission. H. Oser and T.D. Guyenne (Eds.), *Life Sciences Research in Space*, Proceedings of the Fifth European Symposium held 26 September - 1 October, in Arcachon, France. ESA SP-366. European Space Agency, 451.

on Spacelab, with experiment packages to be launched by the USA Space Shuttle, and with so-called multi-user facilities, such as Biorack, able to support a dozen biological experiments on those missions. Typically, the experiments were under the responsibility of the scientists, while the integration and mission phase were managed by ESA. Both mission definition and experiment selection rested with the ultimate decision of the responsible ESA Program Board.[55]

The active ESA's space biology program started in 1985, with the launch of space Shuttle Challenger hosting Biorack on board of Spacelab D1, a scientific mission funded and managed by Germany. The flight opportunity on Spacelab D1 for the envisaged ESA-developed Biorack facility was offered by Germany to ESA.[56] Despite the unique chance to conduct experiments in weightlessness, ESA scientists had emphasised the downsides of conducting biological experiments on board a manned spacecraft. René Demets underlined, “As the Biorack experiment packages were manually handled by the crew, the hardware had to comply with elevated safety standards. This had its toll on the experiment design and on the preparation of the safety documents. On top of that, the stringent non-toxicity requirements prohibited the use of particular chemicals and biological sample materials, affecting the scientific core of some experiments”.[57]

It was, by then, agreed that Biorack would fly again as part of the IML-1 Payload,[58] the First International Microgravity Laboratory (IML-1) in a series of missions with payloads dedicated to life sciences and microgravity science research. The re-flight of the Biorack facility was foreseen on the joint ESA/NASA IML-1 mission, in May 1987.[59] However, after the Challenger accident on January 28, 1986, all Shuttle activities were stopped and the Microgravity Program was delayed consequently. It was given the go-ahead for extension of the Phase-2 Program in 1988. The main objectives of this extension were: to bridge the gap in flight opportunities caused by the Challenger accident by using Shuttle-

[55] Oser. (1984).

[56] Krige, J., Russo A., Sebesta, L. (2000). *A History of the European Space Agency,* Vol. II, European Space Agency, 597.

[57] Correspondence with R. Demets, March 26, 2023.

[58] Brillouet, C. A. (1994). *The Second flight of the ESA Biorack facility on the spacelab IML-1 Mission*, ESA SP-366.

[59] Krige, Russo, Sebesta. (2000), 597.

independent flight opportunities, and to keep the interest and safeguard the competence of the European microgravity science communities.[60]

Indeed, in the aftermath of the USA Space Shuttle Challenger's tragic explosion, the national delegates at ESA's microgravity program board "urged ESA to look for new flight opportunities for all areas relevant to microgravity, and particularly for opportunities with the Soviet Union".[61] Consequently, ESA flew its microgravity experiments three times with the Soviet Cosmos program, starting in September 1987 with *Bion-8* (still called Cosmos 1887 at that time), then followed by *Bion-9* in 1989 and *Foton-7* in October 1991, just before the USSR political and geographical entity begun to fall into pieces.[62]

Although Shuttle flights resumed in September 1988, the re-flight of Biorack had to wait for another few years. Following the loss of the Challenger, biological missions in Earth's orbit were suspended until 1992. A restricted program continued between 1986 and 1992 on sounding rockets.[63] The Biorack facility was included in the IML-1 and scheduled for flight aboard Space Shuttle Discovery in January 1992. Other ESA-developed facilities were flown on the German D2 mission in April-May 1993, and on the IML-2 mission in July 1994. The former carried a sophisticated facility for experiments in human physiology (Anthrorack) and an Advanced Fluid Physics facility. The latter carried four multi-user facilities: the Biorack, a Bubble Drop and Particle Unit (BDPU), a Critical Point Facility (CPF) for the study of transparent fluids, and an Advanced Protein Crystallisation Facility (APCF) for studying the growth of single protein crystals.[64]

It is in this context of ESA's life sciences development that Soviet-European cooperation on the *Bions* fit in the middle of the 1980s. However, the first steps towards convergence in this area date back to an earlier period.

[60] ESA Bulletin (1988), 55, 46.

[61] European Space Agency – Microgravity Program Board Third Meeting Held in Paris on May 23, 1986 (ESA/PB-MG/MIN/3; Annexes ESA/PB-MG/OJ/3). The document was kindly provided to the author by Antonio Verga.

[62] Zhdanovich. (2012), 327

[63] Correspondence with R. Demets, March 26, 2023.

[64] Krige, Russo, Sebesta. (2000), 602.

Laying the Groundwork for Rapprochement

In the field of bio-medical research, before *perestroika,* there had been sporadic contacts between ESA and the Soviet Union for several years. As Evgeny Ilyin recalled, the initiative for joint research came from the Soviet scientists, who showed interest in the research activities of their European colleagues. "Soviet scientists had a question as to why ESA was not taking part in the *Bion* program. We made a proposal to the Intercosmos Council of the Academy of Sciences that Intercosmos formally invite ESA to participate in this research".[65]

In his next interview, Ilyin corrected the officially accepted version about the beginning of cooperation as it seems, to a more probable one. "The decision to invite foreigners", he recalled, "came not from scientists, but from above. It was a decision of higher authorities. They determined foreign policy: with whom to co-operate and in what directions. We were only asked whether we were interested or not. We did not make the decision. The decision was made by higher authorities".[66]

In fact, initial contacts were established through the Scientific Attaché of the Soviet Embassy (Elchin), following an exchange of letters between ESA's Director General and Vladlen Vereshchetin. As a counterpart for these talks, Oleg Gazenko, director of IBMP, was nominated. The pioneer of Soviet-European relations on ESA's side was Heinz Oser. He visited the Soviet Union many times and became a direct participant in Soviet-European co-operation, being actively involved in the early establishment of relations and in their further strengthening and development. It was his task to get the maximum amount of information from the first ESA visit to Moscow in June 1979. As Oser reported, by the time of his first trip to the Soviet Union he had already personally met Gazenko (at the 50th Annual Scientific Meeting of the International Rocket and Space Medical Association in Washington DC on May 1979), and some other Soviet space biologists (during IAF Congresses in Prague, 1977, and Dubrovnik, 1978). It is therefore possible that he was chosen due to his personal connections with Soviet scientists.

Oser defined the main goal of his mission "to explore the present situation in Soviet Space life science activities and to sound out possibilities for

[65] Interview with E. Ilyin, March 15, 2023.

[66] Interview with E. Ilyin, April 17, 2023.

potential cooperation with and participation in the Soviet Space Flight Program". The visit to Moscow had a purely informative goal and was "entirely unofficial", coinciding with the 6th International Conference on Space Biology and Rocket and Space Medicine in Kaluga. There were approximately 180 participants, including four USA researchers, one from FRG, one from Sweden, two from France, and other scholars from socialist bloc countries.[67] Moreover, he visited the Department of Biology at the Lomonosov Moscow State University (MGU) and IBMP. However, he was not granted access to any of the laboratories associated with this Institute.

During the meeting, Oleg Gazenko made two proposals of cooperation: on board of the Salyut-6 manned space station and in the Biosatellite program. Oser did not assess station opportunity but concentrated his analysis on the possible collaboration on the *Bion* spacecraft. He put ahead negative sides of potential future cooperation. First of all, he emphasised that the satellites used for the program were "old" satellites, though were refurbished after approximately 20–30 days orbiting. Moreover, increased overloads associated with mechanical stress during take-off and landing had a devastating effect on the cells, tissues, or higher organisms. The spin-stabilised orbital attitude of *Bion* and *Foton* and the absence of specific accelerometers capable of measuring the mechanical loads during all the flight's phases appeared to Oser as to constrain biological experiments to the field of radiobiology.

At the same time, several positive factors may have encouraged ESA's interest in cooperation: the ability to monitor experimental data in real-time via telemetry, the full recovery of specimens and hardware, no additional costs for integration and launch, and the opportunity to conduct ground-based studies simulating zero gravity to study its effects on biological systems. In conclusion Oser added a personal remark: "I received a very warm welcome with tremendous hospitality and noticed a fine distinction between personal and political issues".[68]

Several months later, in September-October 1979, Roy Gibson travelled to the USSR, where, among other Soviet scientific centres (Novosibirsk Institute of Atomic Physics, IKI, Star City), he also visited IBMP and met with Oleg Gazenko. In his report Gibson underlined Gazenko's dramatic willingness to establish collaborative relationship with

[67] Oser, H. *Report on Mission to URSS 4-8 June 1979.* In ESA-28129, HAEU.
[68] Ibid.

ESA. "As usual, he was tremendously enthusiastic but, unlike previous occasions, very anxious to have a collaboration. He had been pleased with the visit of Dr. Oser and would be glad to receive specialists for deeper discussions 'at any time'".[69]

In comments on the Director's report, Jean Arets envisaged a continuation of cooperation in life sciences in the form of further visits of European specialists to the USSR maintaining generally "informal character" of contact with the Soviets.[70] Thus, at that time, the circumstances that could serve as a trigger for launching Soviet-European projects in space biology and medicine did not yet stand. Moreover, it was rather the existence of inhibiting political factors that had a major impact on the development of the relationship. The obvious reasons for the failure of these first contacts in space biology can most likely be attributed to the sharp deterioration of the international situation in 1979, due to the Soviet invasion of Afghanistan, as well as the focus of ESA's biological program at that time of using Spacelab for its experiments.

Step by Step: Slowly But Surely Forward

In November 1984 Heinz Oser was back in Moscow, at IBMP, to attend the presentation of the results of the next-flown *Bion* mission. Oser listened to the speeches of individual researchers and scientists, presenting their findings and results of the latest Biosatellite Cosmos-1514 flight (*Bion-6*, Dec. 1983). As Oser reported, "the meeting was well organised; approx. 250 people attended out of which 54 were foreign nationals (GDR, Hungary, Bulgaria, Czechoslovakia, France, United States (including NASA), and West Germany)". He revealed how surprised he was to learn how many foreign investigations and co-investigations were carried out on that mission, predominantly coming from the USA and France. He underlined that the Soviets contended that, without foreign assistance, they may not have been able to achieve their goals. Oser was amazed at how open the Soviet counterparts were in this respect, in particular they did not hide the enormous support they got from CNES and NASA, both in onboard instrumentation and ground facilities supporting the mission.

[69] Gibson, R. *Report of visit to U.S.S.R.* September 28 –October 6, 1979. October 8, 1979. In ESA-28129, HAEU.

[70] Arets, J. *Commentaires sur le rapport de la visite en URSS du DG.* Octobre 10, 1979. In ESA-28129, HAEU.

"The Soviets did not 'disguise' any one of their approaches, particularly the hardware they have obtained from outside the Soviet Union (e.g. the advanced computer tomography scanner from General Electric, or the sophisticated French technology in computers and data management), and argued that without this Western technology they would not be able to conduct their research". At the same time, Oser reported that "Soviets clearly demonstrated that they had now an advanced system in their hands (sophisticated on-board data acquisition, tele-commanding from the ground, soft landing capacity, field laboratory for immediate post-flight sample analysis, etc.)".

He noted the changes that had taken place in his reception compared to his previous visit to Moscow in 1979. "I received a very special treatment and personal tours through various institutions". He was accompanied at IBMP and at the Soviet National Cardiology Centre. According to Oser's information, the former "had approx. 26 different institutions with approx. 1800 people and the later was an extremely advanced and most modern facility and could easily be compared with high Western standards. It was certainly unique for the Soviet Union".[71] He was positively surprised by the possibility to have had open discussions with everyone he was introduced to on any questions or topics, without reservation.

Even the mission results, according to Oser, were presented in a very open way, with frank discussions, critical analysis, and high quality (e.g. excellent slides and perfect simultaneous translations). However, the *Bion-6* mission was not very successful: although the topics addressed in their investigations were of a highly relevant nature (primarily gravitational biology), the mission had to be aborted prematurely after only 5 days in orbit (7–8 days were planned) due to the fact that one of the two primates on board got out of his harness. Also, one of the two primates died approximately 60 hrs after recovery. Still, *Bion-6* was important because it marked the start of the monkey missions, breaking the long Soviet tradition of using dogs as their primary research animal.[72]

Despite the relative failures with primate experiments, Oser had to admit that "all other investigations in the field of embryology, organogenesis, geotropism etc. did reveal some interesting results, but were difficult

[71] Oser, H. *Report on Mission to U.S.S.R.* Moscow, November 21-24, 1984. In ESA-28042, HAEU.

[72] Harvey, Zakutnyaya. (2011), 444.

to interpret, mainly due to the short flight duration". He added that they "rather contributed to a steadily increasing data base".

Oser had interest in and admiration for the conference, but he was merely a guest and a passive listener to the papers presented by other scholars. In his post-mission report, he made an attempt to analyse the reasons of ESA's refusal to collaborate with Soviet scientists in the field of life sciences. As he put it, "the enormous complexity and high risk of flying instrumented animals with only limited results underlined the importance of the ESA Life Sciences Working Group. It recommended since quite a while not to embark on this research approach at this stage but rather go for non-invasive human investigation as planned with the Anthrorack facility, at least what concerned human physiology".[73]

New Circumstances, New Opportunities

The Soviet-European cooperation in the field of medicine and biology received a new impetus in 1986, when the Soviet side once again invited ESA colleagues to co-operate on the *Bions*. The new invitation was no coincidence. The Soviet Union was already aware of this fruitful moment in active collaboration with ESA: they could take advantage of the USA difficulties following the fatal Shuttle accident that occurred early that year. Spotting an opportunity, the Intercosmos Council approached ESA, offering the *Bion* capsule for life sciences experiments.[74] The invitation for cooperation from Intercosmos was made via Karl Hecht of Charité Hospital, the medical university in Berlin responsible for space biology in the GDR.[75] Antonio Verga reconstructed the inception of the collaboration. He stated that, "It all began in the GDR, which was part of the communist bloc. The space program of the GDR was then under the leadership of Karl Hecht, who also served as the director of the Charité Hospital in East Berlin. Hecht initiated contact with Heinz Oser, one of our employees at ESA/HQ, due to shared interests in the effects of space on human and animal biology. This marked the starting point. Information was conveyed to ESA that the Soviet Union had this program".[76] By then, the

[73] Oser, H. *Report on Mission to U.S.S.R.* Moscow, November 21-24, 1984. In ESA - 28,042, HAEU.

[74] Harvey. (2024), 21.

[75] Ibid.

[76] Interview with A. Verga, March 27, 2023.

main common interests had already been identified in basic research, e.g. gravitational biology, human physiology, and exobiology.[77]

In its official letter, Intercosmos informed the ESA leadership about the main objectives of the Soviet program and the conditions for the participation of foreign scientists in it. The objectives of these investigations were: assessment of the adverse effects of microgravity on the human body, development of scientifically sustained principles and methods of medical support in space missions of various duration and solutions to fundamental problems of general biology, such as the role of gravity in the growth, development, ageing, and evolution of living systems. Foreign participants were expected to be involved in the preparation of biological objects, development of flight qualified equipment, and the conduct of experiments.[78]

The next symposium, organised by IBMP in June 1986, was well attended by the Soviet scientists responsible for the experiments aboard Biosatellite-1667 (*Bion-7*),[79] which had flown on a seven-day mission in July 1985. Heinz Oser not only gave a presentation, but also had the full authority to sound possibilities and identify research areas, though without any official mandate for negotiations which, in case of a favourable Soviet attitude, had to be discussed in future Soviet-ESA meetings at a later stage.

As Oser reported, despite numerous invitations sent out to Western countries, only a few attended from: France (6 people), Denmark (1), Switzerland (1), and the USA (2), as opposed to the previous meeting in November 1984. Some scientists from socialist bloc countries were present. The program essentially dealt with results obtained on invasive studies on the cardiovascular, neurosensory, muscular-skeletal, and endocrine systems of both primates and rodents. According to Oser, "The level of the presentations was good, and the Soviets have considerably advanced in their knowledge in comparison with his first visit on 1979". He underlined a substantial number of new younger scientists, "well educated and knowledgeable in their field of research". He regretted the impossibility to attend all panels and discover all outcomes of the mission. For next chances

[77] Relations with the U.S.S.R. IRAC, October 1, 1986. In ESA-28130, HAEU.

[78] Interkosmos to J.F. Danisse, February 26, 1987. In ESA-28130, HAEU.

[79] Cosmos 1667, also known as Bion-7, was a 1985 biomedical research mission satellite involving scientists from nine countries (Bulgaria, Czechoslovakia, East Germany, France, Hungary, Poland, Romania, Soviet Union, and United States).

to attend at similar future events, he suggested that "ESA should send at least one or two more experts since this is often the only opportunity to gain insight into their [Soviets'—O.D.] experiments, science and results".[80]

Oser intended to present his own paper about the SLED[81] and the Biorack experiments flown on the Spacelab mission D1 (1985), and he also asked Oleg Gazenko to "set aside a time for himself and some of his collaborators, to discuss the potential of future ESA cooperation with the Soviet Union on mission opportunities and experiments in the field of life sciences in space".[82] Oser's proposal to the Soviet side included the collaboration, both on board of the Soviet space station and within the Soviet Biosatellite program. He centred his proposal on using existing, flown, and space-proved ESA hardware at various and increasing levels of complexity, supporting research in fundamental biology and physiology. Regarding the main common interest, Oser defined it "a basic research, e.g. gravitational biology, human physiology and exobiology".[83]

To start with the collaboration, Oser outlined to the Soviets a range of conditions and prerequisites "for the meaningful conduct of ESA experiments". First of all, he insisted on the "strong involvement of ESA experimenters in the preparation, training, conduct and evaluation of experiments". Other conditions were "on-site delivery and monitoring of experiments", late access possibilities pre-launch (particularly for sensitive biological samples) including suitable laboratory facilities and conditions, ground-based control experiments and rapid post-mission analysis of samples or chemical treatment for shipment to a home laboratory. He also highlighted that no experimental hardware should be left on board the space station.[84]

Oser did not count too much on the fact that the terms of cooperation with the Soviet Union put forward by him would be fully accepted by the Soviet side. Given his previous experience with the Soviets, he suggested that "it will be worth trying to let them know upfront that we want to go

[80] Oser, H. *Report on mission to the Soviet Union*, Moscow, June 18-23, 1986. In ESA-28042, HAEU.

[81] An ESA experiment launched on Spacelab-1 on October 30, 1985. SLED investigated space sickness by monitoring eye movement as the seat moved horizontally and tilted while cold or hot air was blown into the astronaut's ears.

[82] Oser, H. to Gazenko O., May 30, 1986. In ESA-28042, HAEU.

[83] Oser, H. *Report on mission to the Soviet Union*, Moscow, June 18-23, 1986. In ESA-28042, HAEU.

[84] Oser, H. *Mission to U.S.S.R.* June 16, 1986. In ESA-28042.

a different avenue than all their other arrangements they made with other countries, and that we do not want this to be a 'political' issue but rather would like to conduct serious science in space". Furthermore, Oser raised the bid of sending a European astronaut aboard a Soviet orbital station for experiments in the field of physiology, since "the Soviets normally do not agree to do this on their own crew".[85]

According to Oser, the Soviet reaction was more than welcome. He even added that "it seemed to run through open doors". Moreover, the Soviet side stressed the need for a longer-term cooperation agreement with ESA. However, the Soviets had to reject most of ESA's requirements outlined by Oser. As he put it, "this approach was mainly based on the limitations and boundary conditions laid down by their space program policy, to which even IBMP had to abide to". In fact, Soviet conditions contradicted the requests of ESA: no access for ESA experimenters to the launch site; delivery of experimental hardware to Moscow; preparation of the experiment by ESA-experimenter on Moscow-based laboratories; both for manned and unmanned missions, the experiment conduct, supervision, and operation of ground controls were being made by the Soviets; sample/data return to ESA experimenters after notification, analysis, and publications of results commonly done by ESA and Soviet scientists. At the same time the Soviet side agreed to deliver data obtained outside a common experiment protocol when using ESA-provided hardware, a fact that turned out to be of high relevance in future cooperative agreements.

Despite these constraints and boundary conditions, Oser still highlighted the only but very important deference to earlier discussions: the Soviet eagerness to have such cooperation. He concluded that now ESA "might be in better negotiating position with respect to the limitations and constraints the Soviets have put until now".[86] For ESA, it was extremely important to "make it clear to the Soviets that we will select our experiments through our normal science peer reviews and that ESA/ESTEC staff and selected ESA experimenters should play a key role in the experiment integration, and not just deliver ideas and hardware". Oser believed that "it would prove to be the most difficult obstacle to be overcome".[87] Nevertheless, despite all the difficulties, already in the process of

[85] Ibid.

[86] Oser, H. *Mission opportunities for Life Sciences experiments on Soviet Spacecraft*, October 23, 1986. In ESA-28042, HAEU.

[87] Ibid.

preliminary discussions on future cooperation in June 1986, the Soviet side demonstrated to be very open and available for cooperation, offering the first flight opportunities already with the next Biosatellite in Autumn, 1987.

It was suggested that no time be wasted in formulating these oral agreements into a formal script. The initial ESA intent to seek for occasional mission opportunities on a modest interface basis during the meeting was turned around by the Soviets in suggesting a longer-term cooperative program in the life sciences domain. Soviets considered the ESA-Intercosmos Agreement of 1972 as too generic and proposed to sign a separate agreement as it was done for cooperation on *Phobos*, X-ray telescope, *Vega* and *Giotto* mission and ISTP. They suggested ESA's Director General addressing the letter to the Chairman of the Intercosmos Council and vice-president of the Academy of Sciences Vladimir Kotelnikov. Soviet officials advised that the letter should contain all aspects, including the ESA astronaut flight but be of a more general nature and mentioning cooperation for a longer term.

As for the flight of a European astronaut on board the Soviet space station, due to the complexity and political dimension of this issue, its discussion was beyond the competences of both Oser and IBMP. In fact, the European astronaut flight was considered by IBMP's delegates as a political issue that was "immediately deferred to other arrangements at highest political level". The Intercosmos delegates suggested to address this question to the recently formed organisation Glavkosmos which was "responsible for more technical and crew-related matters".[88] Subordinate to the MOM, i.e. the ministry responsible for the entire military and civil space industry in the Soviet Union and fundamental component of the VPK, Glavkosmos was the only window of access to Soviet manned cosmonautics. Therefore, IBMP officials, although involved in the development and conduct of the manned program, had no access to the decision-making process. They were unable to assume responsibility or respond positively to the request to utilise Soviet cosmonauts as test subjects for research in human physiology.

Proof of the keen interest in cooperation with ESA was a letter from the Director of IBMP Oleg Gazenko, sent in July 1986 personally to Reimar Lüst, Director General of ESA. He referred to his visit to ESTEC in April

[88] Oser, H. *Report on mission to the Soviet Union*. Moscow, June 18-23, 1986. In ESA-28042, HAEU.

1986 and to his talks with Oser in Moscow to emphasise that "the cooperation in space biology and medicine research between IBMP and West European scientists was very promising". Based on his "important experience gained from the cooperation with scientists from many countries, including the USA (NASA) and France (CNES)" he stressed that "the international cooperation was useful and beneficial for every side involved" in it. However, Gazenko did not bother to pay attention to the fact that "biomedical research programs did not occupy an appropriate place in ESA plans". So, he suggested to Lüst to take an "opportunity to give a spectacular impetus to the pertinent investigations in future ESA programs".[89] Lüst replied to Gazenko only in September informing him that he requested Philip Goldsmith, Director of Earth Observation and Microgravity, to be in touch with IBMP and "to organise further contacts as soon as possible".[90] A letter from Goldsmith to Gazenko immediately followed, where he offered a concrete date to visit Moscow "with a small ESA delegation in order to discuss further the question of common interest in the field of life sciences".[91]

After Lüst's briefing of IRAG (International Relations Advisory Group), the ESA leadership was encouraged to pursue the possibilities of life sciences experiments to be flown on Soviet spacecraft and to prepare a potential cooperation agreement. In response to Lüst's and Goldsmith's letters, Gazenko invited European experts to hold a meeting in Moscow in November 1986, in order to discuss all details of the future cooperation. Before travelling to Moscow, ESA's scientists considered it necessary in their interests to hold an information meeting with CNES representatives and French scientists, due to their active cooperation with the Soviets in life sciences, in order to obtain information on their experience, to avoid duplication or interference with their efforts, and to streamline ESA's attitude in a coordinated approach, vis-à-vis the Soviets.[92]

[89] Gazenko, O. to Lüst, R. Moscow, July 23, 1986. In ESA-28042, HAEU.

[90] Lüst, R. to Gazenko, O. Paris, September 11, 1986. In ESA-28042, HAEU.

[91] Goldsmith, Ph. to Gazenko, O. September 17, 1986. In ESA-28042, HAEU.

[92] Oser, H. *Mission opportunities for Life Sciences experiments on Soviet Spacecraft*. October 23, 1986. In ESA-28042, HAEU.

Collaboration Unveiled: Understanding the Why

Back in 1979, at the moment of ESA's first unofficial visit to Moscow, there was an awareness of urgent need among the European scientific community to increase the ability to launch their experiments into space. At that moment, as Oser mentioned, there was "only one manned mission (SL-1)" foreseen for European space biologists. Therefore, he outlined "the need and urge for more frequent missions, easier access to mission, less complicated procedures and lower costs for developing and launching the scientific payload". Besides the USA, only the Soviet Union could offer flight opportunities on manned spacecraft, hence the visit to Moscow had taken on an obvious and crucial relevance.[93]

Seven years later, as a result of events discussed above, the "uniqueness" of the Soviet space capability to put automatic scientific satellites into orbit and to carry experiments on board an orbital station became even more unique. Thus, the primary advantage for ESA laid in the use of Soviet biosatellites and their access to the orbital station for long duration experiments, including their crews as operators and test subjects on a somewhat limited basis. ESA officials stressed the "extreme usefulness" for the Agency of this type of cooperation with the Soviet Union in the field of life sciences "given Soviet unique facilities and their advanced program in this field".[94]

More precisely, these unique capabilities, which were not available at ESA or NASA, were long duration experiments or studies in physiology on an invasive basis (e.g. with primates and rodents). Oser also disclosed an extreme need to collaborate with the Soviets on board *Bions*: "the Western world did not have such a capacity of a retrievable, unmanned Biosatellite, allowing for great number of – even sometimes quite complex—experiments, producing the required database on a comparably cheap and quick basis" (Fig. 7.7).[95]

The ultimate goal of such experiments by the European officials was the adequate preparation of the scientific community for the Columbus-ISS era.[96] In fact, despite the difficult negotiations ahead with the Soviet side, Oser considered it extremely important to start discussions on

93 Oser, H. *Report on Mission to URSS.* June 4-8, 1979. In ESA-28129, HAEU.

94 *Relations with the U.S.S.R. IRAC*, October 1, 1986. In ESA-28130, HAEU.

95 *Mission Report to U.S.S.R.*, December 19, 1988. In ESA-28042, HAEU.

96 Oser, H. *Report on mission to the Soviet Union.* Moscow, June 18-23, 1986. In ESA-28042, HAEU.

Fig. 7.7 *Bion* and *Foton* re-entry capsules in stock at TsSKB-Progress. Source: A. Verga

cooperation, since "we would be able to provide for needed additional mission opportunities and thus not be ill-prepared for the space station in the life sciences".[97]

As far as the Soviet side was concerned, in the early 1980s before *perestroika*, Oser primarily emphasised the "geopolitical nature" of motivation on the Soviet side. He noted the difficulties that had recently appeared with France and the United States—the main partners of the Soviet Union in the field of space biomedicine. In fact, the "current political climate" in the first half of the 1980s did not promote rapprochement. "I left with the impression that their collaborative efforts with NASA had practically ceased and France's contributions, although continuing, were limited. Some bilateral planning efforts with the Netherlands, Sweden, Switzerland,

[97] Oser, H. *Mission to U.S.S.R.* June 16, 1986. In ESA-28042, HAEU.

West Germany proved", according to Oser's sources, "to be impractical". Thus, the Soviet Union was looking for a "new partner".[98]

However, with the coming to power of Gorbachev and the corresponding warming of Soviet-American and Soviet-French relations, the situation changed in Oser's eyes. In 1986, during his visit to Moscow, where he took part in a discussion on the results of the *Bion-7* flight (see above), he drew attention to the fact that IBMP officials also had separate meetings with NASA (Ames Research Centre), the French delegation and with Augusto Cogoli from ETH Zürich. Yet, according to Oser, the Soviets and NASA did not come to a renewal of their former cooperative Agreement in the life sciences, which had ceased around 1980, despite many efforts and negotiations in the meantime. He believed that it would require a governmental Agreement on both sides beforehand. Meanwhile, the Soviet-French cooperation was developing—both on Biosatellites and the Mir-space station. Both countries continued to build up their opportunities through an established governmental Agreement already dating back several years.[99]

Oser also commented that the Soviets preferred to deal with ESA's proposal to overcome "tedious and numerous negotiations with single ESA member states".[100] This signalled a change in the attitude of the Soviet leadership as a whole towards the European Community. Even Evgeny Ilyin recognised that the arguments of the scientists directly involved in the cooperation should be complemented by a purely political motivation, which he believed was the real driving force behind the cooperation between ESA and IBMP. The political recognition of the integration processes of the European Community by the Soviet Union, as well as Gorbachev's principles contained in the concept of a "Common European Home", served as a reliable political framework for a concrete Soviet-European rapprochement and consequently to the establishment of relations directly with ESA, particularly in the field of life sciences.

It is pertinent to note a seemingly obvious, yet occasionally overlooked fact: ESA and the European Economic Community (EEC) constituted distinct entities during the 1980s. At that time, ESA comprised 13 member states, while the EEC encompassed 12. Significantly, the membership

[98] Oser, H. *Report on Mission to U.S.S.R.* Moscow, November 21-24, 1984. In ESA-28042, HAEU.

[99] Oser, H., *ESA/Soviet Cooperation in Life Sciences and Mission Report to the IUPS/CGP Congress*, Tokyo, November 14, 1986. In ESA-28042, HAEU.

[100] Ibid.

compositions of these two organisations did not entirely coincide. Thus, although a degree of membership overlap existed, their respective activities and objectives remained largely independent.[101] As Antonio Verga recalled, "even the Soviet colleagues and authorities were mesmerised and sometimes they did not understand why they had to deal with two different European institutions. ESA's policy might have also differed from the EEC one and even from those of individual member states, adding yet more confusion".[102]

Ilyin, for his part, and on behalf of IBMP, focused his analysis of motivations on the technical aspects of cooperation, leaving out its "geopolitical" component. Summarising the question of the Soviet motivation for participating in international cooperation, both in the case of the USA and ESA colleagues, Ilyin identified four main reasons. Firstly, Soviet scientists aimed to "expand the effectiveness of their research by conducting new experiments that foreign specialists could offer. Let us say that Soviet Union proposed twenty experiments, and the Americans offered another twenty of their own. If IBMP felt that it would not interfere with its own research and would improve the efficiency of the launch, the proposal was accepted. In this way, more information could be obtained from the same animals in one flight".[103] As for their European colleagues, there was also an element of enthusiasm for Soviet scientists, manifested in purely scientific interest. As Ilyin put it, "We were interested, knowing that ESA specialists had conducted numerous experiments on Space Shuttle missions. We realised that ESA were worthy partners who understood the essence of the problems and had experience in preparing and conducting biological and radiobiological experiments. This gave us additional strength and eagerness".[104]

Secondly, Ilyin emphasised the importance of sharing research methodologies. As he argued, "Different countries use different research methodologies. We knew that there were techniques in the USA and Europe that we did not use, but that seemed useful and important for our research. That is, it was a question of mastering research methods that our side, for

[101] In the late 1980s EEC included Belgium, France, West Germany, Italy, Luxemburg, Netherlands, Denmark, Ireland, UK, Greece, Portugal, and Spain. ESA was composed by Austria, Belgium, Denmark, France, Germany, Ireland, Italy, Netherland, Norway, Spain, Sweden, Switzerland, and U.K.

[102] Interview with A. Verga, February 20, 2024.

[103] Interview with E. Ilyin, March 15, 2023.

[104] Ibid.

one reason or another, did not possess". In analysing the lack of knowledge among Soviet scientists, Oser drew attention to the scientific rather than the methodological component. He believed, the Soviet side's interest was in fundamental biology and physiology due to the fact that as the Soviets recognised themselves, "they had in the past almost all their efforts in operational biology and physiology, neglecting fundamental aspects of these sciences".[105]

Thirdly, as previously mentioned in the reports of ESA scientists, the Soviet side was interested "in using modern parts of equipment, which were developed abroad and not available to IBMP. We were interested in flying and using European scientific instrumentation for biological experiments that had been developed by ESA for the Space Shuttle and Spacelab missions. These were cutting edge pieces of scientific instrumentation. We had analogues, but they were somewhat outdated in some of their capabilities for conducting such experiments. I will not deny that we had an interest in using European equipment together to obtain accurate, advanced knowledge in the course of these studies".[106]

At the same time, Ilyin paid special attention to the fact that this equipment was used "for joint research. It is not that we gave Europeans or Americans an opportunity to put a free device on our satellite so that they could get scientific information for themselves. We did this under the agreement that the results of the research would be jointly shared, and the authorship would belong to both parties. That is, we strived to use modern devices that we did not have, but which could enrich our science".[107] Furthermore, while offering their automated spacecraft for European research, Soviet officials in turn expressed an interest in participating in ESA projects such as SLED, Biorack, or Anthrorack on Spacelab missions. They also were inquiring about collaboration on the Eureca (European Retrievable Carrier) facilities, its Botany facility, and Exobiology/Radiobiology Assembly (ERA).[108]

Finally, Ilyin mentioned the cultural factor: "We were just interested in communicating with foreign scientists from the point of view of understanding their mentality, getting acquainted with their traditions, culture,

[105] Oser, H. *Report on Mission to U.S.S.R.* Moscow, November 21-24, 1984. In ESA-28042.

[106] Interview with E. Ilyin, March 15, 2023.

[107] Ibid.

[108] Oser, H. *Report on mission to the Soviet Union*. Moscow, June 18-23, 1986. In ESA-28042, HAEU.

behaviour. Purely human factor. And how are they? Who are they? I think it was also interesting for foreigners to look at Soviet citizens and realise that they are not barbarians, but similarly educated, intelligent people with whom you can cooperate and be friends".[109] In this, Ilyin's recollections largely agree with René Demets' testimony: "Both parties (ESA–IBMP) were very curious about each other's lives and lifestyles. The Soviet Union had been full of secrets. Vice versa, most Russians that we met had no idea, or a false idea, about life in the Western world. Meetings took place in Moscow and in Noordwijk (ESTEC). IBMP colleagues were invited into the homes of ESA people, and vice versa".[110]

Hence, during their visits to the Netherlands, Soviet specialists had the opportunity to acquaint themselves with the scientific and production facilities of their counterparts and explore various premises dedicated to the preparation of experiments and the manufacturing of scientific equipment. A report on their visit to ESTEC in 1988, for instance, illustrates this engagement: "at the end of their Noordwijk journey the five Soviet specialists visited ESTEC facilities where they saw various testing facilities, Biorack equipment, Mini-Gravitational-Biology facility, experiment growth chambers and a mock-up of the Columbus Attached Pressurised module".[111]

Thus, both sides were guided simultaneously by purely scientific goals (sharing experience, mutually beneficial filling of missing knowledge, joining efforts), as well as through pragmatic considerations, finding a favourable partner while reducing financial costs and providing the greatest benefit. If at the turn of the 1970s and 1980s ESA did not feel an acute need to cooperate with IBMP, with the change in the political situation, in the absence of a real alternative at that time and against the background of unchanged interest from the Soviet side since the late 1970s, the Europeans moved towards rapprochement and formalisation of relations (Fig. 7.8).

[109] Interview with E. Ilyin, April 17, 2023.

[110] Correspondence with R. Demets, March 26, 2023.

[111] *Protocol of Third ESA-U.S.S.R. Working Meeting on Space Biology and Medicine.* Noordwijk, Netherlands, April 25-29, 1988. In ESA-28042, HAEU.

Fig. 7.8 Dam square in Amsterdam, April 1988. Soviet delegation with ESA colleagues. Source: R. Demets

From Negotiations to Lift-off: The Journey of *Bion*-8

In January 1987, a delegation from ESA, consisting of Ph. Goldsmith, J. Arets, W. Ockels (ESTEC, ESA astronaut), G. Seibert (head of ESA's Microgravity Department), and H. Oser went to Moscow for negotiations. As Gazenko would recall afterwards, this visit represented "the first real step towards the establishment of mutually beneficial cooperation".[112] From the Soviet side, an official reception was given in accordance with the high status of guests and the importance that the IBMP and Intercosmos scientists and managers attached to the upcoming cooperation with ESA. So, during the first meeting ESA's delegation was "received and cordially welcomed" by Vladimir Kotelnikov. Both sides demonstrated the interest in the establishment of long-term scientific contacts and the conduct of joint biomedical investigations in three fields: on board manned and unmanned ESA and Soviet space vehicles, as well as ground-based

[112] Gazenko, O. to Lüst, R. January 8, 1987. In ESA-28042, HAEU.

simulation studies. Moreover, ESA and Soviet officials specified areas of research to be performed on a joint basis. Besides the classical declaration to assist researchers in performing their scientific activities in the specified areas of potential cooperation and commitments to increase regular exchange of scientific information and literature, ESA and Intercosmos representatives agreed to establish a Joint Working Group on space medicine and biology.

The even more practical proposal of the participation in the 8th Biosatellite was made by the Soviet representatives. The next *Bion* had to be launched in Autumn 1987, thus ESA scientists did not have much time to prepare experiments. Accordingly, it was decided to accelerate the legal formalisation of cooperation: an agreement had to be prepared "in the very limited time frame". During these talks, a foundation was also made for the future development of bilateral relations: "more complex type of cooperation, including the active participation of ESA in the Soviet manned space flights and the possibility of embarking Soviet life sciences experiments in the framework of ESA programs".[113] Since the issues of manned flights were not within the competence of IBMP (see above), representatives of the Moscow Institute declined to discuss the possibility of an ESA astronaut flying aboard the Mir orbital station. As already mentioned, this issue was under the jurisdiction of Glavkosmos. To the disappointment of ESA officials during that visit, it was not possible to establish contact with representatives of the recently created Glavkosmos.

Returning after the meeting, Goldsmith hastened to assure Gazenko that he "sensed a real will for a close and mutual beneficial scientific cooperation by all the scientists taking part in the discussions" and he informed his Soviet counterpart that at ESA's Management Board meeting the "results of his visit to the Soviet Union were quite enthusiastically received and professor Reimar Lüst, ESA's Director General, was particularly satisfied with the outcome".[114]

The preparation of ESA experiments for flight aboard *Bion*-8 commenced immediately following the completion of official agreements. However, it is important to note that animal studies were not part of ESA's agenda. As Antonio Verga remembers, "The reason why both CNES and NASA were involved in the Biocosmos program is that they were very interested in the studies that IBMP was doing on primates and

[113] *Protocol ESA and U.S.S.R. meeting*. Moscow, January 6-9, 1987. In ESA-28042, HAEU.
[114] Goldsmith, Ph. to Gazenko, O. January 16, 1987. In ESA-28042, HAEU.

other mammals (rodents) whereas ESA was not for ethical reasons and still is not. Conversely, ESA was very interested in molecular and cellular biology and then extended to the biology of micro-organisms (microbes, bacteria) but certainly not mammals or animals of this class".[115] Indeed, research on small organisms and microbes started in 1985 with Biorack on board of Challenger's Spacelab D1 (see above).[116]

As Biorack was designed as a crew-operated apparatus, it could not be used on board of automatic *Bion* satellites. Instead, *Bion* offered the possibility to place experiment packages not only inside the capsule, but also outside in containers called KNA, attached to the external shell of the pressurised module.[117] ESA had been informed by IBMP that it would be possible to fly experiments under low-shielding conditions in the outside-exposure KNA facilities.[118] These KNAs were used on *Bion* capsules for the first time. Indeed, both missions, *Bion-8* and *Bion-9*, differed from the previous ones by the presence of the KNAs; each *Bion* carried four of them. Ilyin drew attention to the fact that "we had not conducted such experiments before, neither with NASA, nor with CNES. For the first time we started to conduct experiments not only inside the satellite, but also outside. That was the novelty of the experiments that we first started to carry out with ESA specialists. This significantly enriched the scientific program and allowed us to obtain a number of new interesting data".[119]

Once the satellites were in orbit, biological samples were exposed to the factors of open space: to vacuum, large temperature differences, and a high level of ionising cosmic radiation. Remarkably, the KNAs were launched with their lids spread open, the samples only protected by the fairing. During the descent phase, with a very simple action of mechanical straps and springs, the lids of these containers were closed to make the bio-objects safely return to Earth. Throughout atmospheric re-entry, the KNAs were protected by thermo-insulating shells from the scorching of aerodynamic heating, typical of a spacecraft descending from orbit (Fig. 7.9). Antonio Verga claims that "KNA were the forerunners of Biopan, ESA's exposure container that made its first flight in 1992 on *Foton-8*. Biopan was somewhat larger (diameter 38 cm, height 23 cm)

[115] Interview with A. Verga, March 27, 2023.
[116] Harvey (2024), 21-22.
[117] Correspondence with R. Demets, March 26, 2023.
[118] Demets, R. (1996). *ESA SP-1190 Bion-8 and -9*, ISBN 92-9092-359-8, 11.
[119] Interview with E. Ilyin, March 15, 2023.

Fig. 7.9 Post-flight inspection of *Bion*-10's four KNA facilities. Source: European Space Agency - ESA.

than the KNA (diameter 30 cm, height 8.5 cm) and was equipped, unlike the KNA, with a suite of sensors to monitor the harsh space environment, which the bio-samples were exposed to".[120]

Dick Mesland, ESA project scientist of Biorack, was sent to Moscow in April 1987 to explore what ESA could achieve on *Bion*. During his visit the scientific, technical, and methodological aspects of the experiments were discussed. ESA proposed a program of joint biological, radiation-physics and radiobiological research on the next Soviet biosatellite (Fig. 7.10).

With a very short lead time, it was decided to keep joint experiments simple, without power and data interfaces.[121] Thus, three autonomous experiments were provided by, and selected by, the ESA's and Soviet

[120] Interview with A. Verga, March 27, 2023.
[121] Harvey (2024), 21-22.

Fig. 7.10 Dick Mesland and René Demets, 2019. Source: R. Demets

Principal Investigators for the *Bion-8* flight in 1987.[122] Named SEEDS, CARAUCOS, and DOSICOS, each one was a joint ESA/IBMP experiment. Thanks to the tandem accommodation options (inside and outside of the re-entry module), a combination of biology specimens and dosimeters could be implemented.[123]

It was also important for ESA staff to discuss the moment when the Soviet side informed them about the date of the satellite launch. Due to the fact that the cycle of preparing bio-assembly for the flight experiment was about 40 days, it was decided that the Soviet launch authority would inform the ESA specialists about the launch date in advance by telex.[124]

ESA specialists were requested to hand over the filled containers to the Soviet scientists, no later than 5 days before the launch. Separately, it was stipulated that the Soviet side took upon itself the transportation of all instruments to the launch site and back to Moscow 1–2 days after the landing of the biosatellite. Moreover, the Soviet team was due to

[122] Seibert, G. et al. (2001). *A World without gravity. Research in Space for health and industrial processes*, ESA, 449.

[123] Mesland, D. *Progress Report – ESA/Biocosmos Experiment.* May 7, 1987. In ESA-28042, HAEU.

[124] *Protocol ESA and U.S.S.R. meeting*. Moscow, April 21-24, 1987. In ESA-28042, HAEU.

hand-over to the ESA delegates the part of the biomaterials intended for them, no later than two days after landing.[125]

The formal wording of the agreements and protocols was enlivened and filled with emotional colouring thanks to the memories of those directly involved in the project. Thus, Evgeny Ilyin recalled: "We worked in a spirit of reciprocal respect, common understanding, and mutual support. We helped the European specialists prepare their experiments; we loaded their bio-samples. They did some part of the work together with us, because we had a distribution of responsibilities on *Bion-8* and *Bion-9*. European specialists prepared biomaterials for research, supplied by the European side. In addition, they placed these biomaterials in scientific equipment. But here we were also involved. And then, transporting the bio-samples to the launch site, working with them at the launch site, control over the course of experiments in the flight control centre, then their landing and their delivery from the landing site in Moscow—all this was done by Soviet specialists. ESA staff finally returned the flown specimens to the respective European institutes where the research was conducted. That is, we helped ESA specialists in preparing the experiments, and they helped us with the equipment".[126]

Among other things, it was the collaborative nature of these experiments that Ilyin drew attention to in his recollections. "The main condition for cooperation was that the experiments had to be jointly conceived and designed. This meant that both parties acted as equal partners, the exchange of scientific results from joint experiments and joint publication of research were also due. The wording predicated that all research was conducted on the basis of mutual scientific and technical interest. There was no financial gain for either party. It was exactly an exchange of experience".[127]

As for the financial side of the issue, in the final version of the contract it was agreed that "both parties would each bear the costs of discharging their respective responsibilities, including travel and subsistence of their own personnel. With regard to the transportation of the experiments, ESA would bear all the costs to bring them to Moscow for the flight preparation and to return them from Moscow after the flight. Intercosmos would bear the costs for transporting the experiments from Moscow to the

[125] Ibid.
[126] Interview with E. Ilyin, March 15, 2023.
[127] Ibid.

launch site and from the landing site back to Moscow".[128] The agreement was entered into force by the exchange of letters between ESA Director General Reimar Lüst and the Vice-President of the Soviet Academy of Sciences, Vladimir Kotelnikov, on September 8, 1987.

Bion-8 had not yet returned to Earth delivering the results of the experiments on board, when Goldsmith had already sent a long letter to Gazenko with detailed proposals for further cooperation, not only on board the biosatellites, but also on the Mir orbital station. Goldsmith indicated areas for cooperation that foresaw both the Mir "on crew" and "crew interaction" experiments and Biocosmos satellites. He asked Gazenko to provide him with "some background information of sufficient detail", in order to be able to inform the scientists accordingly and to build appropriate hardware. Goldsmith explained his anticipated need for information by the ESA rules that obliged him to proceed with a call for experiment proposals to the scientific community in ESA member states.[129]

A whole series of bilateral cooperation issues was discussed during the visit of ESA delegation to Moscow and talks in the frame of the ESA-USSR Working Meeting held at IBMP in November 1987 (Fig. 7.11). Both sides discussed the preliminary results of the *Bion-8* flight and control experiments on ground, emphasising their successful completion. Moreover, they agreed upon the time schedule of data exchange in support of joint experiments. Soviet specialists also accepted ESA's proposal to carry out joint experiments on board the biosatellite *Cosmos-2044* (*Bion-9*) scheduled for 1989, using ESA hardware. It was agreed that the total size of ESA's hardware delivered to *Bion-9* would be the same as previously on *Bion-8*. Moreover, the sides agreed that ESA would supply equipment for an IBMP-based laboratory to support joint pre- and post-flight studies, as well as control simulation studies. Consequently, IBMP was required to provide a designated room for this purpose.

The Soviet side also expressed its interest in using incubators of ESA Biorack equipment in joint biology and radiation biology experiments and the Mini-gravitational biology facility to be flown on *Bion* satellites and the Mir station after the year 1990. As Oser put it, "the Soviets appreciated

[128] Lüst, R. to Kotelnikov, V. July 21, 1987. In ESA-28042, HAEU.
[129] Goldsmith, Ph. to Gazenko, O. September 29, 1987. In ESA-28042, HAEU.

Fig. 7.11 November 1987. ESA delegates in Star City. Source: R. Demets.

this project and proposed the potential inclusion for this flight after the year 1990. This project could include cell and developmental studies on a higher organism level (tissues, plants, small animals, insects etc.)".[130] Furthermore, ESA officials proposed to carry out three joint human physiology experiments on the Mir station and promised to provide necessary flight and laboratory equipment for these experiments.[131] Since these proposals required a crew, both as operators and test subjects, they could potentially justify the presence of an ESA astronaut aboard the Mir space station. The Soviets showed high interest in European proposed studies and expressed their willingness to fly them as soon as possible.[132] Finally, both sides agreed to the usefulness of a joint ESA-Soviet working group

[130] Oser, H. *Report on mission to the Soviet Union*, 11 December 1987. In ESA-28042, HAEU.
[131] *Protocol ESA/U.S.S.R. Second Working Meeting on space life sciences cooperation.* Moscow, November 23-27, 1987. In ESA-28042.
[132] Oser, H. *Report on mission to the Soviet Union.* December 11, 1987. In ESA-28042, HAEU.

on space biology and medicine to be established during the next year, 1988. The scientific results of the joint ESA-IBMP experiments on *Bion-8* were published over a four-year period (1988–1992) with a peak in 1990, three years after the mission.[133]

During the Discussion Medilab: From Vision to Reality

Let us take our minds off the *Bions* for a moment and recall another interesting Soviet project—Medilab. In fact, there was another proposal made by IBMP representatives to ESA scientists that dated from the same period. This was an invitation to participate in the Medilab project.[134] It was originally planned that a special biomedical module, Medilab, would be attached to the Mir orbital station in the early 1990s. The module was designed to be 4.1 m in diameter and 11 m in length, with a pressurised volume of 100 m^3, having integrated computer support, an animal holding facility, and a man-rated centrifuge. It was planned that Medilab would have been used continuously by one clinical physician and one physiologist. There were also plans to perform survivable animal surgery and to develop anaesthetic techniques on board Medilab. One of the goals was to identify the optimal duration of human exposure to spaceflight, as it was considered a critical step in enabling future long-duration space missions such as a Mars expedition.[135]

In September 1987, the Soviet scientist in biomedicine Sergei Kholin from IBMP presented the Medilab at the 3rd European Symposium on Life Sciences research in Space, held in Graz (Austria). The new module dedicated to life sciences was supposed to be docked onto the Mir-station in the near future, and ESA was invited to consider its utilisation and to come forward with cooperative proposals. As reported by Oser during the Symposium, the Medilab resembled somewhat of a Spacelab long module, but was double the size. He underlined that "it would be primarily dedicated to basic research which the Soviets seem to have 'neglected' in this

[133] Demets. (1996), 19.

[134] See more in Baranov, U. et al. (1991). Concept of "Medilab" orbital bio-medical laboratory. *Acta Astronautica*, 23, 299-306.

[135] Campbell, M. R. (2017). The Medilab Mir Module. *Aerospace Medicine and Human Performance,* 88 (11), 1056.

area over the past decades".[136] Kholin was authorised to admit that ESA's potential expressed interest in using Medilab could expedite the realisation of the project within the Soviet Union.

During ESA's delegation visit to IBMP in November 1987, the Soviets updated ESA specialists on the development of the Medilab project and confirmed their invitation to explore the possibility of performing joint studies on Medilab and developing facilities of the Anthrorack and Biorack type for joint studies. In reply, ESA delegates expressed their interest in participating in the Medilab project, thus it was agreed that before June 1988 ESA would send its preliminary proposals to IBMP.[137] In April 1988, the Soviet side reminded ESA to submit its proposals for Medilab equipment by July 1988. However, ESA could not respond to this invitation until July 1, 1989, due to the length of its decision-making process.[138]

The Medilab Module was an ambitious concept that unfortunately was never realised due to the collapse of the Soviet Union and the resulting financial crisis.[139] Nevertheless, the ambition of Soviet scientists in the field of space biomedicine, their willingness to involve European colleagues in their own projects, and the mentioned interest in the joint development of the project, with co-financing, were noteworthy and possibly paved the way to future joint ventures.

Bion-9: Overcoming CoCom Hurdles and Launch Setbacks

The successful experience of jointly conducting experiments on *Bion-8* allowed ESA and Intercosmos to continue their cooperation and start discussing a number of experiments on board the next *Bion-9* spacecraft (Fig. 7.12). In addition to Soviet scientists, specialists from Austria, Hungary, DDR, Canada, Poland, Romania, the USA, France, Czechoslovakia, as well as ESA took part in the scientific program. 81 experiments were carried out

[136] Oser, H. *3rd European Symposium on Life Sciences research in Space*. September 21, 1987. In ESA-28042, HAEU.

[137] *Protocol ESA/U.S.S.R. Second Working Meeting on space life sciences cooperation*. Moscow, November 23-27, 1987. In ESA-28042, HAEU.

[138] *Protocol of Third ESA-U.S.S.R. Working Meeting on Space Biology and Medicine*. Noordwijk, Netherlands, April 25-29, 1988. In ESA-28042, HAEU.

[139] Campbell. (2017).

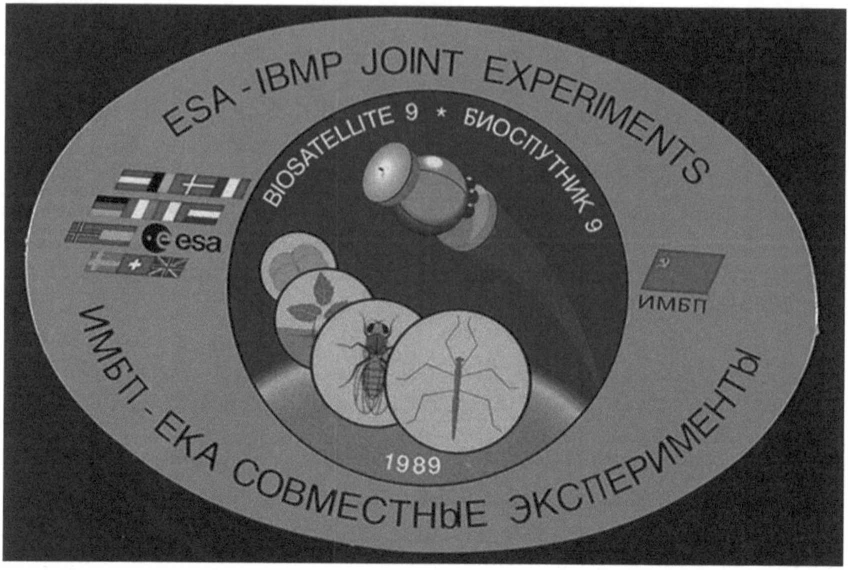

Fig. 7.12 *Bion*-9 mission logo, designed by Claus Alfermann (ESA). Source: European Space Agency - ESA

on board,[140] in such fields as motion sickness, reproduction and regeneration, immunology, and re-adaptation to a normal gravity environment. A number of different biological specimens were used, including an assortment of rodents, newts, fish, fruit flies, ants, worms, cells, and seeds. The prime animals were two macaque monkeys. The 2.3-m diameter descent sphere was successfully recovered after 14 days, but a failure in the thermal control system resulted in the deaths of some of the specimens.[141]

It was the second time that ESA participated in such a mission. Five experiments were carried out jointly by scientists from the Moscow IBMP, Kiev Institute of Botany and Kazan Institute of Biology[142] and from ESA. ESA scientists came from institutes in Norway, Denmark, FRG, Switzerland, France, and Spain. Studies were done on plants, seeds, plant protoplasts, *Drosophila melanogaster* (fruit flies), *Carausius morosus* (common stick insect), and radiation dosimetry techniques, in and outside the

[140] Nikitin, S. (1990). Biosputnik “Kosmos-2044”. *Priroda*, 1.

[141] Bion-9. Description. Retrieved February 26, 2025, from https://nssdc.gsfc.nasa.gov/nmc/spacecraft/display.action?id=1989-075A.

[142] Demets. (1996), 37

re-entry module.[143] *Bion-9* carried five joint ESA-IBMP experiments: CARAUCOS, DOSICOS, and SEEDS (as on board of *Bion*-8) to which were added FLIES and PROTODYN.[144]

The third ESA/USSR working group met at ESTEC in April 1988. General issues of biology and medicine cooperation were discussed along with future joint missions on *Bion-9*. The Soviets informed ESA about the scientific objectives for *Bion-9* and *-10* and indicated for *Bion-9* a launch date around September 1989 and for *Bion-10* in 1991.[145]

To support the joint ESA-Soviet experiments aboard the satellite, ESA decided to install some ground laboratory equipment at IBMP. According to the preliminary agreement, this equipment had to be available in May 1989 for pre-mission, ground control, and post-mission experiment activities. Both parties decided that ESA equipment would be available to IBMP also for future joint experiments, while still remaining property of ESA.

During the discussion of the legal side of the supply of equipment from Western Europe to the USSR, the issue of technology exports to the socialist bloc countries, regulated by CoCom, arose. Although ESA was a pan-European organisation, in this case the exporting part was the country where ESA's main technical premises (ESTEC) was located. Although procured from different suppliers, such equipment needed to obtain an export licence from the Dutch authorities, since it had to be transported to the Soviet Union from ESTEC, located in the Netherlands. Thus, the export procedure had to be conducted according to Dutch laws and regulations. Therefore, it was the responsibility of the national authorities to check the list of the equipment to be exported and its compliance with the CoCom mechanism. As Gabriel Lafferranderie, ESA legal adviser, put it "ESA status (privileges and immunities) had nothing to do with this".[146]

An ESA officer had to navigate the complicated procedure of obtaining authorisation to export scientific equipment to the Soviet Union,

[143] Mesland, D. *Invitation to Biokosmos 9 and 10 Meeting at Madrid*. February 12, 1990. In ESA-28043, HAEU.

[144] Demets. (1996), 23-41.

[145] *Protocol of Third ESA-U.S.S.R. Working Meeting on Space Biology and.*, Noordwijk, Netherlands, April 25-29, 1988. In ESA-28042, HAEU.

[146] Lafferranderie, G. to Jansen, W. February 8, 1989. In ESA-28042, HAEU.

providing a detailed justification for the necessity of such an operation. In order to get an export licence, ESA stated that all exported items were for the common use of the researchers involved from Western Europe and the Soviet Union, while any ground equipment had to be located in a laboratory based in Moscow and left at the disposal by IBMP. This was necessary because the biological material used in the experiments needed to be prepared as close to the launch as possible and analysed as soon as possible after landing. ESA staff sought to reassure Dutch authorities that "this ground equipment was a property of ESA, was registered as an inventory item in ESTEC, Noordwijk, and would be placed temporarily at IBMP in Moscow for the specific purposes as described above".[147] Thanks to a favourable decision, ESA was able to send the necessary equipment to Moscow. With one month delay to the due date (June 21), the first batch of ground-support equipment was transported by truck from ESTEC to Moscow. Additional transport followed on July 21, 1989, by truck and on August 17 by plane.[148]

Both sides had also agreed to organise three working meetings, in order to prepare the joint experiments. It was foreseen that up to eleven people had to be sent to Moscow in the period between 7 days before and 7 days after the *Bion-9* flight. It was decided to exchange written preliminary reports (including raw data and preliminary analysis) at around 6 months after the flight, and about one year after the flight—final reports and drafts of joint articles ready to be submitted for publication.[149] In the end, three ESA staff members were assigned to support *Bion-9* ground operations in Moscow: W. Jansen (project manager), D. Mesland (project scientist), and A. Franz (technical support).[150] As it was during the implementation of the *Bion-8* mission, ESA specialists did not have access to the launch site or the landing site, nor contact with mission operators. ESA's role in Moscow was confined to loading samples into flight hardware, handing it over to the IBMP and receiving it back afterwards.[151]

In the cooperation agreement for the launch of *Bion-9*, a separate article highlighted the financial side of the issue. According to Article. 6 of the "Letters on the flight of joint experiments on Biocosmos-9" "both parties

[147] Jansen, W. *Moscow GSE*, January 25, 1989. In ESA-28042, HAEU.

[148] Demets. (1996), 27.

[149] Summary of the meeting between U.S.S.R. and ESA representatives in the joint experiments aboard the biosatellite, November 29-30, 1988. In ESA-28042, HAEU.

[150] Demets. (1996), 27.

[151] Harvey. (2024), 22.

would each bear the costs of discharging their respective responsibilities, including travel and subsistence of their own personnel. With regard to the transportation of the experiments, ESA would bear all the costs to bring them to Moscow for the flight preparation and to return them from Moscow after the flight. The Intercosmos Council would bear the costs for transporting the experiments from Moscow to the launch site and from the landing site to Moscow".[152]

The *Bion-9* mission has a remarkable history with significant and far-reaching consequences. The spacecraft was launched on September 15, 1989, and came back to Earth 14 days later on August 29. The launch, originally scheduled for September 8, 1989, was cancelled when a short circuit was discovered in one of the spacecraft batteries. The error was detected during the final spacecraft checkout with the fully integrated rocket on the launch pad.[153] As Ilyin recalled, "When the biological samples were already loaded into the spacecraft and the rocket was already at the launch site, just a day before the launch, malfunctions in the launch system were detected, and the launch was postponed for a week. In this regard, the rocket was moved from a vertical position to a horizontal one, it was again transported to the MIK (Assembly and Test Building) where the spacecraft was removed from its fairings and parts of equipment were unloaded from it, including the containers with biomaterials and with animals".[154]

As René Demets explains, "In the evening of September 7, less than 1 day before the scheduled launch, an emergency telephone call from Plesetsk was received in Moscow, announcing a one-week launch delay due to launcher problems. It was very difficult for IBMP to make contact with ESA personnel from within the launch area. The call announcing the launch delay was actually made from outside the security zone, from a post office in the town of Mirny, at a distance of about 40 km from the launch pad. Why ESA was not simply notified via the flight control centre in Moscow remains unexplained". Demets emphasised the fact that this was also the first time such a situation had occurred for Soviet scientists since

[152] *U.S.S.R., Institute for Biomedical problems (IBMP): exchange of letters.* Mars 29, 1989. In ESA-28042, HAEU.

[153] Demets. (1996), 27.

[154] Interview with E. Ilyin, March 15, 2023.

the use of *Bions* for bio-medical research began, thus "no contingency plans had been prepared".[155]

Ilyin continued: "We began to discuss with our European colleagues what we should do. We decided not to touch the rats and run them as planned. But it was decided to replace some of the bio-samples that were prepared by our European colleagues, because they were selected by age, by stage of development for a specific date of the flight. This shift by a week would have spoiled the entire scientific program of the research".[156] René Demets confirmed that "a quick assessment revealed that a one-week period of waiting was incompatible with the developmental studies CARAUCOS, FLIES and PROTODYN, because after that week the biological materials would have grown beyond the required developmental stage. First of all, therefore, the flight hardware had to be returned from Plesetsk to Moscow immediately and next, fresh samples had to be acquired, prepared and loaded. That was easier said than done because some investigators were already on their way back home".[157]

In a few words, Ilyin managed to describe a week-long force majeure situation, which ultimately had a happy ending: "We urgently sent containers with bio-objects from the cosmodrome to Moscow. In Moscow they were handed over to our European colleagues, and they urgently sent them to Europe. New preparations began there. They were prepared, re-loaded into their containers, and then sent to Moscow so that they could be re-installed inside the spacecraft in two days. This made the situation more complicated. It was a great stress and astonishment for everyone, including us, because we had to do a lot of extra work. We changed our biomaterials as well, but not to a large extent. But the Europeans had to change all the samples with few exceptions".[158]

Despite the trouble that happened with the rocket launch and the urgent replacement of the biomaterials, the scientific outcome was sufficiently valuable to be published in international science journals.[159] The scientific findings from the joint ESA-IBMP experiments on *Bion-9* were released over a four-year period (1990–1994) with a peak in 1992, 3 years after the mission.[160]

[155] Demets. (1996), 28.
[156] Interview with E. Ilyin, March 15, 2023.
[157] Demets. (1996), 28.
[158] Interview with E. Ilyin, March 15, 2023.
[159] Ibid.
[160] Demets. (1996), 40.

Moslab: Advancing Collaboration to the Next Level

The force majeure circumstances that the ESA science team urgently had to overcome during the *Bion-9* mission, requiring additional time, financial, and human resources, led to the adoption of an important decision: the establishment of an in-house laboratory at IBMP. As Ilyin put it, "ESA, taking into account the experience of *Bion-9*, when the bio-samples had to be exchanged at the last moment, decided to create a bio-laboratory at the premises of our Institute, in order to prepare the biomaterials as close as possible to the launch date. If done abroad time for transport is to be accounted for, and in case of a launch postponement, extra time and extra costs arose for transport, logistics, and so forth".[161] The ESA laboratory, located directly in Moscow, enabled the management of potential contingencies that inevitably arise in cosmonautics, given the inherently risky nature of this field of research (Fig. 7.13).

Fig. 7.13 ESA laboratory Moslab. Source: R. Demets

[161] Interview with E. Ilyin, March 15, 2023.

During a meeting held in Moscow in September 1990, even if discussions centred on the issue of joint experiments to be flown aboard the satellite, the question of ground support equipment arose. Moreover, the preliminary implementation study performed and presented by ESA showed that the *Bion-10* joint experiments required considerably more laboratory space for European instrumentation and preparation works than what was made available for *Bion-9* by the IBMP administration.[162]

In ESA Special Publication issued in 2002, Wim Jansen gave his less formal explanation for the Lab construction: "Moscow lacked the minimum infrastructure that was so desperately needed for the job we had to do – which is the reason why we built our own house there, Moslab". As he explained, "Before Moslab, ESA and its science partners had no chances to breed cell cultures at IBMP, and cell biology experiments became of high importance to ESA's life science program. Therefore, a strategically located bio-laboratory as Moslab was, gave us the instrumental infrastructures and instrumentations to begin the path of cell biology research onboard *Bion*. This endorsed the development of Biobox".[163] Biobox was a new automated incubator for cell culturing experiments. Based on the previous Biorack, it was simplified and scaled down to be fit between the two monkey seats in the *Bion* capsule. Biobox received power from the spacecraft's batteries and could be operated from the ground by telecommand.[164]

ESA then grabbed the idea of the installation of a pre-fabricated moveable laboratory space ("Prefabs") and outlined a possible configuration. According to initial calculations, the Lab configuration would require 72 m^2 of flat surface and a central connection for electricity and water. A sewage system and telephone line were also planned.[165] During the Joint Working Group meeting on August 19, 1991, in Leningrad, the question of the future European laboratory on the territory of IBMP was raised again and its objectives were more clearly defined. This laboratory aimed to conduct synchronous ground control experiments, to carry out launch preparations, and to process scientific results from flight

[162] *Protocol of the ESA-IBMP working meeting on the future joint experiments aboard biosatellite 10.* Moscow, September 3-7, 1990. In ESA-28043, HAEU.

[163] Demets, R., Jansen, W., Simeone, E. (2002). ESA SP-1208 'Bion-10'. *ESA Publications Division Noordwijk, the Netherlands.* European Space Agency, viii.

[164] Harvey. (2024), 22.

[165] *Protocol of the ESA-IBMP working meeting on the future joint experiments aboard biosatellite 10.* Moscow, 3-7 September 1990. In ESA-28043.

experiments.[166] Unfortunately, the meeting was interrupted because of the GKChP coup attempt and its foreign participants urgently left the Soviet Union.[167]

Amidst political turbulence in the Soviet Union, the wobbling colossus, the communist regime, was living out its last months. Despite these upheavals, scientific contacts continued to proceed as usual. The laboratory issue was revisited a month later during the subsequent meeting in Moscow in September 1991, when ESA specialists inspected the first prefabricated laboratory units and the IBMP infrastructure. As Jansen reported, the Joint Group gave the prefab Lab project the name Moslab (Moscow Laboratory of ESA) which by its vagueness and avoidance of such terms as "joint" or "*Bion*" stressed its versatile applicability beyond the limits of the *Bion-10* Implementing Arrangement (Fig. 7.14). In this

Fig. 7.14 René Demets and Wim Jansen, 2019. Source: R. Demets

[166] Seibert, G. to Grigoriev, A. *Comments to your Draft protocol of the interrupted joint working group meeting on 19 August 1991 in Leningrad*. In ESA-28043, HAEU.
[167] About return flight see Harvey. (2024), 22.

way, future projects like the Biopan test flight (see below), Biopan operational flights, Kashtan of material science type missions, and the Working Group meeting room and office facilities could be carried out inside this laboratory.[168]

The joint scientific program planned for the *Bion-10* mission was carried out on a no exchange of funds basis between ESA and IBMP, with compensation in kind for the engineering and operational services provided to ESA by IBMP and its industrial partners. Thus, Moslab, its lab instruments, and office equipment became a logical part of the compensation package which was finally agreed upon by both parties.[169]

Several trailers came from Europe, bringing the blocks of the new laboratory. The contribution of the Soviet side was to build the foundation for this laboratory. Based on this foundation, blocks in the form of several laboratory rooms were constructed, capable of accommodating both engineers and biologists. A vivarium was also included, allowing for the accommodation of animals, such as rodents, on-site. Incubators were installed too, in which it was possible to conduct ground control experiments and replicate the temperature conditions in which biological samples would be in space flight. As Ilyin stated, Moslab represented a new and more advanced stage of collaboration.

Still, he highlighted a less positive side of the laboratory: "Of course, on the one hand it was better, but on the other hand it led to the fact that a large number of European specialists had to stay in Russia for long periods of time. Sometimes the business trips were as long as five weeks. That is such a long stay". Preparation of bio-samples before the flight, then in parallel with the flight there were control experiments with modelling of temperature conditions that were observed in the flight, then post-flight control experiments. However, in Ilyin's opinion, this was not very difficult for ESA staff. First of all, "Moscow is such a centre, where there is a lot to see and go to. In this way, our European colleagues had an opportunity to get better acquainted with our everyday life, history, and culture". And secondly, long stays in Moscow led to the establishment of "closer human relations with specialists from Belgium, France, Spain, Italy, who participated under the auspices of ESA".[170] However, as it

[168] Jansen, W. *Prefab Lab Protocol*, October 3, 1991. In ESA-28043, HAEU.
[169] Demets, Jansen, Simeone. (2002), 14.
[170] Interview with E. Ilyin, March 15, 2023.

turned out, Europeans were not always enthusiastic about their business trips to Moscow, and the picture they painted of late Soviet life did not always align with the expectations of their Moscow colleagues.

Daily Life in the Late USSR: Perspectives from ESA Staff

The dry protocols and reports of the meetings did not fully convey the atmosphere in which the cooperation unfolded. To a much greater extent, this is manifested in private correspondence among ESA employees, where they detailed their frequent "logistic" and everyday problems—issues that had no place in official reports of the intermediate results of cooperation. Their impressions also testified to the conditions in which they and Soviet scientists had to work in the early 1990s, realising long-planned projects, despite financial and political difficulties.

One of the few ESA staff who managed to directly observe events in the Soviet Union was Wim Jansen. As he recalled, "Because of my work, I was a regular in Moscow in the late eighties and early nineties [...]. I still think it was a great privilege to sit in the front row during this whole revolutionary period—the heyday of *perestroika*, the rise, fall, and return of Yeltsin, the coup against Gorbachev, the collapse of the Union, the shelling of the White House, and the first years of the new Russia".[171]

In his regular reports to ESA management, Wim Jansen detailed the challenges he encountered on a daily basis. Specifically, he highlighted the difficulties in reaching the hotel, which had been booked for him by IBMP personnel, located at a considerable distance from the Institute where the meetings were held. "A normal working day at IBMP tended to be very irregular, and driving along the beltway for 2–4 h/day was no fun, especially in the dark, without any illumination along the road". Losing tyres and reloading became a problem since "there were no services stations along the beltway with guaranteed supplies".[172]

Recalling his impressions from more than thirty years ago, René Demets was also not thrilled with his first trip to Moscow in 1990, expressing some reservations or concerns about the experience: "Russia was

[171] Jansen, W. (2002). *Relax! - Ons Nederlands toen, nu en straks, Het Taalfonds,* 117. Thanks to René Demets for providing the English translation.

[172] Jansen, W. to Barbance, K. *Bion-10 Implementing Arrangement.* March 19, 1991. In ESA-28043, HAEU.

different—no advertising, no colours, no shop windows—just a simple sign with 'Bread' or 'Shoes' to indicate that behind an anonymous door a shop was hidden. Sheremetyevo-2, the International Airport, looked gloomy, even sinister in those days. People in the streets never smiled. The buildings and roads in Moscow were in a miserable state, everything wanting maintenance and repair. Having said that, the city also had its charm. There was a lot to be admired in Moscow, like the wonderful Stalin-era metro stations".[173]

Another difficult problem to overcome was insufficient per diem travel allowance, which, according to the framework agreement, the Soviet side had to allocate to European specialists who visited Moscow on working visits. As Jansen put it in his official report "a rouble subsistence for ESA personnel in Moscow, no matter at what exchange rate, didn't cover even a fraction of our per diem expenditure, since all basic services were nowadays payable in hard currency only".[174] An absurd situation arose in which accommodation of ESA personnel by IBMP in hard currency hotels and a per diem in hard currency was far more expensive to the Institute than providing this support to their own personnel in ESA member states where hotels were three times less expensive.[175]

The everyday life of ESA personnel was exacerbated by the peculiarities of the socio-economic conditions of the late USSR. Thus, "taxis, international phone calls, etc., could only be obtained by paying in hard currency. Some rouble restaurants (like 'Praga') required table reservations two weeks in advance. Others (like the 'Pekin' and 'Baku') had introduced hard currency dining rooms for foreigners, excluding our access to their rouble sections. As a rule, restaurants now accepted table reservations only with guaranteed credit card payment, as soon as your appearance or accent reveals you're a foreigner".

Jansen's overall description was complemented by a sketch depicting the life of a foreigner in the late Soviet Union: "If you manage to intimidate the manager and insist on payment in roubles, the final bill will largely exceed any subsistence we might expect from IBMP".[176] A private dinner invitation at the home of Soviet colleagues did not appear to offer a

[173] Correspondence with R. Demets, March 26, 2023.

[174] Jansen, W. to Barbance, K. *Bion-10 Implementing Arrangement.* March 19, 1991. In ESA-28043, HAEU.

[175] Ibid.

[176] Jansen. W. to Barbance, K. March 19, 1991. In ESA-28043, HAEU.

solution to the difficult financial situation, as "a bouquet of flowers for the hostess, purchased at the central market, was equivalent to the IBMP rouble per diem".[177]

Relations with IBMP were complicated by the difficult situation in which the Institute found itself during the transitional period. Demets recalls the building's wistful appearance to visitors: "unfinished buildings, used ones needing repair and cleaning, dirty, primitive, cockroaches, toilets not working, even the crockery in the canteen was cracked".[178] Expressing empathy for his Soviet colleagues, he also added that "Times were hard, the income of employees at IBMP did not catch up with the increasing prices".[179] As Jansen reported, "we seemed to be dealing with an organisation which was struggling for survival, leading to chaos on all possible ways". Among disorganisation examples he mentioned bad meeting preparation, total lack of coordination with IBMP's industrial partner KB Foton (Kuybyshev TsSKB, see below), unexplained absence of key people during the discussions, IBMP's science involvement in Biobox "which was approaching zero", etc. Jansen concludes about total unreliability of the Institute "in this turmoil" period.

In 2002, Wim Jansen published a book, in which he devoted two chapters to relations with Soviet partners. He cited several episodes illustrating misunderstandings between the parties due to cultural differences. The lack of knives in the canteen, the inconvenience of the Akademicheskaya Hotel intended for foreign delegations, and the Europeans' enterprise and initiative in finding a new hotel and renting a car for traveling freely around Moscow all caused misunderstandings with their Soviet colleagues.

"They never answered an implicit question and did not understand our excitement about it. Consider the following incident, which took place exactly as described, in a hall on the ground floor of a laboratory on the outskirts of Moscow. Together with a Russian key holder, we were standing there with a large load of equipment that needed to go to the second floor (the first floor above the ground floor). I asked, 'Is there an elevator or a freight elevator?' The response was, 'Yes, at the back of the hall,' accompanied by a vague hand gesture towards the darkness. A moment later, our conversation partner had disappeared, leaving us to start lugging the equipment ourselves. We dragged the equipment through the hall and

[177] Ibid.
[178] Harvey, B. (2023). Interview: René Demets. *Quest*, 30 (2), 54.
[179] Correspondence with R. Demets, March 26, 2023.

piled it up in front of the elevator door. I pressed the button, but no light came on. Worse still, the entire elevator did not work. The key holder, who had reappeared after our lugging, looked at us with pity. I had asked about the presence of the elevator, not its functionality! It was deeply ingrained: information was not released without asking, because it could cause trouble.

There was much that our Russian partners did not understand, and little of it was just a language problem. We had somehow overcome the language barrier that separated us by improvising. The real barrier was one of culture and mentality. Seventy years of Soviet indoctrination on top of an already not-so-enlightened history had created a distance between them and us that was much greater than that between the Dutch and Russian languages".[180]

René Demets attributed Jansen's critical attitude to the fact that "he had to resolve everything on his own. Moreover, Wim's superiors within the ESA hierarchy had little understanding of Soviet culture and were insufficiently aware of the challenges he faced during his work in Moscow. He had a lot of responsibility in resolving a huge number of difficulties in Russia. He had studied Russian before he entered the *Bion* project, but this advantage was not always sufficient to resolve the communication problems".[181]

Of course, the political and economic situation in the late USSR had an impact on the Europeans' perception of the country as a whole and the cooperation itself. Solidarity and empathy were influenced by the personal characteristics of each participant in the cooperation, while the establishment of personal relations depended on communicative abilities, willingness, reciprocity, and favourable circumstances. The emotional state and attitude of the characters in the narrative are always very personal factors and crucial for historians, as they add colour to the dry presentation of facts. However, it is essential to recognise their highly subjective nature, influenced by a multitude of factors that, for obvious reasons, sometimes escape the historian's analysis.

[180] Jansen, W. (2002), 118-119.
[181] Interview with R. Demets, July 4, 2024.

The Interference of Glavkosmos

Glavkosmos, which will be addressed in greater detail later, attempted to interfere with the seemingly well-established professional and personal relationships between IBMP and ESA representatives. On the one hand, as we have already seen, the Institute's scientists themselves advised the ESA officials to resolve certain issues directly with Glavkosmos. The establishment of direct contacts between ESA and Glavkosmos, bypassing IBMP, occurred with the explicit consent of the scientists. On the other hand, the subsequent actions of the Glavkosmos staff could be compared to the scenario described in the old proverb about throwing a cat among the pigeons. This inevitably caused perplexity on both sides for and IBMP. The latter faced the risk of losing the privilege of direct communication with European partners, particularly as ESA was expected to shift towards a more commercial relationship regarding the use of *Bion* and its twin companion, *Foton*. The commercial use of the Soviet space infrastructure was entirely managed by Glavkosmos.

IBMP was involved, directly or indirectly, in two additional areas of interest for ESA: materials science experiments conducted aboard the *Foton* spacecraft and biomedical experiments implemented on the Mir orbital station. Both activities were overseen by Glavkosmos. Against the backdrop of sustained, primarily scientific cooperation with IBMP—albeit with ESA's financial investments in the production of scientific equipment, so-called in kind compensation—the dynamics of the relationship between ESA and Glavkosmos, regarding the execution of European experiments on board *Foton* satellite and Mir differed from those of *Bion* managed by IBMP.

Concerning cooperation in the field of materials science, according to Antonio Verga's recollections, "initial contacts on this issue between ESA and Glavkosmos commenced in June 1988 during a somewhat scientific meeting in Essen, Germany. The impetus for this interaction was that, at that time, the forward-thinking German company, Kayser-Threde, had already recognised the potential for collaboration with the Soviet space programs".[182] Seeking to conduct experiments on *Foton*, which was manufactured at the same facility as *Bion*—namely, TsSKB-Progress in Kuybyshev—ESA program managers had to address this matter through Glavkosmos that oversaw the external contacts with TsSKB. The inherent

[182] Interview with A. Verga, March 27, 2023.

nature of materials science research compelled ESA to engage in closer collaboration with the engineers in Kuybyshev—a level of cooperation that essentially an academic institution Intercosmos (the official curator of the Biocosmos program) was unable to facilitate. Antonio Verga elucidates the significance of direct contact between the ESA specialists and the *Foton* satellite designers: "While initially on the first *Bion-8* and *-9*, ESA's experiments were simple and passive, in the sense that they had no electrical interfaces with *Bion*, ESA had in its bosom the idea of developing other instruments especially in the field of physical science research, where the importance of on-board electrical connections and data recording was essential. Therefore, it was necessary that the Soviet/Russian technical counterpart of the ESA engineers had the competence to discuss these technical and engineering issues, as well as the full knowledge of the *Foton* and *Bion* spacecraft subsystems. The biologists were not engineers and therefore could not hold technical and engineering discussions with ESA or with TsSKB personnel".[183] During that period, only Glavkosmos could facilitate direct access to the satellite manufacturers and engage in detailed discussions about the project with them.

Moreover, experiencing difficulties in collaboration with IBMP, ESA looked for alternative possibilities for the flight of its scientific facilities, i.e. Biobox and Biopan. Thus, this period was characterised by a growing number of contacts between ESA and Glavkosmos, indicating a substantial shift in the very essence of Soviet-European scientific cooperation. Furthermore, ESA actively facilitated its own interactions with Glavkosmos, not solely relying on the Biopan manufacturer Kayser-Threde, to handle all technical matters entirely (Fig. 7.15).

ESA was also exploring avenues for closer collaboration with TsSKB in the realm of materials science and the commercial utilisation of Soviet satellites. In this pursuit, akin to the precedent set by national agencies that had previously pioneered Soviet rocket and space technology, ESA was following a similar trajectory, exemplified in the instance of *Bion*. Valery Abrashkin reminisced about the initial interactions with ESA in the following manner: "Prior to formal engagement with ESA, there were preliminary attempts by foreign partners, among which German and French national space agencies, to assess the feasibility of collaboration with us. These efforts involved independent, smaller-scale experiments conducted on our satellites. Upon realising our commitment to fulfilling

[183] Ibid.

Fig. 7.15 Installation of Biopan on the exterior of the Foton re-entry capsule. Source: ©ESA – S. Corvaja

all contractual obligations and terms, these initial endeavours paved the way for a more comprehensive and substantial contract with ESA. The progression with ESA also began with smaller, individual experiments before evolving into a full-fledged, large-scale collaboration".[184]

Another area of cooperation in which ESA could not engage directly with IBMP was the execution of ESA science experiments aboard the Mir space station, along with the remote, yet highly desired prospect of a European astronaut participating in a mission aboard the station. Indeed, Glavkosmos endeavoured to assume control of collaboration in the biomedical field, focusing primarily on manned flights. European representatives were then scheduled to engage in discussions with MOM officials regarding the organisation and execution of scientific experiments aboard the Mir station. Those discussions were planned in Moscow by January 1987, but as mentioned earlier the first official European delegation failed to meet with Glavkosmos representatives. Then, in December of the same

[184] Interview with V. Abrashkin, March 25, 2023.

year Oser managed to clarify some issues of such a cooperation with officials from Glavkosmos. He specified that Glavkosmos was responsible for technical and engineering features of the spacecraft (Cosmos biosatellites and Mir), its interfaces to payloads, as well as programmatic and financial matters related to the cost of reimbursable payloads. Nevertheless, the main purpose of the meeting with Glavkosmos was primarily to obtain detailed technical information and payload test requirements.[185]

Glavkosmos' officials were clearly hinting at a shift from the previous exclusively scientific, non-commercial nature of cooperation in the field of life sciences to new market relations. This shift would involve the utilisation of not only the technical capabilities of the orbital station but also those of the satellites. The Soviet counterpart delivered the information upon cost refundable flights: the flight opportunities on Cosmos satellites and Mir had to be booked approximately 18 months prior to launch; the price charged by Glavkosmos was of the order of $/kg 15,000, which is of potential interest for fluid physics and materials sciences experiments. The costs for utilisation of Mir on a cost reimbursement basis could be as high as $/kg 60,000, taking into account crew time and sample return. Glavkosmos also offered to fly a European astronaut for a cost of about $10 million, or reimbursement in terms of hardware (e.g. additional experiment models).[186] In line with the new policy of *perestroika* in the space sector, the Soviet side emphasised that the foreign experiments would be treated as "black boxes". This designation granted the paying customer sole ownership of experimental results and any potential property rights.

At a subsequent meeting in Noordwijk in April 1988, during discussions on the scientific component of the cooperation aboard *Bion-10*, the issue of payment for Soviet space services became the focal point of discussions between ESA and IBMP representatives. ESA specialists presented and handed over summaries of their proposals concerning joint experiments, both aboard unmanned spacecraft and on a manned orbital station. The Soviet side had to identify those ESA experiments in which they had an interest in cooperation. ESA presented the Mini-Gravitational Biology Facility (MGBF), indicating that it was very important to find a flight opportunity for MGBF soon, so that the development work on it could be

[185] Oser, H. *Report on mission to the Soviet Union*. December 11, 1987. In ESA-28042, HAEU.

[186] Ibid.

started. The Soviet side took time and promised to inform ESA about their own interest to fly the MGBF on *Bion-10* as a joint project, as well as about the terms and conditions of joint utilisation and the modalities of flight cost compensation, if applicable. In addition, the two sides discussed a preliminary list of Ground Support Equipment necessary for the support of joint experiments.[187]

Furthermore, ESA provided a summary of the scientific objectives intended for a manned orbital station, and the Soviet side committed to informing ESA about the first potential flight opportunity on Mir. ESA's representatives also broached the topic of potential investigations in the field of immunology for potential inclusion on board the Mir, alongside the Biorack incubator. Whereas the Soviet side refrained from an immediate expression of interest, attributing it to the necessity for additional time to reach a decision, the question of compensation in kind assumed a central role in this context. The Soviets introduced the concept of "flight compensation" within the framework of the collaboration aboard the orbital station. In addition to discussions about Biorack flights, Soviet specialists briefed the ESA delegation on the possibility of conducting ESA experiments on Soviet space vehicles on a commercial basis. However, IBMP officials promptly distanced themselves from direct discourse on this matter, proposing that intermediary agencies, with Glavkosmos serving as the representative for the Soviet Union, handle such arrangements.[188]

As Oser observed during the meeting, it became apparent that contrary to earlier cooperation policies, Glavkosmos then took the lead in determining the conditions for cooperation. At the same time, it was confirmed that cooperation via Intercosmos and at the institute level still remained valid, but Glavkosmos would need to approve the conditions, particularly in terms of compensation. This might involve ground support equipment for joint experiments, rather than a cash payment.[189]

Philip Goldsmith commented on the new Soviet approach. As he put it, "the Soviet site made it quite clear that contrary to the 1987 ESA/Intercosmos Cooperation Agreement, which was based on the principle of no exchange of funds, all future cooperation agreements would in all

[187] *Protocol of Third ESA-U.S.S.R. Working Meeting on Space Biology and Medicine.* Noordwyk, Netherlands, April 25-29, 1988. In ESA-28042, HAEU.

[188] Ibid.

[189] Oser, H. *Mission report. Third meeting ESA/Interkosmos cooperation in life sciences; ESTEC,* April 25-29, 1988. May 10, 1988. In ESA-28042, HAEU.

likelihood have to be concluded with Glavkosmos on the principle of compensation. This means that both Intercosmos/IBMP and ESA would have to pay jointly the flight costs". The Soviet side explained that the ESA portion of the flight costs could be covered by "compensation in kind". For example, ESA could provide experiment ground support equipment or other general biochemical laboratory equipment for joint use in a laboratory at IBMP in Moscow. Goldsmith clarified that, at that time, the Soviet side was unable to provide the exact terms of "compensation in kind", but they indicated that such terms would vary depending on the specific case.[190]

In view of the above, Goldsmith proposed his vision of future collaboration with the Soviet Union regarding MGBF and the Simple Protein Crystallization Facility (SPCF) built by ESA. He suggested taking into account firstly, that as a consequence of the Shuttle accident, there were no new NASA-offered flight opportunities for experiments requiring retrieval before 1993–1994. Secondly, he cited prior experiences of national agencies that had engaged in commercial cooperation with the USSR. NASA and the CNES used Soviet-offered flight opportunities very extensively in the past years, a French astronaut would fly for four weeks on Mir in 1989, several USA and French laboratory facilities were installed at IBMP, while Germany was in negotiation with the Soviet Union; the Netherlands, Belgium, and Sweden had frame agreements with the Soviet Union and Austria accepted a Soviet offer to fly an Austrian astronaut plus a 500 kg payload on Mir in 1992 (with a cost of $10 M, or compensation in kind). Thirdly, Goldsmith highlighted that the costs offered by the Soviets were very attractive, compared to the costs per kilogram requested by the Chinese Academy of Space Technology (CAST), which were more than three times higher. Additionally, the costs for Spacelab experiments were five to ten times higher. Fourthly, he underscored the importance of the experiments planned to be performed within the MGBF, which could be regarded as fundamental research, while those planned for the SPCF encompassed both fundamental and applied aspects. Thus, he proposed to consider the opportunity to make the MGBF fly on a cooperative basis (50% compensation in kind) with Intercosmos/IBMP and the SPCF flying on a full "compensation in kind" basis as a "black box" experiment,[191]

[190] Goldsmith, Ph. *Cooperation between ESA and the U.S.S.R. in Microgravity under New Conditions*. May 27, 1988. In ESA-28042, HAEU.

[191] Ibid.

endorsing the concept of employing Soviet hardware for conducting scientific experiments in materials science on a commercial basis.

In consideration of Glavkosmos' conditions, ESA officials proceeded with negotiations with their Soviet counterparts on the questions of joint experiments in the field of human physiology on board of the Mir orbiting station. During the meeting held in Moscow in November 1988, it was agreed that in the near future, ESA should draw up and submit to the Soviet side precise descriptions of their experiments, including a detailed description of the scientific research apparatus necessary for carrying them out.[192]

The transition of cooperation to a commercial basis, the increasing complexity of ESA experiments and their technical interaction with the Soviet spacecraft interface, and the gradual opening of the rocket and space sector to international cooperation under the close supervision of Soviet state authorities, with Glavkosmos acting as their "spokesman", resulted in IBMP gradually losing its primary role in cooperation with ESA. In the domain of biological experiments in collaboration with IBMP, the European Space Agency sought to uphold its position of adhering to the original agreements on non-reimbursable cooperation.

IBMP's Resistance

In June 1988, Goldsmith wrote to Anatoly Grigoriev a letter conveying the ESA's response to the revised Soviet financial policy. He stressed that ESA "wanted to continue the excellent cooperation with the Academy of Sciences, Intercosmos, IKI and IBMP and it should like that this cooperation be based on the principle of noexchange of funds". Subsequently, he directly addressed the collaboration in life sciences, organised in a manner where ESA developed and provided experimental equipment. This equipment would be utilised for the execution of joint experiments, with the Soviet side managing the launch and retrieval processes. Goldsmith communicated to his Soviet counterpart that European scientists were planning to develop, among other smaller experiments, an MGBF for joint Soviet/ESA experiments. He drew Grigoriev's attention to the substantial development costs associated with such an ESA facility, emphasising the Agency's "willingness to contribute for the benefit of Russian scientists".[193]

[192] *Minutes of the meeting*. Moscow, November 29-30, 1988. In ESA-28042, HAEU.
[193] Goldsmith, Ph. to Grigoriev, A. June 29, 1988. In ESA-28042, HAEU.

Both ESA and IBMP representatives shared a similar stance concerning Glavkosmos' efforts not only to assume control of scientific collaboration in space biomedicine but also to transform it into a commercial venture. Participants in ESA and IBMP cooperation exhibited evident annoyance on both sides. Glavkosmos attempted to exert influence on the previously concluded agreements between ESA and IBMP regarding the execution of experiments aboard *Bion-9* and *-10*. As stated by Evgeny Ilyin, during a phone conversation with Oser, "our agreements reached earlier (i.e. cooperation based on no exchange of funds) were still valid" and he confirmed "no interference of Glavkosmos". Moreover, he asked to transmit to him "detailed technical and engineering discussion in the Mini-Gravitational Biology Facility on Cosmos (*Bion*), irrespective of Glavkosmos 'conditions'".[194] The MGBF was under development by ESA, and efforts were undertaken by IBMP specialists to integrate the MGBF onto one of their forthcoming satellites. Despite the readiness of IBMP scientists for collaborative efforts with ESA on the installation of the MGBF aboard Soviet spacecraft, the prospect of deploying this facility on a non-exchange of funds basis appeared remote. Günther Seibert believed that in such a scenario, IBMP would be required to cover mission costs to Glavkosmos.[195]

In November 1989, the validity of this assumption was substantiated during the ESA delegation's visit to IBMP. The purpose of the visit was to discuss the outcomes of the *Bion-9* flight and meticulously plan collaborative life sciences experiments for the upcoming *Bion-10* mission. IBMP highlighted a shift in their circumstances within the Soviet Union, emphasising that, as part of the Ministry of Health, they were now obligated to cover the expenses for the launch and services from various Soviet organisations. Despite this change, IBMP acknowledged that the previously agreed terms and conditions for ESA remained unaltered for the forthcoming *Bion-10* mission. Specifically, the collaboration would persist on a scientific basis through Intercosmos.[196]

The dedication of IBMP officials to the agreed terms of joint work was also influenced by the ESA's firm stance on the matter. ESA emphasised to both the IBMP and TsSKB representatives that the conditions for ESA's

[194] Oser, H. to Seibert, G. and Barbance, K. *Moscow Mission, Information on phone call to Dr. Ilyin*. November 14, 1988. In ESA-28042, HAEU.

[195] Main points and actions arising from the 14th meeting of the Microgravity programme board held on December 9, 1988. In ESA-28042.

[196] Oser, H. *Mission report on a Meeting with IBMP and Photon (Company) at the Institute of Biomedical Problems in Moscow*. November 23-24, 1989. In ESA-28043, HAEU.

involvement in this mission strictly adhered to cooperative principles as outlined in the ESA/Intercosmos agreement, specifically, on a no exchange of funds basis. During that meeting, ESA representatives once again raised the issue of ESA investigators and personnel access to launch and landing sites. This aspect held particular significance for life sciences activities, due to the stringent requirements for late access and the early retrieval of biosamples.[197] Nonetheless, ESA personnel were not granted access to the launch site until 1992, despite the Agency having scientific payloads launched from Plesetsk starting in 1987.[198] At the same time, as René Demets suggests, "CNES and Kayser-Threde GmbH from the FRG visited Plesetsk earlier than ESA to watch the launch of a *Foton* spacecraft".[199]

Overall, the IBMP staff strived to assume a central role in space biomedicine cooperation with ESA. This commitment was evident, among other things, in the decision to establish an independent Working Group and in efforts to address logistical issues related to the work of ESA personnel in the USSR. Thus, Ilyin expressed a strong inclination "towards creating a joint Life Science Working Group independent of political decisions, i.e., free from Glavkosmos interference".[200] In 1988, a draft Charter for a joint ESA-Intercosmos Life Science Working Group was developed. The primary objective of this Joint Working Group (JWG) was to review the ongoing activities conducted by Intercosmos and ESA in the life sciences field and to identify and evaluate potential areas for cooperation between ESA and Intercosmos in this domain.[201] This initiative further emphasised the IBMP officials' determination to distance themselves from Glavkosmos, maintain autonomy in decision-making, and promote direct engagement with international partners.

IBMP management remained the focal point for ESA officials, seeking resolution to various issues that had arisen from the launch of two biosatellites and the accrued experience of collaboration. Goldsmith endeavoured to enhance the efficiency of joint projects and streamline bureaucratic complexities. In a personal letter directed to Evgeny Ilyin, with whom

[197] Ibid.

[198] Demets, Jansen, Simeone. (2002), 12.

[199] Correspondence with R. Demets, February 7, 2024.

[200] Oser, H. to Seibert, G. and Barbance, K. Moscow Mission, Information on phone call to Dr. Ilyin. In ESA-28043, HAEU.

[201] Draft. Charter for a joint ESA-Interkosmos life science working group, November 1988. In ESA-28042, HAEU.

ESA and its science teams collaborated directly on *Bion* experiments, Goldsmith delineated several problematic aspects and proposed measures to enhance interactions.

The first aspect pertained to the imperative of efficient communication, a pivotal factor for the success of joint endeavours. Goldsmith advocated for the establishment of an operational telephone link between the ground control laboratory and the launch site, alongside easily accessible telephone and telefax links with European staff in ESA member states. He underscored that ESA leadership deemed these enhancements vital for the success of any future joint missions. Notably, the ESA team in Moscow faced a limitation in accessing their experiments after handing over materials to the Soviets for transportation to the launch site; they were not permitted to travel to the launch site, nor to oversee the final installation.[202] The second aspect addressed the challenging bureaucratic procedures for the entry and exit of European personnel working at IBMP for an extended duration. Goldsmith inquired about the feasibility of issuing permanent visitor badges to European specialists to streamline the process, thereby avoiding unnecessary delays. His final request pertained to obtaining multi-use entrance visas to the Soviet Union for ESA staff who regularly visited IBMP. Goldsmith believed that this would enhance flexibility for both the Soviet and European sides in planning meetings and visits.[203]

Ilyin responded positively to nearly all of Goldsmith's requests. He affirmed that the matter of obtaining multi-use entry visas for ESA staff regularly visiting IBMP could be readily addressed. He also requested the preparation of a list of specialists interested in such visas. Additionally, he committed to overseeing the establishment of an operational telephone connection with the launching site. Furthermore, plans were underway to install a telefax connection, facilitating a quick and reliable exchange of information between ESA and IBMP. However, the third request regarding the reduction of bureaucratic control over the entrance of ESA personnel into the institute was not fulfilled, due to the stipulation that foreign specialists participating in joint research works were only permitted entry in the presence of institute officials.[204]

Gradually, it became apparent to ESA staff that they would persist in their scientific, no exchange of funds cooperation with specialists from

[202] Harvey. (2021), 60.
[203] Goldsmith, Ph. to Ilyin, E. January 8, 1990. In ESA-28043, HAEU.
[204] Ilyin, E. to Goldsmith, Ph. May 21, 1990. In ESA-28043, HAEU.

IBMP, otherwise addressing the technical requirements for the *Bion-10* and other missions directly with TsSKB through Glavkosmos. Additionally, they were to discuss financial aspects of the situation with the latter. Thus, despite the dissatisfaction of ESA management with Glavkosmos' interference in its relations with IBMP, interactions with the former continued to progress. The exclusively commercial relationship between ESA and the USSR regarding the *Foton* satellites and the Mir station would become one of the main projects of Russian-ESA cooperation in the subsequent post-Soviet era. Meanwhile, the final *Bion-10* mission, in which Soviet and ESA scientists collaborated in preparation for launch, will be discussed below. This period coincided with the closing years of the Soviet Union.

Bion-10: Preparing for Launch

Antecedent to the dissolution of the Soviet Union in December 25, 1991, and well in advance of the mission's launch, its key features were delineated, setting it apart from the preceding two *Bions* that ESA and IBMP had collaborated on. Three primary categories of differentiation can be identified: techno-scientific, legal, and structural. The initial category encompasses the utilisation of Biobox by European scientists for onboard research and the Moslab laboratory for the preparation of their scientific experiments (Fig. 7.16).

With regard to scientific and technological innovations, between 1989 and 1992, ESA developed two new multi-user platforms for conducting biological experiments on the unmanned *Bion* capsules. One was to study the effects of microgravity on cell cultures (Biobox), the other (Biopan) was designed to test the influence of the harsh space environment on small biological organisms.[205] If Biopan would fly on board the *Foton* satellite, Biobox was destined to be launched on board of *Bion-10*.

In his memoirs, Evgeny Ilyin distinctly referenced Biobox. As he recalled, "*Bion-10* was a new stage of co-operation, not only because the Europeans built Moslab to facilitate their work and improve the quality of biomaterials preparation, but also because we decided to use the European-made device Biobox in flight. This advanced thermostat box provided the ability to keep various biological samples, different cultures

[205] Goodbye! HE space "legend" René Demets retires - 30 years of space biology at ESTEC. *HE spaces*, May 26, 2020. Retrieved February 26, 2025, from https://www.hespace.com/newsroom/news/rene_demets.

Fig. 7.16 *Bion*-10 mission logo, created by René Demets (ESA). Source: European Space Agency - ESA

and tissues requiring specific temperatures in strictly controlled conditions. Moreover, this device allowed fractional fixation of bio-objects at different stages of flight, so that it was possible to evaluate the influence of the factor of duration of stay in flight on certain indicators. At the end of the flight, before the descent from orbit the fixation of biological samples was also carried out. Thus, the experiments on *Bion-10* were a landmark, as they were conducted for the first time with such a fractional fixation of the flight automatically. When the experiments were conducted on the Shuttle's Spacelab, an astronaut could perform experiments by pressing buttons or injecting drugs into vials and syringes. In contrast, on *Bion*, everything was done automatically. This was such a significant achievement in conducting experiments in unmanned flights without any crew's aid".[206]

[206] Interview with E. Ilyin, March 15, 2023.

At the meeting in Moscow held in February-March 1990, the total experiment package for *Bion-10* was defined. In Biobox, three experiments were foreseen while the other four were to be flown as single autonomous experiments. Moreover, the Soviet side proposed to conduct a joint experiment inside Biobox named "Fibroblast" within the frame of a program of experiments in bone cell cultures. In addition, the Soviet specialists invited ESA to take part in the experiments "Population" (demographic studies on the fruit fly *Drosophila*) and "Beetle" (study of circadian rhythms in desert beetles). At that stage, the ESA delegates appreciated the invitation and promised to seek co-investigators (Fig. 7.17). The Biobox was to be transported to the launch site by passenger train, and the ESA specialists were requested to ensure that the dimensions and weight of the equipment being transported from Moscow complied with the train's limitations.[207] Ultimately, the satellite carried ten joint experiments from ESA and Moscow's Institute of Biomedical Problems.[208]

The legal innovations associated with this mission were closely linked to the political developments unfolding in the Soviet Union at the time. Although discussions regarding the European Space Agency's participation in the *Bion*-10 mission began as early as 1989, the legal framework governing the cooperation remained unresolved for an extended time period. Seibert hurried Anatoly Grigoriev, citing the importance of the project and the lack of a formal document, which could undermine its implementation. He even proposed to avoid including the USSR Academy of Sciences "due to its uncertain future, formally speaking".[209]

Ultimately, *Bion-10* was destined to be encompassed within an overarching arrangement between ESA and the Soviet Union: "Agreement between the European Space Agency and the Government of the Union of Soviet Socialist Republics concerning Cooperation in the Field of the Exploration and Use of Outer Space for Peaceful Purposes" (see Chap. 5). This document, known as the Cooperation Agreement and signed on April 25, 1990, outlines, among other things, cooperation on joint projects involving the design, development, launch, and operation of payloads. It also covers the conduct of joint experiments in orbit across various fields of space activities, including space biology, space medicine, and

[207] *Protocol of the ESA-IBMP working meeting on the future joint experiments aboard satellite 10.* Moscow, 26 February- 2 March 1990. In ESA-28043, HAEU.

[208] Demets, Jansen, Simeone. (2002), 12.

[209] Seibert, G. to Grigoriev, A. In ESA-28043, HAEU.

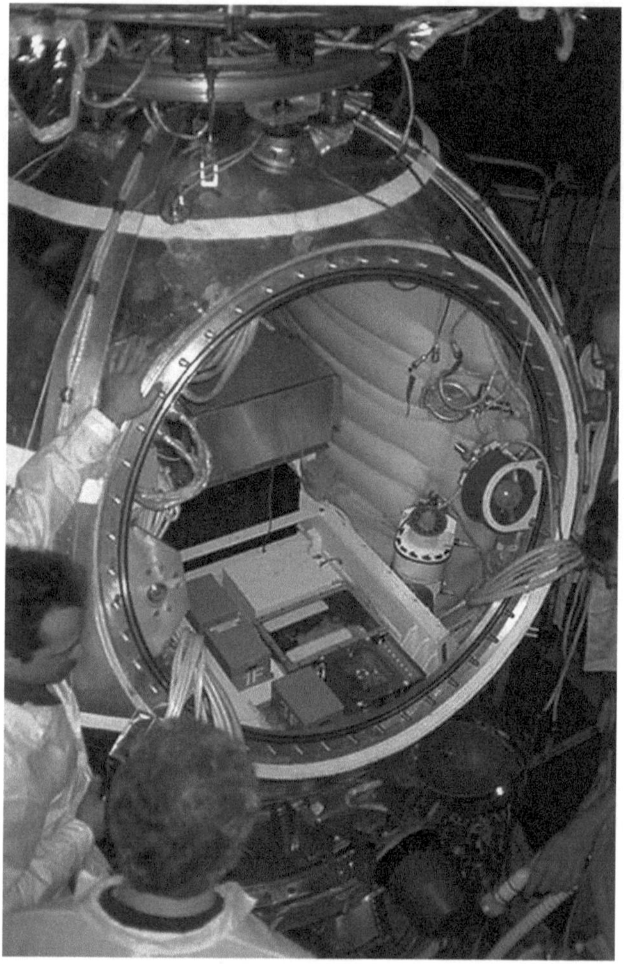

Fig. 7.17 *Bion-10* during installation of the ESA experiments. Source: European Space Agency - ESA

fundamental research in microgravity. The Cooperation Agreement also reiterates that, as a rule, no exchange of funds is foreseen, unless agreed otherwise by the parties, that compensation measures may be worked out, and that projects implemented on a refundable basis may be included.[210]

Subsequently, a draft *Bion*-10 Arrangement was developed by ESA and agreed upon by IBMP and the Intercosmos Council in 1991. No exchange of funds was anticipated under the Implementing Arrangement, meaning that each party was responsible for covering the costs associated with fulfilling its program responsibilities. ESA was responsible for covering the transportation charges of all ESA-provided experiment and facility hardware from any ESA member states to Moscow, as well as the costs for their return from Moscow. IBMP would finance the transportation charges for this hardware between Moscow and any site within the Soviet Union, as well as the costs for its return to Moscow. ESA was also responsible for covering the transportation charges of ESA-provided ground laboratory equipment between ESA member states and Moscow. In the event that this equipment required repair, IBMP would bear the related transportation costs from Moscow to the ESA member states. ESA in turn would cover the costs of repair or replacement, as applicable, as well as the costs for the equipment's return to Moscow.[211]

Essentially the same as *Bion-9*, but to an even greater extent, ESA participated on the basis of barter: instead of paying launch costs, ESA provided IBMP with laboratory equipment, radiation readers, electrical equipment, temperature loggers, as well as office appliances.[212] It is imperative not to overlook the principal scientific and technical contribution from ESA to the collaborative project, that was Moslab. The *Bion-10* Arrangement between ESA and RKA (Russian Space Agency) was signed on October 12, 1992, less than three months before the launch of the *Bion-10* satellite.[213]

The allowance of incorporating elements of compensation and reimbursement into the implementation arrangement facilitated the extension of cooperation to encompass the *Bion* satellite designer and payload integrator, namely the Design Bureau Foton (KB Foton), a constituent of the

[210] Demets, Jansen, Simeone. (2002), 2.

[211] *Draft of Implementing Arrangement between the ESA and the Academy of sciences of the U.S.S.R. for the conduct of joint experiments on the Bion-10 mission*, 7 November 1991. In ESA-28043, HAEU.

[212] Harvey. (2024), 23.

[213] Demets, Jansen, Simeone. (2002), 3.

Central Specialised Design Bureau, TsSKB. The industrial counterpart was eligible for reimbursement, proportional to its contributions associated with the integration of the joint payload.[214]

The formal avenue for direct engagement with the manufacturer of space equipment triggered a structural transformation in Soviet-European relations within the domain of space biology and medicine. Initially, direct communication channels were established with the, until recently, classified production and launch facilities, which held significant importance as part of the overarching military manufacturing complex of the USSR. Over time, connections with the Institute of Biomedical Problems for experiments involving automatic spacecraft experienced a weakening, culminating in its cessation. Next, we will delineate the process of structural change in Soviet-European cooperation, with a specific emphasis on its primary causes and ensuing consequences.

The Evolution of Cooperation with IBMP

Despite the political and economic difficulties of the late *perestroika* period, IBMP staff aimed not only to maintain, but also to deepen cooperation with ESA. During the meeting held in Moscow in February 1991, Soviet specialists informed ESA again about the possibility of flying experiments outside the framework of the Biocosmos program. Specifically, it was mentioned that proposals from ESA to perform joint experiments in existing experiment hardware, under the conditions applied for Biocosmos, would be welcomed and would have a high probability of being accepted for flight. The Soviet side intended the hardware to be flown on the *Resurs* satellites.[215] In fact, a few European instruments were already flown on *Resurs*, like Cosima, built by Intospace, a German company based in Berlin (*Resurs-F #5*, 1989, and *Resurs-F #6*, 1990), the Flugdatenrekorder, built by Kayser-Threde (*Resurs-F #5*, *Resurs-F #6* and *Resurs-F #9*, 1990), and Casimir, jointly built by Intospace and the French company Carrar (*Resurs-F #9*) (see Chap. 10). One year later, Grigoriev proposed to involve ESA scientists in joint experiments on Soviet facilities and Soviet

[214] Ibid., 2.

[215] *Protocol of the ESA-IBMP working meeting on the future joint experiments aboard satellite 10.* Moscow, 26 February- 2 March 1990. In ESA-28043, HAEU.

scientists in ESA flight opportunities (Spacelab, Eureca, and Sounding Rockets).[216]

Notwithstanding, the IBMP's eagerness to collaborate, and its exploration of potential avenues to sustain joint experiments involving automatic stations, the interest from ESA gradually diminished. Over time, it became clear to ESA officials that IBMP was no longer receiving the necessary funds from the Soviet government to continue research on *Bion* as before. This meant that the IBMP would lose the possibility of affording dedicated satellites. Therefore, for ESA officials, it was already obvious that IBMP was losing control of the situation, not receiving the previous government support and not being able to independently conduct experiments onboard dedicated satellites.

In this new context, ESA specialists had two options for collaborating with IBMP. They could either continue their cooperation with IBMP, with key aspects such as funding, in-kind contributions, and management structures yet to be defined, or they could establish direct commercial agreements with Soviet industries. Karin Barbance emphasised that the second modality was preferable for the ESA's requirements. But at the same time, she believed that care should be taken not to close the door to cooperation with IBMP, which remained the key institute in its scientific field.[217]

This new approach was vividly demonstrated when ESA harboured serious doubts about launching Biobox on IBMP's promised *Bion-10*. Despite its initial reluctance, ESA ultimately had to engage with Glavkosmos for the launch of Biobox, which had originally been intended to operate aboard *Bion*. If in 1991, the flight of the Biopan aboard a *Foton* satellite was already a matter of course, then the next flight of the Biobox aboard the *Bion* remained in doubt. In this situation, ESA tried to "jump" on board the *Foton* or the *Resurs-F #11* and *Resurs-F #12* spacecraft, having found out in advance what the new space vehicle would turn into for them. This issue had to be discussed only with Glavkosmos. Thus, in November 1991, Seibert asked Vladimir Podsadnik from Glavkosmos to let him have "a price proposal for the re-flight of Biobox".[218] Obviously, during that period, ESA had already acquiesced to the shift towards

[216] Seibert, G. to Oser, H. August 27, 1991. In ESA-28043, HAEU.

[217] Barbance, K. *Mission en Urss*, Aout 25, 1991. In ESA-28043, HAEU.

[218] Seibert, G. to Podsadnik, V. November 19, 1991. In ESA-28043, HAEU.

commercial cooperation, making its primary objective the launch of Biobox into space, even if payment was required for the use of a Soviet spacecraft.

Even if the Biobox ended up flying aboard *Bion-10*, in the realm of biological experiments conducted in weightlessness during the 1990s, ESA's primary counterpart became TsSKB, utilising *Foton* as spacecraft for its experimental hardware. Despite the provision of alternative cooperative possibilities, researchers from IBMP endeavoured to uphold their esteemed collaboration with ESA. Subsequent to the launch of *Bion-10* in 1992, wherein ESA shifted its focus from the *Bion* program to *Foton*, the collaboration endured in various other research domains. This ongoing partnership extended to areas such as human psychology in the space environment and the adaptation of the human organism to space conditions.

Interest in the psychological aspects of extended human presence in space, which eventually paved the way for Russian-European cooperation in this domain during the 1990s, began to emerge in the later years of *perestroika*. Illustratively, in June 1990, Oser participated in the 9th National Conference on Space Biology and Medicine of the Soviet Union in Kaluga, attended by 300 Soviet participants and several Western scientists representing NASA, Canada, FRG, and ESA. Oser directed his attention towards the primary focus of the conference—a manned mission to Mars. Notably, he highlighted the pronounced scepticism among the participants regarding the feasibility of such a mission from a physiological standpoint.[219]

While the prospect of a Mars expedition failed to generate considerable enthusiasm among Europeans, the attainable goal of sending an ESA astronaut into space aboard the Mir station became a tangible objective through collaboration with IBMP. Taking into account all of the above, it did not come as a surprise that the Soviet-European cooperation in the biomedical field switched from biological topics to the study of the psychological factors of a persons' long stay in space. This can be verified by analysing the projects proposed in 1991 by the Soviet side and the agreements on joint experiments, signed in the 1990s. Of the thirteen memoranda signed with ESA in the 1990s, nine were related to the psychological and physical conditions of humans during long-duration space missions.[220]

[219] Oser, H. *Mission to U.S.S.R.*, June 18-22, 1990. In ESA-28043, HAEU.
[220] Thanks to Tatiana Agaptseva from IBMP for providing the documents.

Therefore, in the early 1990s, a paradigm shift was underway ushering in entirely novel conditions. On the one hand, the IBMP lost the capability to furnish ESA with flight opportunities for research conducted by European scientists. Yet, IBMP owned accumulated expertise and persisted in endeavours related to medical support for space flights and the training of astronauts for space conditions. ESA, cognisant of the potential launch of its own astronaut aboard the Mir station, expressed interest in collaborating with the Institute under these circumstances. Nonetheless, in order to foster further collaboration, the European leadership needed to address the adverse impressions left by ESA participants following their experiences on the *Bion* missions.

Discussions on Further Cooperation

The aforementioned proposal by Karin Barbance to continue the cooperation with IBMP elicited diverse responses among ESA staff and proponents of space science. At times, it encountered opposition from some European scientists actively engaged in national cooperative programs. As Seibert put it in his letter to Oser, the inclination of ESA scientists to engage in collaborative efforts with Soviet specialists in joint experiments was markedly constrained because of the "viscous cooperation due to bad communication links, lack of appropriate equipment in Soviet labs, local travel difficulties, etc.", as well as the "low level of contribution and involvement of Soviet scientists in joint experiments". Additionally, as Professor of Physics at Université Libre de Bruxelles Manuel Paiva emphasised, "the low scientific standard of Soviet instruments for human physiology, when compared to USA or west-European technical level".[221]

Wim Jansen, who played a pivotal role in all the cooperation initiatives, also criticised the Soviet side for the lack of real scientific participation in joint projects on Biobox and on ESA-IBMP payload on *Bion-10*. Moreover, he argued that the cooperation was almost reduced to "a disparate search for high-tech equipment from ESA and trips to ESA member states", which caused the reduction of duration and number of meetings between two parties.[222] He recommended exploring alternative flight opportunities and minimising, whenever feasible, joint endeavours with IBMP. Indeed, he held a particularly critical stance towards collaboration with Soviet

[221] Seibert, G. to Oser, H. In ESA-28043, HAEU.
[222] Jansen, W. to Barbance, K. March 19, 1991. In ESA-28043, HAEU.

partners. His reservations were primarily rooted in disparities in scientific approaches, working culture, and everyday conditions prevalent in the USSR. Regarding working culture, Jansen drew attention to the specific Soviet working style—very different from those European. At the same time, he suggested not to "impose on our partner" the European style to "which they were only gradually adapting".[223] Otherwise, he believed, it could have been counterproductive in the long run.

Oser agreed not to force things, given the difficult political and economic situation in which Soviet colleagues had to endure and their readiness to accept the proposals and changes put forward by ESA. "What lasts here one hour lasts there two. Just take the translations, or writing up of protocol/corrections into account. In their current difficult time, we should not press them too hard and apply an ESA/NASA working style with all its advanced amenities. After all, it is the start and we can try to gradually "educate" them, and I am sure they are more than willing to learn".[224]

Jansen presented his vision for further cooperation with IBMP on *Bions* that he defined as "uncertain". He underlined that the Institute had no more financial resources to own *Bion-11*. According to his information, IBMP was abandoning its former manufacturer, Kuybyshev Progress plant, and was already in negotiations with NPO Lavochkin[225] concerning the design of a totally new satellite for microgravity research, to be launched on dismissed strategic missiles. Moreover, Jansen reported that IBMP established direct contacts with foreign partners, first of all with NASA in order to create an international consortium or joint venture, to finance the development and operation of this new satellite system. However, this did not worry Jansen too much, since he "saw very little or no project-oriented cooperation with IBMP (joint missions) in the immediate future after *Bion-10*". Furthermore, he doubted that ESA's "immediate future was in joint missions with Russian institutions, at least not in joint mission of the Biobox complexity".[226]

Oser acknowledged the current instability and insecurity regarding ESA's future legal partner and "what our negotiations will be worth

[223] Jansen, W. *Working Group Meeting Moscow*. November 7, 1991. In ESA-28043, HAEU.

[224] Oser, H. *Working group meeting Moscow*. November 7, 1991. In ESA-28043, HAEU.

[225] *Bion-11* was built by TsSKB-Progress and launched in 1996 with financial support from NASA.

[226] Jansen, W. to Barbance, K. March 19, 1991. In ESA-28043, HAEU.

tomorrow". Nonetheless, referring to the agreement taken by ESA's managers, he suggested taking the risk of further collaboration because "if we do not do it now, we will kick ourselves out when 'better times' will occur on the horizon but ESA again will be late. Our intentions are so highly focused (except the ESA astronaut) that I believe it does not matter whenever we go there unless there are no flights anymore. Also, ESA will not have ground based common projects in the life sciences other that pre-, in- and post-flight control experiments".[227]

Additionally, the establishment of the ESA laboratory at the Institute served as a compelling rationale for the persistence of collaboration with IBMP. Oser did not shy away from leveraging the progress in the laboratory's initiatives to rekindle discussions about future cooperation with IBMP. He underscored the significance of sustaining professional relations with the Moscow institute, especially in the context of several projects associated with Moslab. According to Oser, these considerations were persuasive enough to warrant the continuation of cooperation with IBMP, despite encountering several significant challenges. Indeed, notwithstanding the often-negative attitudes towards some aspects of the joint project, based on official reports authored by ESA participants, European scientists expressed satisfaction with the scientific advancements, achieved through their work on Soviet automatic spacecraft.

The representative of the Soviet side had different recollections. For instance, Evgeny Ilyin did not notice many changes in scientific approaches with his European colleagues. He attributed the mutual understanding in work to the accumulated experience of cooperation with foreigners. "We started international cooperation in 1975, and with ESA in 1987. By that time several biosatellites had already flown by. We already had such a rich experience! We worked with leading scientific institutes in the GDR, Poland, Czechoslovakia, Romania, Bulgaria, Hungary. There were numerous proficient scientists and scientific centres, including the contributions of American and French researchers. In essence, when we initiated collaboration with ESA, we had already undergone extensive experience in cooperating with European scientists. Consequently, we did not perceive a substantial distinction in approaches and scientific perspectives. The shared emphasis was on a rigorous, scientific approach to establishing controlled conditions. It was important that the conditions of the experiment were strictly controlled: humidity, temperature, oxygen content, and so

[227] Oser, H. *Working group meeting Moscow*. November 7, 1991. In ESA-28043, HAEU.

on. We provided all of that. I should note that the Europeans were very pedantic about this. Everything had to be perfect, strict, everything had to match. What was the temperature during the flight on a certain day, it was necessary to reproduce the same temperature in laboratory conditions on the ground control experiment. No more and no less. A very strict, meticulous approach. We have taken the same approach as always and we are proud that we have managed to conduct experiments in strictly controlled conditions on biosatellites".[228]

René Demets, for his part, claims "that biology experiments in space must be conducted much more strictly than experiments on ground. On ground, a biology experiment can immediately be repeated if not satisfactorily executed the first time around. This luxury does not exist for a space experiment. ESA biologists usually get just a single chance to fly their experiment in space, and ESA is fully liable for their implementation and technical execution".[229]

As mentioned above, ESA was gradually moving towards establishing direct contacts with TsSKB for technical and interface matters. Valery Abrashkin emphasised the proximity in approach and the swift mutual understanding that Soviet, and subsequently Russian engineers, were able to establish with their ESA colleagues. "There was a seamless integration of work styles during that period. Visiting engineers, much like ourselves, noted the absence of significant disparities in our qualifications. The familiarity with computer-based work and documentation development further facilitated a smooth collaboration. The absence of difficulties in these technical aspects fostered enduring friendships that persist to this day".[230]

Two caveats likely warrant consideration at this point. Firstly, the remarks regarding collaboration with Soviet scientists originated from the contemporaneous observations of ESA's staff, recorded in real-time during the deployment of cooperation. Secondly, Evgeny Ilyin's and Valery Abrashkin's comments were provided 30 years after the events in question, introducing a temporal gap that may influence the perspective presented. Notably, Ilyin translated the divergence in working styles into the meticulous fulfilment by the Soviets of their obligations in creating flight conditions. Evidently, this constituted a paramount requirement for ESA, serving as an indicator of the efficacy of the work conducted by Soviet

[228] Interview with E. Ilyin, March 15, 2023.
[229] Correspondence with R. Demets, February 8, 2024.
[230] Interview with V. Abrashkin, March 4, 2023.

engineers and scientists. In his memoirs, Ilyin further highlighted the "pedantry" and "scrupulousness" exhibited by his European colleagues. They displayed a keen emphasis on precise formulations in protocols and held themselves and their Soviet counterparts to a stringent standard, demanding the meticulous fulfilment of all commitments within the agreed-upon timeframe. The "rigorous approach" exhibited by the European colleagues, as recalled by Ilyin, stood in stark contrast to the recurrent delays and postponements characteristic of the Soviet side's work, a point consistently underscored by ESA officials in their reports. For the Soviets, as emphasised by Ilyin, cosmonautics inherently represented an unpredictable realm of human activity, necessitating a flexible approach to deadlines and commitments.

In turn, Abrashkin highlighted the elevated level of bureaucracy that Soviet engineers encountered in their collaboration with ESA. "In contrast to the relatively straightforward collaboration with German and French specialists from their respective national space agencies, working with ESA proved challenging due to bureaucratic complexities. Contracts with ESA were characterised by intricate structures and numerous clauses, in stark contrast to the simplicity of a half-sheet contract signed with the Germans, for instance. Kayser-Threde's inquiry about the feasibility of creating specific equipment was met with an agreement from our side. Once both managers signed the letter, work commenced. The collaborative relationship with German specialists was marked by trust and mutual support, extending to assistance with equipment repairs and on-site aid at the cosmodrome when needed".[231]

In this context, Antonio Verga explained the position of the ESA staff and the difference between the work of ESA scientists and engineers and their colleagues from national agencies. "What our Soviet and then Russian counterparts failed to grasp is the fact that ESA acted as an agency on behalf of the European communities of scientists and industrial partners that delegated to ESA the role of monitoring and controlling the correct implementation both of the scientific programs and of the engineering and operational requirements, set forth by the specific design of ESA's payload. To this extent, ESA's liability vis-a-vis its member states was enormous and the investments at stakes were too high for leaving any odd and detail at bay of shear chance. Contrary to national endeavours, ESA was not the design authority of the boarded instruments and, therefore, last

[231] Interview with V. Abrashkin, March 25, 2023.

minute repairs or changes were totally out of question. For later projects, when the cooperation and the joint tasks with TsSKB-Progress became routine, ESA was entitled to entrust the participation of its scientists and its industrial partners to test and launch campaigns at TsSKB's premises in Samara and at the launch site, respectively. The latest *Foton* projects were conducted indeed by what we can call *integrated teams*".[232]

Summarising the outcomes of the collaboration on *Bions*, Evgeny Ilyin underscored that approximately "twenty scientific papers were published. Half of these papers were collaborative efforts, while the remaining half originated from singular perspectives. The collaboration successfully yielded the desired outcomes for both European and Soviet scientists, culminating not only in a fruitful partnership but also in the joint publication of the research findings".[233]

Without assuming an arbitrating role, it is important to highlight the inherent differences in approach that both sides had to navigate. This makes the results of the cooperation even more commendable, given the significant efforts made to achieve them. Despite ESA management's decision to end collaboration on the *Bion* program, Soviet (later Russian)-European cooperation continued with the *Foton* missions. As previously noted, a key factor in ESA's disengagement from joint efforts with the Institute of Biomedical Problems on *Bion* missions was its decision to establish direct arrangements with the spacecraft manufacturer, thereby bypassing the Moscow Institute—at least for all interface-related matters, both at the physical and operational levels.

New Partner

Let us recall that, from the inception of the collaboration, the primary counterpart for life sciences experiments on *Bion* missions was the Institute of Biomedical Problems. Over time, the entity referred to in ESA documents as the "Foton company" (formally KB-Foton, or Foton Design Bureau, of TsSKB) gradually assumed an equivalent role in Soviet-European cooperation. Notably, this design bureau occupied a unique position within the Soviet rocket and space sector. KB-Foton was responsible for the design and development of several space systems, including the *Soyuz* launchers and unmanned spacecraft such as *Foton*, *Bion*, and

[232] Interview with A. Verga, March 15, 2024.
[233] Interview with E. Ilyin, March 15, 2023.

Resurs. All these satellites were recoverable, with typical mission durations ranging from 14 to 30 days.

As it follows from the reports of ESA officials returning from Moscow, they had meetings with representatives of the KB-Foton company, a special unit responsible for the commercialisation of the hardware manufactured by TsSKB, Kuybyshev Central Specialised Design Bureau, headed by General Designer Dmitry Kozlov. TsSKB worked in close cooperation with Kuybyshev Progress plant, forming in 1996 the unified State Research and Production Rocket and Space Centre TsSKB-Progress (today the Progress Rocket Space Centre), one of the most important Russian producers of space hardware.[234]

The Foton Design Bureau, which served as ESA's primary point of contact, was established in the late 1980s. Its main objectives included fostering and managing relationships with foreign customers while overseeing the plant's commercial activities. Valery Abrashkin recalls that "KB-Foton was an entity embedded within a larger organisation, specifically functioning as a structural department of TsSKB. Its inception dates to 1987, when it originated as a modest structural unit comprising three departments. Notably, during this period, the organisation gained authorisation to engage in collaborations with foreign entities—a privilege previously restricted to Glavkosmos. This change in permissions was prompted by the TsSKB management's recognition of the potential for *Bions*, *Fotons*, and other space vehicles to contribute to international space endeavours. The design bureau was under the leadership of Vladimir Kozlov, son of Dmitry Kozlov".[235]

The dynamic between ESA and TsSKB personnel underwent a swift transformation over the few years of ESA's collaboration with the Soviet Union on *Bions*. Representatives from the Bureau, which had been previously inaccessible to foreign contacts, progressively engaged in meetings between scientists from IBMP and ESA. As Abrashkin confirmed, "We were not involved in work with the foreigners through IBMP; rather, it primarily centred around the installation of equipment on the satellite. Direct contacts were not permitted at that time".[236] The situation underwent a gradual transformation, particularly with the commencement of collaborative efforts

[234] The information was sourced from the RKT Progress website. Retrieved February 20, 2023, from https://www.samspace.ru/about/history/.
[235] Interview with V. Abrashkin, March 4, 2023.
[236] Ibid.

with ESA on *Bions*. By the time of cooperation on *Bion-10*, TsSKB engineers were actively engaged in direct discussions and project tasks with their European colleagues. This marked a significant shift, highlighting the evolving nature of the working relationship between TsSKB and ESA. Employees of KB Foton were invited to contribute insights into technical aspects and provide clarifications, for instance, on issues related to the integration of the planned ESA facility, Biobox, aboard *Bion-10*.

In Autumn 1990 for the first time, an ESA delegation went to KB Foton in Kuybyshev (Fig. 7.18). As Wim Jansen put it, the ESA delegation was the second one of Westerners to be allowed into the hitherto closed city on the Volga River. He considered it as a "great internal victory of KB Foton and IBMP over their own apparatus". The direct presence of European specialists, according to Jansen, "contributed to a further relaxation in the relations" with Soviet engineers and "set an important precedent for the future". He noticed that "the preparation of the meetings

Fig. 7.18 A street seller of *kvas* in Samara in the early 1990s. Source: A. Verga

was considerably more professional than with IBMP", intending "punctual transport by company plane to and from Moscow, easy hotel accommodation,[237] well-prepared agenda, visit to the Soyuz launcher assembly hall, language interpreter provided all the time, etc."[238] Antonio Verga confirms: "The workforce of TsSKB was estimated to be 15,000 individuals, indicating a robust organizational structure and ample financial support. This strength stemmed from its active engagement in manned and military programs".[239]

In general, Wim Jansen highly appreciated the work with the Soviet spacecraft manufacturer, emphasising the fact that "KB Foton was an unavoidable and indispensable extra, because they built the satellites, integrated the payloads, and were responsible for all satellite-payload interface engineering; their involvement simplified matters, rather than complicating them". Returning from his first trip to TsSKB, Jansen was quite optimistic about further collaboration with Kuybyshev's plant: "This was the encouraging message from our trip. I am confident that, after the lengthy and detailed technical discussions on Biobox/*Bion-10*, any future discussions on flying scientific hardware on their spacecraft will be much easier".[240]

During the meeting at TsSKB, a supply with additional equipment was negotiated as well. Though during the next *Bion-10* mission a "User Centre KB Foton" was supposed to be responsible for downloading the telemetry data from ESA's Biobox facility, although the question about the reception, processing, and utilisation of the telemetry data remained open. To guarantee fast communication with the foreign specialists, visual display and documentation of the results of telemetry measurements, it was necessary to supply the Kuybyshev User Centre with additional equipment. The ESA specialists recommended the presence at KB Foton of a laboratory for the incoming inspection of the payload hardware, and to organise a working area for

[237] Jansen was referring to the fact that, as he stated, "rouble hotels in Moscow, accessible to the Academy of Sciences, were either 'Akademicheskaya' or the 'Uzkaya'. The former accommodation was below any reasonable standard of hygiene; the latter one was at 1–1.5 hrs driving distance from the ESA/IBMP Ground Control Laboratory. Moreover, IBMP had never guaranteed single room accommodation".

[238] Jansen, W. to Barbance, K. *Bion-10 Implementing Arrangement*. March 19, 1991. In ESA-28043, HAEU.

[239] Interview with A. Verga, March 15, 2024.

[240] Jansen, W. *Protocol ESA-IBMP-KB Foton*. November 6, 1990. In ESA-28043, HAEU.

the verification of mechanical and electrical interfaces between the payload hardware and mock-ups of the spacecraft before delivery of the payload hardware to the spacecraft manufacture's plant. In addition, it was supposed that this working area would be used for de-installing payloads from the spacecraft re-entry vehicle after its landing and delivery to Kuybyshev for post-flight servicing of the flown hardware, and its preparation for return to ESA-ESTEC.[241] The centre could also accommodate foreign teams of payload developers for the supervision of synchronous ground control experiments, the analysis of received telemetry data, and the preparation of telecommand uplinks. At the meeting held at ESTEC in April 1991, KB Foton confirmed that a special laboratory would be installed in the User Centre for incoming inspections and post-flight analysis of payloads.[242]

As Antonio Vergia explained, "That centre, called EIK-3,[243] was finally located within an uncompleted building that belonged to the Rocket and space University of Samara. Unfortunately, the dispatch of the down-linked telemetry data was done with rather primitive means, such as telefaxed fragments of paper rolls, requiring a significant effort by ESA's team for their interpretation. Despite the good intents of specialists of both KB-Foton and ESA, the analysis of telemetry data often led to interpretation errors and raised unnecessary concerns upon the actual performance of the flying hardware and its embarked experiments".[244] The aforementioned Biopan became a symbol of cooperation between ESA and TsSKB. Its development and integration into the Soviet, and later Russian, *Foton* spacecraft resulted from close collaboration between the scientific and technical teams of ESA and the Kuybyshev Design Bureau.

Biopan: A Catalyst for Continuity in Space Research

The conclusion of the Soviet era marked the inception of a new endeavour that would evolve into a pivotal project in Russian-ESA scientific collaboration, particularly in the domains of biology and materials sciences. This initiative relates to the Soviet (Russian) *Foton* automatic spacecraft and an

[241] *Protocol of the working meeting between representatives of the ESA, the Design Bureau "Foton" and IBMP concerning the "Biobox" project, Kuibyshev*, Oct.29-Nov.21990. In ESA-28043, HAEU.

[242] Minutes of meeting, ESTEC, April 15-19, 1991. In ESA-28043, HAEU.

[243] Eksperimental'no-Ispytatel'nyj Kompleks (Experiment-Test Complex).

[244] Interview with A. Verga, February 26, 2024.

ESA's facility integrated as part of its payload. By early 1991 the ESA scientists and engineers started work on a new project: Biopan. Until that time, the Soviet KNA exposure containers were used on *Bion* spacecraft to conduct experiments involving the exposure of biological materials to the space environment. ESA then decided to develop a full-European and technically more advanced successor than the KNA.[245] As Valery Abrashkin recalled, " some Soviet specialists suggested to foreign partners the idea of creating a Biopan container for attaching scientific experiments outside the SA[246] and returning them to Earth. To develop their analogues of our KNA container, ESA's project team used the most advanced instruments and electronics".[247] Moreover, "they asked us to produce a heat-shield coating, because they had no experience in the manufacturing of appropriate materials (Fig. 7.19). This thermal protection was made at our experimental production centre".[248]

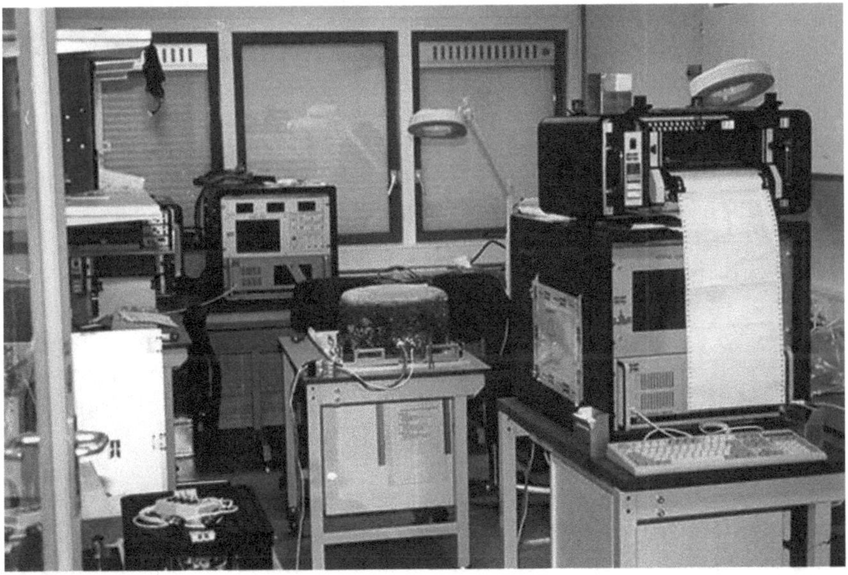

Fig. 7.19 Biopan-0 after Foton-8 flight. Source: European Space Agency - ESA

[245] Harvey. (2024), 23.
[246] Spuskaemyj Apparat (Re-entry module).
[247] Abrashkin. (2017). 144.
[248] Interview with V. Abrashkin, March 4, 2023.

Antonio Verga confirmed that "the idea of having a pan-shaped container that opens its lid once in orbit was indeed a TsSKB idea. They suggested developing the Biopan with higher performance than the KNA in terms of both volume and thermal control, absent in the KNA's design, causing that several samples were found completely roasted, upon their ground recovery. KNAs had serious limitations; Biopan, on the other hand, had its own thermal control: its two trays were kept at a temperature at which the cells could survive".[249]

Biopan was designed and manufactured for ESA by Munich-based Kayser-Threde. The selection of the prime contractor was deliberate. As it was earlier mentioned, before beginning collaboration with ESA, Kayser-Threde had already established strong connections with engineers from TsSKB, had visited the Plesetsk launch site, had opened an office in Moscow, and owned a genuine *Foton* re-entry capsule that had orbited the Earth in 1985. Somehow, Kayser-Threde paved the way for ESA to fly European hardware on *Foton*.[250]

As Valeriy Abrashkin recalled, "the start of collaboration on *Foton* projects began when European counterparts approached us with a proposal. Our satellite stood out as the only one worldwide capable of conducting the specific types of experiments that ESA envisioned. Hence, our European partners sought our expertise and capabilities, leading to their proposal for joint work on *Foton* missions".[251]

ESA officials initially explored the prospect of launching Biopan into space with the then-major biomedical partner, i.e., IBMP. In July 1990, ESA and IBMP engaged in discussions regarding potential collaboration within ESA's Biopan project, which was then in the developmental phase and viewed as a prospective avenue for extended cooperation between ESA and IBMP. During the meeting, representatives from both sides introduced the hardware manufacturers, respectively, Kayser-Threde and KB-Foton. IBMP specialists acknowledged that the Biopan project represented a significant enhancement of the hardware employed for experiments in open space. It could have been used for research in radiobiology, radiation physics, biology, exobiology, and for the study of the influence of ultraviolet radiation in different biological sample materials, as well as for experiments investigating the combined effects of space flight factors.

[249] Interview with A. Verga, March 27, 2023.
[250] Harvey. (2024), 23.
[251] Interview with V. Abrashkin, March 25, 2023.

The instrument could also be used for technology and materials science experiments. As Wim Jansen reported, the Soviet side confirmed the technical and operational feasibility of frequent Biopan missions on a range of satellites between 1992 and 2000.[252]

Initially, it was envisioned that Biopan could potentially be installed on various Soviet satellites, including *Bion*, *Foton*, *Resurs-F #11*, *Resurs-F #12*, and *Nika-T*. However, specialists from TsSKB proposed the two *Resurs* satellites as the primary carriers for installing Biopan. In a meeting held in Moscow in November 1990, Kayser-Threde, the manufacturer of Biopan, presented extensive materials detailing the design of the Biopan instrument that took into account the requirements and the recommendations put forth by the Soviet side. On the basis of these preliminary technical characteristics of the facility as presented by Kayser-Threde, the accommodation of Biopan on the satellites *Resurs-F #1*, *Resurs-F #2*, and *Foton* was considered possible while the accommodation on the *Bion-10* was not because of its limitations in terms of electrical power supply. KB-Foton confirmed its readiness to carry out a test flight and an operational flight of the Biopan facility in the years 1992–1993.[253] Ultimately, the competition for the privilege of launching European scientific payloads into space cast its preference for *Foton*. This marked the inception of a mutually advantageous collaboration, primarily focused on the Russian *Foton* spacecraft for a span of sixteen years (1992–2008).

The inaugural *Foton* spacecraft was launched in 1985 and spent 13 days in orbit. From 1985 to 2007, fourteen *Foton* satellites were successfully launched and put into low-Earth orbit, with one unsuccessful attempt in October 2002, due to the crash of its Soyuz launch vehicle 29 seconds after lift-off. The duration of those flights ranged from 13–18 days, and the descent vehicles employed a soft-landing system. Throughout this period, the *Foton* spacecraft facilitated a vast research program encompassing fluid physics, semiconductor and optical materials, biotechnology, cell biology, organic chemistry, and the impact of space factors on the environment.[254]

Although *Foton*, like *Bion*, was manufactured by the TsSKB-Progress plant, it was not under the jurisdiction of IBMP. This was one of the

[252] Jansen, W. *IBMP-ESA Protocol Biopan*, 30 July 1990. In ESA-28043, HAEU.

[253] *Protocol of the meeting between representatives of IBMP, RCSRS, KB Photon and ESA on the Biopan project*. Moscow, November 20-23, 1990. In ESA-28043, HAEU.

[254] Novyye tekhnologii v usloviyakh mikrogravitatsii: kosmicheskiy apparat «Foton». In Yu. Izyumova. (2017), 147.

reasons for ESA's shift from cooperation with the Moscow Institute to direct contacts with TsSKB. The prominent role of TsSKB-Progress was underlined also by IBMP, whose specialists suggested involving KB-Foton as the only organisation capable of providing support to the further elaboration and development of the Biopan project.[255] At the same time, the Institute gradually began to lose its priority in cooperation with ESA. As Wim Jansen put it, all efforts from the ESA side between July-November 1990, to incorporate Biopan as an element of cooperation between ESA and IBMP, failed because of the inability of IBMP to find and offer to ESA a joint launch opportunity.[256] Thus, ESA was gradually moving away from "programmatic" cooperation with IBMP on *Bion* missions and shifting towards TsSKB, the direct manufacturer of unmanned spacecraft.

To summarise, the structural set-up of the Biopan project associated with the *Foton* spacecraft, which presumed direct cooperation with the TsSKB, suited ESA better than the cooperation with IBMP also for the flight of Biobox aboard *Bion*, even despite its "commercial" nature. Firstly, "KB-Foton acted professionally" and reassured more stable and clear communication with its European partners. Secondly, ESA was not "lumbered with the burden of joint science management". Thirdly, according to Wim Jansen, the Institute failed to win the competition with TsSKB Progress even though ESA "throughout 1990 fed them (IBMP) with technical and operational data, and aligned the scientific methods of the design with their inputs. Until November 1990 we gave them all the hints that they needed to create a cooperative framework for this new facility, but they were totally outplayed by KB-Foton".[257]

In January 1991, the first meeting of the Soviet-ESA Working Group on Fundamental Research in the field of microgravity was held at ESA-HQ. It was agreed that KB-Foton would perform, as subcontractor of the Biopan main contractor Kayser-Threde, all the required engineering tasks, including all interface testing with the spacecraft, the Biopan's heat shield production and coating, and its test flight. Three flights of Biopan were foreseen in the period of 1993 to late 1994, either on two or three separate missions of the Soviet *Resurs* or *Foton* retrievable satellites. In return for the provision of the operational flights, ESA reimbursed the

[255] *Protocol of the meeting between representatives of IBMP and ESA on the Biopan project*, Moscow, July 23-25, 1990. In ESA-28043, HAEU.

[256] Jansen W. to Barbance, K. March 19, 1991. In ESA-28043, HAEU.

[257] Ibid.

Soviet partner 1.8 MDM (in 1991 DM, Deutsche Mark) that corresponded to a flight cost of 30,000 DM/kg.[258]

By 1993, following the *Bion-10* flight, ESA discontinued its collaboration with IBMP on *Bion* missions and fully transferred its efforts towards working relations with TsSKB, mainly on *Foton* missions.[259] However, let us not pre-empt events. *Fotons*, Biopan, and other ESA facilities belong to a distinct era and would take centre stage in future research endeavours. The same applies to Russian-European cooperation on the Mir orbital station, which originated in the Soviet era but progressed further in the 1990s.

Conclusion

In summarising Soviet-ESA cooperation in the field of space biomedicine during the *perestroika* period, it is worth noting manifold characteristic points. First of all, there was an extended phase of mutual observation and gradual warming up, signifying a slow start, before embarking on joint projects. The initial lukewarm cooperation underwent a swift evolution within the four short years that remained before the end of the Soviet era.

The slow initiation was more a reflection of ESA's inertia than of mutual distrust. Soviet officials had extended multiple invitations to European scientists for collaborative efforts on *Bion* since the late 1970s yet encountered a lack of positive response from the ESA side. What prompted the ESA's reluctance or apprehension remains a pertinent question. A major factor was the initial focus of European biomedical research on collaborating with the USA for experiments aboard Spacelab during Shuttle flights. The technical limitations, in terms of power supply, downlink of telemetry data, and uplink of telecommands of the Soviet re-entry capsules, specifically *Bion*, also dissuaded ESA engineers from engaging in the project.

Moreover, the apprehension of transferring advanced technologies used in the production of scientific instruments to the Soviet Union was another deterrent. The conditions of the experiments proposed by the Soviets, wherein Europeans were relegated to the role of an outside

[258] *Protocol on aspects related to the Development and flights of the ESA Biopan experimental facility on Soviet retrievable satellites Resurs and/or Photon*. January 28, 1991. In ESA-28043, HAEU.

[259] Foton-8, launched on October 8, 1992, from Russia's northern Plesetsk cosmodrome, carried the Biopan-0 demonstration version.

observer, relinquishing control over crucial stages of the experiment, further failed to convince ESA staff to participate in cooperation.

Finally, the political context and the intricate decision-making process at ESA should not be overlooked. Achieving full-scale cooperation with the Soviet Union necessitated the agreement of all national delegations within ESA. It took eight years from the escalation of the Cold War in the late 1970s until Gorbachev could successfully persuade the international audience of his peacekeeping intentions. These are the years that span from Oser's initial visit to the Soviet Union in 1979 to the first Soviet-ESA *Bion* flight in 1987.

The cooperation, initiated in 1987, underwent several significant changes in various aspects within a relatively short period of time. There was an enhancement in the technical component of the experiments. The progression was evident, evolving from biomaterials placed in separate autonomous containers on *Bions-8* and *-9* to the development of distinct equipment, such as Biobox, allowing for the simultaneous execution of multiple experiments under strictly controlled conditions. The weight, quantity, and complexity of ESA experiments accommodated in Soviet capsules also experienced a substantial growth.

ESA successfully changed its primary partners in the Soviet Union several times: from IBMP to Glavkosmos, and subsequently to TsSKB Progress. The collaboration, initially based on principles of gratuitous scientific exchange, turned into a commercial arrangement by the end of 1991. Lastly, the level of cooperation, governed on one side by the CoCom regulations and on the other by the secrecy of the strategic VPK, underwent notable changes. This evolution ranged from ESA scientists awaiting the return of biomaterials in Moscow to their later presence at the launch site in Plesetsk.

Regarding the reasons that facilitated the establishment of cooperation, despite the presence of objective factors (such as the postponement of Shuttle flights) and subjective ones (the necessity to adopt advanced technologies and scientific methods in joint work aboard biosatellites), I am contemplating the precedence of purely scientific collaboration for the advancement of science. It would be an error to underestimate the significance of enthusiasm, as expressed by project participants from both sides, as a crucial factor in fostering cooperation: "It was the mutual desire for cooperation and the enthusiastic dedication of the parties involved that made it possible".[260]

[260] Demets. (1996), IV.

PART IV

Commercialisation of the Soviet Space Sector

1.1 Introduction

By the late 1980s, the USSR began its first steps away from the Marxist-Leninist principles of a planned economy, introducing elements of the capitalist model into the crumbling Soviet command system. The shift towards market mechanisms within the Soviet space sector had become inevitable. It was during this period that the Soviet space industry and rocket engineering started to enter the international market for commercial goods and services, while centrifugal forces within the military-industrial complex began to intensify. Although *perestroika* significantly accelerated this process, its roots can be traced to developments that predated Gorbachev's rise to power.

The most reliable and widely recognised commodity that the Soviet space sector could offer on the global stage were its rockets. The commercial promotion of launches marked the beginning of the complex and often ambiguous process of opening up the sector to international cooperation on a commercial basis. In the late 1970s, Soviet space leadership made an initial attempt to sell satellite launch services to ESA as part of the Inmarsat, International Maritime Satellite Organisation. Although this attempt was unsuccessful, it does not diminish the efforts of the first "entrepreneurs" among Soviet officials, who attempted to promote Soviet rockets on the international market and broke into one of the most closed and secretive sectors of Soviet industry. Analysing the negotiation process, the positions of the involved parties, and the reasons behind their decisions sheds light on the international aspect of ESA-Soviet commercial cooperation projects.

The founding of Glavkosmos in 1985 meant a new approach in the USSR to international cooperation in space and to the space sector as a whole. This marked a transition from the traditional practice of conducting joint scientific projects on a non-commercial basis to a purely commercial partnership with foreign entities, under the leadership of the newly established organisation. What exactly was Glavkosmos? What goals were set for its management? What was the scope of its activities, and how were those goals pursued? Answering these questions allows for a comprehensive analysis of a crucial aspect of the space sector's activities during the *perestroika* period: its aptitude, capability, and willingness to adapt to changes dictated by external factors, time, and evolving circumstances.

Foreign partners also played a critical role in the development of the commercialisation process in the late USSR. Potential American and European partners had several reasons for responding in various ways to the emergence of a new, powerful player in the market for space goods and services. The expected heterogeneity of their reactions to Soviet commercial offers—and their influence on the commercialisation process in the Soviet Union—helps explain the outcomes of Glavkosmos efforts and the broader attempts to save the Soviet rocket and space sector. This chapter aims to explore how late Soviet cosmonautics responded to the introduction of market economy elements and to evaluate whether commercialising the sector was an effective strategy to avert its impending collapse.

CHAPTER 8

Marecs and Proton: The Untold Story of a Failed Partnership

At the end of the 1970s, international efforts were made to facilitate the prospective joint launch of an ESA satellite using a Soviet rocket within the framework of Inmarsat. This episode was one of the first Soviet attempts to enter the international space launch market and fight the monopoly of American and European launch vehicles. Possessing a powerful potential and accumulated experience in the field of the rocket and space industry, the Soviet Union could technically compete with private USA companies and the new-born European Ariane. The problem, however, was not the quality of the "product" being sold, but the geopolitical conditions under which the deal was made. This chapter aims to analyse the factors that contributed to the failure of the agreement between ESA and Soviet institutions. The analysis considers the geopolitical context of Soviet-European relations, the influence of third-party involvement in bilateral negotiations, and an assessment of the actions taken by all parties.

Marecs Background

The cluster of four satellites that ESA intended to launch for the needs of Inmarsat held significant importance. Conceived as an experimental system, Marecs evolved to provide Europe with a breakthrough in mobile telecommunications expertise. Several of the ESA member states

O. Dubrovina, *The Soviet Space Program and International Collaboration during Perestroika, 1985–1991*, Palgrave Studies in the History of Science and Technology,
https://doi.org/10.1007/978-3-032-03157-0_8

undertook, in 1973, to fund a satellite program that would guarantee communications between ships and land stations linking into the public networks, at a time when vessels could call only on unpredictable short-wave radio. Marecs then became ESA's first venture into the commercial satellite business.[1]

The Marecs-A and B maritime communications satellites had been developed by ESA from the European Communications Satellite (ECS) series of telecommunications satellites and were then leased by the London-based Inmarsat.[2] Set up in 1979, Inmarsat formally began operations in 1982, with a membership which had grown to more than fifty countries by the end of the 1980s. It functioned as an international co-operative, using a network of nine geostationary satellites spread around the globe which gave virtually worldwide coverage. Two of its satellites, Marecs-A and Marecs-B, were manufactured in Europe, three further satellites were taken over from the Marisat system, which had been operated until the early 1980s by the USA group Comsat General, while Inmarsat also leased space aboard of four Intelsat[3] V orbiters.[4]

Regarding the USSR, it should be emphasised that for the Soviet leadership cooperation with ESA in the sale of launch vehicles represented one of the earliest attempts to establish commercial relations with Western partners in the field of space exploration. Before that, the relationship between the Soviet Union and Western countries was mostly based on a non-exchange of funds principle in the framework of bilateral agreements. It is worth remembering that this was an area that remained highly confidential for an extended period, with access predominantly restricted to representatives from socialist bloc countries and, in rare cases, to Western specialists involved in international collaboration such as the Soyuz-Apollo mission and Franco-Soviet scientific experiments.

The feature of the cooperation between ESA and the USSR on the satellite launch project will not be clear without considering the international circumstances of the design, building, and use of two European satellites, the Marecs-A and B. We will utilise the historical research

[1] See more Marecs. Description. Retrieved March 4, 2025, from https://www.esa.int/esatv/Videos/2005/01/Marecs.

[2] Collins, G. (1990). *Europe in space.* Palgrave Macmillan, 52.

[3] The International Telecommunications Satellite Organisation with headquarters in Washington DC was set up in 1964 with the 11 member countries with the goal of providing global telecommunications links.

[4] Collins. (1990), 108.

conducted by John Krige and his colleagues to reconstruct as comprehensive a picture as possible of the events leading up to the commencement of the Soviet-European negotiations.

As mentioned above, the story of the Marecs satellites began in July 1973, when the second package deal[5] was adopted by the European Space Conference (ESC). This founding agreement for the European space program conceived the creation of ESA, the construction of Ariane, the developing of Spacelab (in cooperation with the American Space Shuttle program) and finally, the development of a maritime navigation satellite, Marots.[6] In June 1975 ESA Council determined on the formation of a Joint Board on Communication Satellite Programs, with responsibility for directing, amongst others, the Marots program. Marots was an experimental maritime communications satellite promoted strongly by the UK (a contract for the development of Marecs-A and B was placed with British Rocket and Space Dynamics). The construction of the satellites was financed by ESA member countries, with the UK (69.89%), Germany (13.29%), and Sweden (6.61%) being the main investors.[7] The launch of the satellites was tentatively scheduled for 1977, which follows from the agreement between ESRO and NASA. However, several modifications and unforeseen circumstances led to a delay in the launch which had to be postponed for several years.

In May 1977, the European Postal Telegraph and Telephone services (PTTS) set up the Eutelsat organisation on an interim basis, which would be responsible for the management of the space segment of European communications satellite systems, including in a first phase the ECS and the Marots satellites.[8] Therefore, initially both ESA Marots satellites were intended for use by Eutelsat. New schedules foresaw Marots-A to be launched into a geostationary orbit above the Atlantic Ocean in October 1980, and Marots-B to be located over the Indian Ocean after a launch in mid-1981.[9] The designs of the two European satellites were subsequently upgraded and the names of both satellites were changed to Marecs.

In 1979, the Inmarsat was finally founded. It was established to operate on a sound economic and financial basis and possessed robust commercial

[5] See more in Krige, Russo and Sebesta. (2000), 83.

[6] Bonnet, Manno. (1994), 20.

[7] Krige, Russo and Sebesta. (2000), 265.

[8] Ibid., 228.

[9] Ibid., 49–50.

principles, being financed by the Signatories to its Operating Agreement. The Signatories were designated by the member countries and, for the most part, they were national telecommunications carriers. Olaf Lundberg,[10] the Director General of Inmarsat, announced that: "the Signatories came from the rich and the developing nations, from East and West. Those Signatories with the largest investment shares were from the USA, Soviet Union, the UK, Norway, and Japan. The investment share was intended to reflect the actual use of the Inmarsat system. The investment covers the costs of the space segment and the operating and administrative costs of the organisation".[11] In the three-tier organisational structure the most relevant body was the Council of Signatories. It was similar to a corporate board of directors, an Inmarsat's policymaking body. In 1984 it consisted of 22 Signatories: 18 with the largest investment shares and four others elected by the Assembly on the principle of merely geographical representation but with due regard for the interests of developing countries. The Council met at least three times a year and each Signatory had a voting power equal to its investment share.[12] Note that ESA lacked voting rights within the Council, rendering it challenging to advance its own interests.

In the end, the ESA Executive proposed leasing the maritime communications capacity of two satellites (Marecs-A and Marecs-B) to Inmarsat for five years (1982–1987), along with spare hardware on the ground. After lengthy negotiations an agreed contract was signed between Inmarsat and ESA in November 1980.[13] The contract specified the availability in orbit of two Marecs satellites against an annual payment equivalent to $12.5 million per year. Terms were also agreed for a reduced service provision corresponding to one satellite in orbit and in line with the leasing concept, no payment would be demanded should no service become available.[14] Marecs-A was launched during the night of December 19/20, 1981, from the Guiana Space Centre aboard Ariane 1. Subsequently, Marecs-B was lost in an Ariane launch failure on September 10, 1982, but

[10] Olaf I. Lundberg has been director general of Inmarsat since its creation in 1979. In 1985 he has been reappointed for a second six-year term as director general of the International Maritime Satellite Organisation.

[11] Lundberg, O. (1984). Mobile Communications via Satellite in the 1990s. *Philosophical Transactions of the Royal Society of London*, 312 (1519), 43.

[12] Ibid., 42–43.

[13] Krige, Russo and Sebesta. (2000), 50.

[14] Letter of Quistgaard, E. to Lundberg, O. October 23, 1980. In ESA-4822, HAEU.

Marecs-B2 was launched successfully on November 10, 1984, on an Ariane 3 launch vehicle—only the second launch for this type of rocket.

The protracted negotiations with the USSR regarding the possibility of participating in the launch operations intriguingly intersected with the fifteen-year narrative of two European satellites. Neither the Western nor the Soviet press of that era, to our best knowledge, made any reference to the Soviet Union's involvement in the prolonged negotiation process. It was only through the examination of documents preserved in the Historical Archives of the European Union (HAEU) that the extent of the USSR's participation in the unfolding events could be comprehensively traced from inception to conclusion.

The USSR Steps In

The Soviet Union took part in the preliminary work of Inmarsat in the early days of its existence. The Convention and Operating Agreement, which served as the formal basis for the future Inmarsat, were opened for signature on September 3, 1976. The USSR designed Morsviazsputnik[15] as entity representing the Soviet Union.[16] In 1977, its quota reached 11%, which was still lower than the USA quota (17%).[17] During the deliberations regarding the selection of launch vehicles capable of deploying European satellites into orbit, the USSR proposed employing a rocket of its own manufacture. Consequently, ESA was presented with a broad spectrum of options, ranging from the domestically constructed Ariane rocket to those offered by competitors, such as the United States and the Soviet Union. Initially, it was assumed that the Soviet launch vehicle would be used for the third Marecs-C satellite but ultimately, the launch of the third satellite into orbit was abandoned and the European-Soviet cooperation project did not reach its conclusion.

[15] All-Union Association of Electric Navigation and Satellite Communications. It was established at the Ministry of Merchant Marine for providing the technical operations of Soviet and international communications satellites for navigation and radio determination of Soviet ships as well as for operational managements for the creation of necessary technic means for maritime satellite systems.

[16] Freeman, A L. (1976). International Conference on The Establishment of an International Maritime Satellite Organisation: Convention on the Inmarsat. *International legal materials* 15 (5), 1051–75.

[17] List of Inmarsat shares for those represented at the meeting of November 10, 1977. In ESA-8046, HAEU.

According to ESA officials, the Soviets first suggested the possibility of providing a launcher for Marots in 1975, but the Europeans were, by then, already planning to launch the satellites on a USA launcher in 1977 (Ariane was not still ready for flight). As the US-ESA negotiations were in the closing stages the Soviet proposal was not taken into consideration and ESA refused it.[18] Initially, two American Delta launchers were procured under the Marots program. In 1975, the NASA/ESRO agreement for spacecraft launching and associated services to be furnished by NASA in connection with the launching of the Marots spacecraft was signed.[19] In the Agreement, it was underlined that "the use of an uprated Delta (3914) launch vehicle was not a standard NASA vehicle but was being developed by the McDonnell Douglas Corporation at no cost to the USA government".[20] The document was updated after the foundation of ESA in 1975 with the indications of the launching time as "fourth quarter of 1977". In 1977, it was found out that the two Delta launchers procured under the program could no longer be used at reasonable cost, since NASA was contemplating closing their launch pads at Cape Kennedy by early 1980, and therefore another launch vehicle would have to be used. Thus, steps should immediately be taken to recover as much as possible of the already incurred cost for the procurement of the two Delta launchers.[21] Ultimately, NASA refunded to ESA $ 44,952.00.[22]

On the one hand, there was a shift in the focus of the treaty: it transitioned from Marots to a modified Marecs model. On the other hand, it is likely that the leadership of ESA recognised the imminent success of the Ariane launch program and deemed it more prudent to redirect efforts away from collaboration with the United States and towards utilising domestically produced launch vehicles. However, the stance of the American side remains ambiguous. According to ESA documents, NASA had not demonstrated substantial interest in launching European satellites.

[18] Galligan, K. (1977). *Informal discussion with representatives of U.S.S.R. to INMARSAT*, May, 12. In ESA-6896, HAEU.

[19] Agreement for spacecraft launching and associated services to be furnished by NASA in connection with the launching of the Marots spacecraft. In ESA-7046, HAEU.

[20] Agreement between the United States National Aeronautics and Space Administration and the European Space Research Organization for spacecraft launching and associated services to be furnished by NASA in connection with the launching of the MAROTS spacecraft. In ESA-7046, HAEU.

[21] Krige, Russo and Sebesta. (2000), 242.

[22] *Marots refund*, June 27, 1980. In ESA-8046, HAEU.

Against the background of ESA-NASA negotiations, consultations with Soviet partners on the issue of launching a third satellite went ahead.[23] In fact, ESA indicated to the Soviets its openness to the prospect of a Soviet launch within the framework of Inmarsat. Subsequently, the Soviets persistently advocated for this plan with ESA representatives during working group sessions convened under the auspices of the pre-Inmarsat Preparatory Committee.[24]

In 1977, through the Soviet embassy in Paris (specifically M. Zhukov) and the Soviet delegation at Inmarsat, the USSR confirmed its readiness to provide, on a commercial basis, a launch vehicle for a European satellite. The head of the Soviet Inmarsat delegation and President of Morsviazsputnik, Yuri Atserov, assured ESA representatives of a lower price compared to any American offer and the costs associated with Ariane. Another advantage of the Soviet proposal compared to the American counterpart was the reliability of the launcher. Contrary to NASA requirements, the contract with the USSR would not include a "clause allowing for ESA's displacement due to shifts in priority by the Soviet government in other areas".[25] In other words, the Soviet launcher for a European satellite would not be cancelled in the event of changing Soviet priorities. ESA officials immediately noted the high level of interest coming from the Soviet side in concluding a contract. As W. Thoma from ESA Directorate of Administration put it: "the Soviet negotiators showed an amazingly high degree of flexibility which proves that they attach considerable importance to the matter".[26]

However, it is evident from the archival documents of ESA how challenging it was for the Soviet side to initiate the negotiation process. According to Zhigarev, science adviser at the Soviet Embassy in Paris, Soviet supporters of cooperation with ESA had to overcome the resistance of the military, who finally agreed to cooperate with the Europeans following the experience of the Soviet-French joint project *Signe-3* already implemented in 1977.[27] The satellite and the scientific equipment of the *Signe-3* mission were produced under the supervision of the French space

[23] Galligan, K. (1977).

[24] Pinglier, A. Arets, J. *URSS and Marecs-C*, September 23, 1980. In ESA-4815, HAEU.

[25] Thoma W. *Report of visit to U.S.S.R.* in January 2–9, 1980. January 15, 1980. In ESA-4815, HAEU.

[26] Ibid.

[27] Arets, J. *Rencontre avec un représentant de l'Ambassade de l'Urss.* Septembre 23, 1977. In ESA-6896, HAEU.

agency, CNES. According to French project participants, the Soviets showed great interest in gamma astronomy and therefore agreed to launch the French satellite, which was ultimately launched from the Kapustin Yar Cosmodrome on June 17, 1977. It was the first time that a French team had been invited to a lift-off from a Soviet military cosmodrome.[28] The fundamental distinction, however, lies in the fact that the French-Soviet project was executed on a grant basis within the framework of a scientific cooperation program. The French satellite was launched using a Soviet Cosmos 3 M launch vehicle in exchange for the reception and processing of the acquired data. Moreover, scientific instruments aboard the satellite were manufactured jointly by French and Soviet institutions.[29]

Prior to submitting the proposal offer to Inmarsat, joint technical groups from ESA and the Soviet Union had to assess all requisite data. The primary challenge lay in determining technical documents, such as interface specifications, that needed to be included in the contract to establish a practical baseline. At the end of 1977, the first meeting between ESA and the USSR was held in Paris to discuss the technical feasibility of launching Marecs-C on a Soviet vehicle. The discussions were also attended by delegates from the newly formed Eutelsat, composed of European PTTs, as well as representatives from Licensintorg, IKI, and Intercosmos. The Letter of Intent was signed by ESA and Licensintorg, representing Morsviazsputnik.[30]

The second technical meeting took place in June 1978 in Moscow, during which both parties arrived at the mutual conclusion that, based on the available information, there were no technical impediments that would render the realisation of the joint project impossible. Despite this, there were several contractual challenges which needed to be overcome to attain complete compatibility of the satellite and the launcher.[31] During this meeting, two possible options for cooperation were ascertained: either ESA paid for the launch of its satellite with a Soviet launch vehicle, or the USSR took part in a Joint Venture providing the launcher.[32]

[28] Ammar-Israël (Ed.). (2015), 82.82.

[29] See also in Harvey. (2021), 42–44.

[30] Letter of intention about the conclusion of an agreement on launching an ESA satellite. In ESA-6896, HAEU.

[31] Minutes. Technical discussions between a Soviet delegation and ESA representatives, Moscow, June 5–7, 1978. In ESA-6896, HAEU.

[32] Arets, J. *Note pour le dossier. Réunion de Moscou*, Juin 12, 1978. In ESA-6896, HAEU.

Finally, the Soviets decided to consider separately their own financial participation in the Joint Venture and the execution of the contract with ESA for the supply of the launcher. Two documents were supposed to be drafted for these purposes. The first would be a contract for launching an ESA satellite by a Soviet launcher to be concluded between ESA and Licensintorg and authorised by Morsviazsputnik. The second would be an agreement elaborating the general principles of cooperation between Morsviazsputnik—the Soviet participant in the Operation Agreement on Inmarsat and the Soviet representative in the Joint Venture—and ESA.[33] It is evident that a distinct division of competence and responsibility between the two Soviet institutions enabled the Soviet side to concurrently pursue commercial interests, facilitated by Licensintorg, as well as cultivate politically oriented relations through Morsviazsputnik.

It is noteworthy to emphasise that three rockets vied for the launch of the European satellite Marecs-C: the Soviet Proton, the European Ariane, and the American Delta. ESA, as indicated in the documentation provided by Inmarsat, should have acknowledged that from a technical standpoint, all three rockets offered an equal chance of success. The difference obviously was the cost of its launch and it was here that the USSR could excel: the Soviets quoted $21.65 million dollars for the launch with a discount for an unsuccessful launch.[34]

As ESA's delegates stressed, the task of promoting the Soviet rocket to Inmarsat relied on the Soviet specialists who knew their brainchild better. In 1979, Yuri Atserov personally informed R. K. Al-Ghunaim, the chairman of the Inmarsat Council, of the ongoing technical negotiations with ESA and the details of the Soviet proposal. He underlined that the offered rocket-launcher was able to put in geostationary orbit a payload of about 1000 kg in "agreed time and with required accuracy".[35] In this appeal, Atserov attempted to enhance the most favourable characteristics of the Soviet launch vehicle, emphasising its reliability, advantageous technical qualities, and the many years of experience of Soviet engineers in the rocket and space field. The official statement of the USSR delegation was made at Inmarsat Council held in London in Autumn 1979. At this meeting, the Soviets informed the Council members of the USSR's willingness

[33] Meeting of a Soviet delegation and ESA representatives in Paris, September 18–22, 1978. In ESA-6896, HAEU.

[34] Letter of Atserov Yu. to Al-Ghunaim, A.R.K. September 19, 1979. In ESA-6276, HAEU.

[35] Ibid.

to launch Inmarsat satellites into geostationary orbit using a Soviet rocket launcher and provided an update on discussions between the USSR and ESA in this regard. Additionally, Atserov emphasised that access to the launch site for foreign specialists had been granted,[36] a fundamental requirement for the European partners. Nevertheless, the project raised numerous questions among ESA officials.

Difficulties Ahead: The Struggles of Soviet-ESA Collaboration

This episode marked the first occasion where ESA officials engaged in a commercial agreement with a Soviet partner, prompting it to insist on early and complete commitment. Notably, ESA lacked a comparable agreement with NASA for the use of the USA Space Shuttle; however, European entities were familiar with the standard contracting and pricing policies adopted by NASA for the USA Shuttle program.[37] Further doubts were the reasons for collaboration with the Soviets. It was fundamental to clearly define the benefits expected by ESA from such a cooperation (gaining experience, diversifying launch offers, expanding cooperation with the USSR, etc.). The consequences and political impact of the decision to launch an ESA satellite using a Soviet launcher were consistently at the forefront of the European agency's considerations. According to the statement, expressed in an internal ESA document, "it was in the European interest in the long term to have a balanced relationship with the two big space powers viz the USA and the Soviet Union".[38]

Questions also arose regarding the conditions under which ESA would agree to cooperate. Poor previous experience highlighted the need to explicitly define key provisions, such as access to the launch pad, obtaining information about the launch vehicle, delineation of areas of responsibility, and various issues related to launch insurance.[39] In addition to purely technical and bureaucratic challenges, further complications emerged

[36] Statement of the U.S.S.R. delegation. October 1979. In ESA-6276, HAEU.

[37] Thoma, W. *Draft Launching contract for Marecs C with Licensintorg*, December 13, 1979. In ESA-7906, HAEU.

[38] Visit of the Morsviazsputnik delegation on November 19–23, 1979. In ESA-4815, HAEU.

[39] Arets, J. URSS. Discussions sur la possibilité de lancer un MARECS avec un lanceur soviétique, Juillet 28, 1978. In ESA-6896, HAEU.

from translation difficulties, conceptual differences, and a general lack of sufficient experience in collaborative work.

While ESA encountered no issues with the interface design of the Marecs-A and Marecs-B satellites launched by the Ariane launcher, interactions with the Soviets presented challenges. The technical specifications of the Soviet Proton were provided to ESA very late, displeasing the European side and casting doubt on further cooperation. The European side's concerns stemmed from the potential necessity of significant adaptations of the satellite for compatibility with a Soviet launch vehicle, thereby risking additional costs that would burden the earlier Inmarsat proposal and diminish its competitiveness. Without doubt, ESA did not want to absorb any additional costs from its own budget.[40]

As a result, in early 1980 negotiations between the Soviets and Europeans achieved the following agreements: it was technically feasible for Marecs-C to be launched on a Soviet launcher; the costs of modifying Marecs-C would be absorbed by the Soviets; the Soviets offered to launch for $23 million dollars and to provide a "re-flight" at 50% of the original cost in the event of a failure. ESA also agreed to include an option for Marecs-C with a Soviet launcher in their offer to Inmarsat.[41]

Particular attention should be paid here to the nature of the relationship between ESA and Inmarsat. Inmarsat acted as a customer of the satellites manufactured by ESA, and ESA in turn was the executor of Inmarsat and made proposals, including those for launch vehicles. In fact, ESA had no formal voice at Inmarsat's meetings and could only submit proposals to that organisation. After the creation of Inmarsat in 1979, ESA also no longer had direct access to the negotiations on the space segment that would be used to assure marine communications.

The Soviet Union, in this context, occupied a most advantageous position. Historically, the Soviet Union had played an important part in the formation of Inmarsat, and it is probably due to their efforts and those of European countries that the organisation was established. In 1979, Soviet investment shares stood at approximately 16%, and the major decisions in Inmarsat were carried out according to the sum of the investment shares of those voting and by a percentage exceeding 66% of the total (see below). By comparison: the USA's non-governmental representative voted about

[40] Discussion avec une Délégation Soviétique. Déclaration préliminaire de l'ESA. In ESA-6276, HAEU.

[41] Pinglier, A., Arets, J. *URSS and Marecs-C*, September 23, 1980. In ESA-4815, HAEU.

22%.[42] Two factors may have positively influenced the selection of the Soviet launcher by ESA management and both were, in one way or another, connected to the principle that Soviet policy adhered to, i.e., *do ut des*. Firstly, the constructive ESA attitude towards the technical possibilities of a Marecs launch by a Soviet launcher would help ensure that the Soviet delegation to Inmarsat would vote in favour of Inmarsat using Marecs[43] and secondly, the Soviet delegation was aware of European aspiration to obtain an observer status, since the relationship between ESA and Inmarsat was non-existent at that time. The Soviet side hinted that there had been opposition to granting observer status in the Council of Inmarsat but simultaneously and according to Soviet representatives, Inmarsat had expressed the intention to contact ESA for more information on the Marecs proposals.[44] In this manner, the Soviet side subtly indicated Inmarsat's interest in European satellites and conveyed the readiness of the Soviet delegation to endorse ESA's candidacy, should ESA opt for a Proton launch vehicle to deploy its satellite into orbit.

Another lever of influence from the Soviet Union was the implicit promise of support for a European candidate, Hans Sondaal, in the elections for the post of Inmarsat Director General in order to "persuade Europe to look favourably on the USSR launch".[45] One gets the impression that the Soviet delegation deftly played its political party according to the rules recognised by all the players, not disdaining to resort to veiled blackmail. Indeed, the Soviet side profited from the situation and tacitly offered itself as an intermediary between ESA and Inmarsat using all political instruments at its disposal. The political stakes in the game involved the selection of the Soviet launch vehicle, which held the promise of several advantages for ESA, not just solely of an economic nature.

One particular risk for the Europeans was the possibility of uniting the efforts of the Soviet and American delegations in the Inmarsat Council against the purchase of ESA satellites (Table 8.1). A joint vote by the Americans and Soviets against the ESA proposal, potentially supported by several other countries, could have adverse repercussions on the outcome.[46]

[42] Ibid.

[43] Visit of the Morsviazsputnik delegation, November 19–23, 1979. In ESA-4815, HAEU.

[44] Report of a meeting held between U.S.S.R. Representatives and ESA for continuing discussions into possibility of the Marecs-C satellite being launched on a Soviet launcher, November 19–23, 1979. In ESA-4815, HAEU.

[45] Galligan, K., Barbance, K. *Launcher questions*, May 18, 1979. In ESA-8046, HAEU.

[46] Ibid.

Table 8.1 The investment shares in the Inmarsat Council in 1979, expressed as percentages (Galligan, K. *Inmarsat Counsil Meeting*, July 23, 1979. In ESA-4822, HAEU)

USA 22.5	Canada 2.9
USSR 15.75	Holland/Belg. 3.22/0.67
UK 11.1	Denmark 1.87
Norway 8.8	Sweden 2.1
Japan 7.8	Spain 2.25
Italy 3.7	

Understanding the complexity of their position, ESA officials developed a strategy that would benefit their organisation. This approach involved presenting two proposals simultaneously to Inmarsat. The rationale behind this strategy was to absolve ESA of any responsibility should the Soviet launch vehicle be rejected. These two formally distinct offers were thus for either the Soviet launcher or Ariane. If, by highlighting the parity between the options, Inmarsat chose ESA's offer, the decision would rest with the ESA Council, with the implicit understanding that Inmarsat would ultimately select the ESA satellite. Jean Arets wrote that, "this procedure would allow us to remain neutral without jeopardizing Soviet support for a decision favourable to Marecs".[47]

Initially, there was ambiguity regarding who held the authority to make the final decision concerning the selection of the launcher, whether it was ESA or Inmarsat. The request to examine the possibility of launching Marecs-C on a Soviet launcher arrived from Inmarsat to ESA in January 1980. At the beginning, Inmarsat's position was key in deciding on the choice of launch vehicle, but later their attitude towards the new satellite had changed: in the Spring of 1980 a lease was requested instead of the sale that was originally foreseen. The prospect of leasing the satellite radically changed the situation, depriving Inmarsat of the right to choose the launcher and completely leaving it up to ESA.[48]

Despite Soviet assurances that there were no obstacles to ESA specialists accompanying the satellite on Soviet territory up to the moment of its launch, the Europeans still encountered the typical Soviet resistance to foreign access to their strategic facilities. According to perceptions of ESA staff in the early 1980s, as they argued there was the "obvious Soviet

[47] Arets, J. *Offre Inmarsat et lanceur Soviétique*, mars 20, 1980. In ESA-4815, HAEU.
[48] Arets, J. *URSS – lancement Marecs C*, Février 26, 1980. In ESA-6896, HAEU.

reluctance to enable ESA to visit the launch site or to give details of facilities".[49] ESA officials insisted on the ability to visit the cosmodrome before signing the contract, while the Soviets promised access only during the final phase of the mission. The Soviet negotiators encountered difficulty in obtaining permission from authorities to allow cosmodrome visits at such an early stage of cooperation.

To prevent possible difficulties in working with Soviet partners, ESA management took steps to gather information from their counterparts and competitors regarding collaboration with the USSR. The Agency endeavoured to ascertain specific challenges encountered by the Americans during their engagement with the USSR Academy of Sciences within the framework of the Soyuz-Apollo project in 1975, as well as with the Indian Space Research Organisation (ISRO) during the launch of their satellite Bhaskara-I in 1979. Furthermore, they naturally expressed interest in the extensive history of cooperation between the Soviet Union and the French space agency, CNES.

As it followed from the answer of USA colleagues, severe restrictions were imposed on the stay of American experts in Baikonur in the mid-1970s, which mainly concerned their movement around the territory and communications.[50] Valerie Hood from ESA International Affairs division pointed out that from her contacts with CNES and ISRO dating back to 1980, "did not appear that conditions had improved in the meantime. ISRO resorted to shift work to have someone with their satellites 24 h/day whilst at the launch site!"[51] Like the Americans, Indians were not issued multiple entry visas, although they had no difficulty at all in obtaining single entry visas at short notice and they were not restricted in their movements. Moreover, they had no difficulty in obtaining either telephone lines or transmission of data to the ISRO control centre.[52]

Meanwhile, disapproving signals were received from the United States regarding the ongoing negotiations between ESA and the Soviets. The Americans believed Europe should step down and use only Ariane. They also suggested that Europe should not force Inmarsat to take a decision on launchers.[53] ESA was also reminded of the need to comply with the

[49] Hood, V. *NASA experience of Baikonur*, Mars 14, 1980. In ESA-6896, HAEU.
[50] Ifft, E. *NASA experience of Baikonur*, Mars 13, 1980. In ESA-6896, HAEU.
[51] Letter of V. Hood to Smith, P. G. 19 Mars, 1980. In ESA-6896, HAEU.
[52] Hood, V. *ISRO's at U.S.S.R. Launch Sites*, April 16, 1980. In ESA-6896, HAEU.
[53] Galligan, K. *Launcher questions*, Mai 18, 1979. In ESA-4822, HAEU.

provisions on the non-proliferation of important technical information used in dual purposes. The US Defence Material Standardisation and Specifications Office prohibited ESA officials[54] from exchanging information contained in the specifications for nitrogen and helium that were similar to the United States MIL-P 27401C (1975) and MIL-P 27407A (1978) specifications. It is curious that this limitation was easily circumvented, due to the fact that the mentioned specifications probably were at the disposal of the USSR Academy of Sciences or could be obtained by the Soviet Academy since it had a cooperative agreement with NASA.[55]

The French delegation to ESA was even less optimistic about utilising the Soviet launch vehicle, primarily due to two main concerns. Firstly, as France was the primary manufacturer of the Ariane, the emergence of a formidable competitor for the European rocket, which had recently passed successful flight tests, was highly undesirable. Consequently, France harboured no desire to assist the Soviets in introducing a long-term competitor to the international market alongside Ariane. Secondly, they believed that ESA made the task too easy for the Soviets by accepting discussions not only on technical matters but also on administrative and financial aspects.[56]

In April 1980, Roy Gibson announced to Boris Kurakin, President of Licensintorg, that "it has not been possible to obtain further financing from the ESA member states for the continuation of the Marecs-C satellite. As a consequence, ESA was in no position to make a formal offer to Inmarsat for a third spacecraft". This decision, however, did not preclude further negotiations with the USSR but rather "ESA intended to hold open the possibility of the acquisition of a third spacecraft by Inmarsat at some later date".[57] The same reason was provided by Erik Quistgaard, who succeeded Gibson as the Director General of ESA, to O. Lundberg: "The leasing of Marecs-A and Marecs-B is financed out of the resources of our Member States. Given the existing burden, these Member States are not in a position to prefinance an additional program in support of a lease offer for Marecs-C".[58] Therefore, Inmarsat's decision not to purchase or

[54] Telegram from Mellors to V. Hood, June 17, 1980. In ESA-6896, HAEU.
[55] Letter of Arets J. to Kurakin, B. July 4, 1980. In ESA-6896, HAEU.
[56] P.J. *URSS*, in ESA-4815, HAEU.
[57] Letter of Gibson, R. to Kurakin, B. April 28, 1980. In ESA-6896, HAEU.
[58] Letter of Quistgaard, E. to Lundberg, O. October 23, 1980. In ESA-4822, HAEU.

to lease the third Marecs led to ESA's abandoning the construction of Marecs-C because its member states were unwilling to risk capital without assurance of the satellite's launch.

Nevertheless, ESA staff continued to consider the possibility of collaborating with the Soviet Union throughout 1980 to ensure their ability to vote in favour of European satellites at any Inmarsat meeting. At the same time, the Soviets remained optimistic about selling their launch vehicle, continuously offering more favourable conditions for the flight (e.g., revision of the launch price, insurance costs below the rates currently being quoted by Western insurance companies[59]). Ultimately, Inmarsat decided to launch only two European satellites, Marecs-A and Marecs-B, using the Ariane rocket, thereby excluding the USSR from participation in the project.

After that unsuccessful first attempt at launching Marecs-B, the Soviets offered again to use their carrier. However, even if Inmarsat had requested that ESA investigate the possibility of a Soviet launch after the failure of the Marecs-B first tentative, and after the causes of the failure had been identified, ESA was confident that Marecs-B would be launched on schedule with Ariane.[60]

Several years later, when the failed attempt surfaced in internal correspondence, ESA officials cited two main reasons for the unsuccessful outcome of the negotiations. According to one version, ESA was to be responsible only for the technical compatibility of the satellite with the requirements of the Soviet launcher, but the final decision on the choice of launch vehicle was determined by Inmarsat. Ultimately, Inmarsat launched the satellites Marecs-A and Marecs-B, abandoning the third Marecs, the launch of which was claimed by the USSR.[61] The analysis of documents presented earlier confirms this version of events. However, there is another, less straightforward version, suggesting that negotiations regarding the utilisation of a Soviet launch vehicle to deploy the European satellite Marecs into orbit, were delayed for several years due to the fault of the Soviet side, ultimately leading to the selection of the Ariane rocket:

[59] Galligan, K. *Telephone call from U.S.S.R. Councillor to Inmarsat*, September 4, 1980. In ESA-4815, HAEU.

[60] Arets, J. *Report of a visit from Licensintorg, U.S.S.R., Representatives in Paris*, April 14, 1981. In ESA-4815, HAEU.

[61] Relations between the European Space Agency (ESA) and the U.S.S.R. in the field of space cooperation, June 1987. In ESA-28130, HAEU.

"the Russians were very slow, and Ariane was ready in time to avoid any choice between an USA or USSR launcher".[62] If in the first case ESA made the leadership of Inmarsat responsible for the failed cooperation, then in the second case, the desire of ESA to support the rocket of its own production and prevent the Soviet launch vehicle from entering the European market became more markedly clear. Moreover, this choice may have been the optimal solution to navigate a predicament where ESA needed to uphold a strategic balance and maintain an independent stance regarding the two major space powers.

The history of the Soviet Union's endeavour to launch a Western European satellite for Inmarsat did not conclude at that juncture. In 1983, as previously mentioned, the Inmarsat board resolved to develop a second-generation satellite for a maritime communication system, incorporating coast stations. These satellites were named "Inmarsat-2". The contract for the manufacture of three samples of the Inmarsat-2 satellite was won by a group of firms led by the UK company British Rocket and Space. In addition to Western European firms, the group included the American Hughes.[63] Inevitably, once again the question arose about the means of launching satellites into orbit.

At that time Inmarsat left the Soviet Union a chance to launch the next generation of satellites. In the call for tenders for new satellites it was specified that the "spacecraft must be compatible with more than one launcher of six listed in the request for proposals. Of the six, one was European (Ariane), one was Soviet (Proton) and four were American (the Space Shuttle, Thor Delta, Atlas Centaur, and Titan)".[64] It is an indisputable fact that Soviet representatives at the Inmarsat Council made concerted efforts to advocate for the inclusion of the Soviet launcher as a viable option for satellite launches.

Ultimately, Inmarsat determined that selecting the Delta and Ariane rockets for its next four launches would be more expedient, despite the lower bid from the Soviet Union. The cost of launching on a Soviet Proton rocket for a medium-sized satellite ranged between $18 and $20 million dollars, approximately half the prevailing rate for Europe's Ariane. However, the decision to utilise the Soviet missile was impeded by two factors: concerns regarding technology transfer and the perceived threat

[62] Arets, J. *Prospects of cooperation with U.S.S.R.*, Mars 24, 1987. In ESA-28130, HAEU.
[63] Goldovsky, D. (1987). *Kosmicheskie programmy zapadno-evropeyskikh stran*, Znaniye.
[64] Lundberg. (1984), 53.

posed to Western companies by dumping prices. Despite promises by the Soviets of "round-the-clock security accompaniment of their [Soviets'—O.D.] precious cargoes"[65] they failed to convince their Western partners of the earnestness of their promises. As Brian Harvey expressed, the actual problem was that the satellite to be launched would be an "export" and, accordingly, required CoCom approval. Another equally important problem for Western rocket manufacturers was that the proposed price undercut the prices of their launch vehicle companies.[66]

Soviet officials did not conceal their feelings of annoyance stemming from the outcome of the competition. During interviews, they openly directed criticism towards the USA side, accusing it of displaying bias against the Soviet proposal and attempting to uphold a monopoly within the space services market. During a meeting with representatives of ESA who visited the Ministry of Telecommunications of the USSR in September 1986, Stepan Bogodyazh from the International Communication Department of Glavkosmos (see Chap. 9) expressed his undisguised "disapproval of the USA which was hindering their [Soviet] cooperation with Intelsat and Inmarsat".[67] *The Wall Street Journal* printed his criticisms of the American position and the unjust decision: the "U.S. stand in relation to the Soviet Proton rocket, which was suggested as a carrier for Inmarsat's ship-to-shore communications network could serve as a classic example of bias".[68]

Conclusion

The unsuccessful endeavour at Soviet-ESA cooperation highlighted the profoundly challenging international landscape within the launcher and space services market during the early 1980s. The story of the European Marecs satellites revealed a complex intricacy of interests of three participants at once: ESA, the USA, and the Soviet Union. At the same time, the Europeans found themselves between two fires, having to choose in the initial phase a launch vehicle proposed by both space powers.

[65] Harvey. (2021), 266.

[66] Ibid.

[67] Block, G.F. *ESA Mission to the U.S.S.R. Ministry of Telecommunications*, September 8, 1986. In ESA-28194, HAEU.

[68] D'Anastasio, M. (1987). Soviets step up efforts to win share in launching of commercial satellites. *The Wall Street Journal*, January 20.

Right from the outset, ESA encountered formidable barriers to entry into the satellite telecommunication services market, despite its concerted efforts to pursue an ambitious foreign policy aimed at fostering lucrative cooperative ventures. Furthermore, their emphasis on promoting Ariane—a project characterised by extensive construction and numerous testing phases—intensified the complexities they encountered. In the decision-making process, ESA management had to consider political motivations alongside other factors, while striving to advance its own objectives. Among these objectives was the pursuit of increased autonomy, achieved through the diversification of proposals under consideration, and the maintenance of an equitable stance vis-à-vis both major competitors.

From the American perspective, significant concerns over technological espionage and the potential erosion of the domestic missile industry's position prompted actions to safeguard USA national interests. Meanwhile, the Soviet Union pursued its own objectives through various means, employing tactics ranging from blackmail to price dumping. By the end of the 1970s, the Soviet Union had initiated discernible efforts to promote its high-quality products in the international market. This endeavour marked the start of acquiring initial experiences in cooperation with Western institutions and companies, both on technical and commercial fronts. Subsequent chapters will delve into further steps taken in this direction.

In summarising the episode detailed above and anticipating the unfolding of the next events, the words of Boris Gubanov are noteworthy. He contended that upon the Soviet Union's entry into the international market, time had been squandered, as positions had already been allocated.[69]

[69] Gubanov. (1998).

CHAPTER 9

Glavkosmos

In 1985 a new department, Glavkosmos, was established within the renowned MOM. Its creation marked the beginning of a new era in Soviet cosmonautics—one defined by the transition of the space sector towards market-oriented relations. Determined that cosmonautics should contribute financially to the national budget, the Soviet leadership established a dedicated body to oversee this objective. This chapter focuses on a key question: who was responsible for introducing elements of capitalism into the command socialist economy of the space industry, what methods they used, and what consequences this bold experiment had.

A Brief Historical Perspective

From its inception, Glavkosmos was led by Aleksandr Dunaev. Western media speculated that Dunaev had "simply taken off his general's uniform" and rebranded an existing agency to civilianise the Soviet rocket marketing effort.[1] As a matter of fact, Dunaev's background was unknown to the Western space community. Aleksandr Dunaev was not a military officer and had no affiliation with any military department. Instead, he

[1] Sawyer, K. (1988). Soviets move aggressively in marketing space services. *Washington Post*, January 1.

O. Dubrovina, *The Soviet Space Program and International Collaboration during Perestroika, 1985–1991*, Palgrave Studies in the History of Science and Technology,
https://doi.org/10.1007/978-3-032-03157-0_9

advanced to senior positions in space industrial management through a career as an engineer and physicist. He had graduated with honours in 1958 from the Faculty of Physics at the Moscow State University Lomonosov and until 1980 he had made a career in the field of radio engineering and radar measurements as well as autonomous control systems for rocket technology at NII 885, under the direction of the space designer Mikhail Ryazansky. Dunaev initiated his political career in 1980 as an official at the MOM. From March 1985, until the conclusion of 1991, he served as the Deputy Minister of MOM, heading the 8th General Directorate, responsible for overseeing multi-year planning programs. As Viktor Nikolaev testifies, Dunaev occupied a special position in the Soviet establishment. On the one hand, "he was under the personal patronage of Oleg Baklanov. Thus, as a deputy minister, he was sent to Baikonur in 1987 to supervise the work on Buran. At the same time, when the Indian IRS was sent there, they started building a hotel at Baikonur, as it was necessary to have some infrastructure to receive foreign guests. Dunaev was also involved in this. On the other hand, he could travel abroad, which was considered an incredible privilege in the rocket and space field at that time".[2] Concurrently he held the position of the head of Glavkosmos.[3] So then, what was Glavkosmos?

To answer this question, we need to look at the political and economic climate that prevailed when Glavkosmos was established. In the middle of the 1980s, following in the footsteps of Peter the Great, who "opened a window to Europe", Dunaev claimed: "It was necessary to open something, open a window".[4] But in his case, it was not about the entire window, but rather just the vent. Simultaneously, both internal and external circumstances played a role in contributing to this endeavour.

The initial efforts to alter the situation and establish a presence in the international market for Soviet space products and services commenced with Gorbachev's policy: the space sector transitioned from being exclusively a military and ideological battleground to gradually opening new horizons for international exchange. Two key trends, both political and economic in nature, facilitated this transformation in the late USSR. The

[2] Interview with V. Nikolaev, June 30, 2024.

[3] Dunaev Aleksandr Ivanovich (1934–2014). Retrieved March 4, 2025 from http://www.biograph.ru/index.php/whoiswho/14-airforces/248-dunaev.

[4] Interview recorded by B. Boyarshinov with A. Dunaev, 2006. Retrieved March 4, 2025 from https://www.youtube.com/watch?v=dq%2D%2Dm8AenEo.

first aspect entailed Gorbachev's proclamation of *glasnost*, openness, international cooperation, and increased freedoms across various aspects of Soviet life, including the economy. These new tendencies were most prominently expressed in Gorbachev's concept of the "Common European Home"[5] (see Chap. 1). Regarding the economic trend, the gradual decline in the USSR's economic growth during the second half of the 1980s led to a tangible crisis by the end of *perestroika*. As a result, it became imperative to urgently explore potential solutions to mitigate this emergency. The decision to actively promote goods and services of the advanced Soviet space industry, in comparison to other categories of Soviet production, was one such opportunity.

Another impetus for the active development of the commercial branch of the Soviet space program was the interruption of both the European and USA launches. The Soviet authorities decided to take advantage of this temporary halt by launching a massive advertising campaign to promote their space technology. The campaign was driven at the highest level by the Chairman of the Council of Ministers Nikolay Ryzhkov in 1986. In June of that year, he sent a letter to the UN Secretary General Javier Perez de Cuellar, in which he wrote that "the Soviet Union declared its readiness to exchange its accomplishments in outer space, to launch peaceful space vehicles of other countries and international organisations by means of Soviet rockets, on mutually acceptable terms".[6] The announcement came at a time when the West was without a commercial launch capacity following the January 28, 1986, explosion of the USA space Shuttle, *Challenger*, in which all seven astronauts on board were killed. Additionally on May 31, 1986, a failure of the French-ruled European Ariane rocket sent the unmanned rocket and a $55 million dollar satellite payload tumbling into the Atlantic,[7] after mission control decided to detonate it nine minutes after lift-off when its third stage failed to ignite. This caused the suspension of the program that was in direct commercial competition with that of the USA.

In January 1987, an article was published in *Pravda* that could be considered as a *perestroika* view on the development of the Soviet space

[5] Malcolm, N. (1989). The "Common European Home" and Soviet European Policy. *International Affairs*, 65 (4).

[6] Soviet Union prepared to launch foreign satellites. *United Press International*, June 13, 1986.

[7] European Rocket Destroyed After Engine Failure. *New York Times*, May 31, 1986.

program. The author of the paper was Nikolay Ryzhkov. In conjunction with the proposition to coordinate international efforts in space, he also broached the subject of commercial opportunities, enumerating the potential advantages of the Soviet space industry and underscoring its readiness to engage in commercial cooperation with all countries. When articulating the benefits of the Soviet space industry, Ryzhkov's words resembled those found in a meticulously crafted advertising proposal:

> As it enters the international market for space technology and services, the Soviet Union cannot, of course, but reckon with the laws of the market. It is only natural that we expect to make some gains. At the same time, we shall be seeking to make the launch terms mutually advantageous for us and our clients. In orbiting spacecraft for developing countries, the Soviet Union is prepared to grant them substantial discounts. Soviet financial organisations can also insure the payloads.[8]

Although Glavkosmos was established in 1985, 2 years before the publication of Ryzhkov's article, it can be argued that Glavkosmos should be seen as a direct embodiment of Gorbachev's broader space policy as outlined in Ryzhkov's publication. It serves as a symbol of the commercialisation of the domestic space program and its entry into international markets. Glavkosmos progressively assumed a crucial role in dealings with international partners. Its principal objective was to leverage the scientific and technical capabilities amassed by the Soviet space industry over more than forty years for commercial purposes. If Intercosmos was the leading organisation, responsible for international scientific cooperation in space, then Glavkosmos was the cornerstone in the gradual entrance of the Soviet rocket and space industry into the international market and for establishing business contacts with foreign partners. Glavkosmos' world-wide activities, including cooperation of a commercial nature, were supported and negotiated by two other long-standing state institutions: Licensintorg in conjunction with the Soviet insurance company Ingosstrakh.[9]

[8] N. Ryzhkov's interview to TASS, January 1987. In ESA-28130, HAEU.

[9] Langereux, P. (1987). L'URSS ouvre ses trois cosmodromes aux lancements commerciaux. *Air&Cosmos*, 1138, Mars 28.

Glavkosmos was established solely as a state institution, operating as one of the departments within MOM. Viktor Nikolaev, who joined Glavkosmos during its initial months, recalls the start of his work in this body:

> When the decision was made to establish Glavkosmos, the question of recruiting specialised engineers arose, and they could not find even one with English proficiency. The point is that NPO Lavochkin and NPO Energya had interpreters, as Lavochkin produced scientific spacecraft for the IKI (e.g., Vega), which cooperated with foreign entities. Energya already had experience working with foreigners during the launch of foreign cosmonauts. However, these were purely translators without engineering education. As a result, Glavkosmos had to recruit translators with English skills from related industries. These individuals were not quite blatant, but they were not specialists in rocketry either.
>
> In the end, we recruited about 30 people across four departments. Our First department was engaged in manned flights; the Second department focused on earth sensing from space; the Third department dealt with the commercialisation of Proton and microgravity research (they were responsible for work on *Fotons*). The Fourth division was the International department. It kept minutes, organised meetings, and handled public relations. In fact, it was responsible for servicing all the other three divisions.
>
> My Manned Space Department had a chief and four staff members: one from Energya, myself from Khrunichev, one from the Ministry (MOM), and one relative of a high-ranking official. When Glavkosmos was created, we were initially located on Miusskaya Square in the MOM building. Since it was clear that we had to communicate with foreigners, the security service initially arranged for us to use the meeting rooms of Obschemashexport, the structure responsible for the production of consumer goods like *Druzhok* bicycles and refrigerators. At first, it was like that, but then security moved us to Krasnoproletarskaya Street. We acquired our own premises with the official signboard "Glavkosmos of the USSR" (Fig. 9.1). We would go to the Ministry for lunch from there.[10]

Western analysts and experts associated the emergence of Glavkosmos with the USSR's attempt to establish its own civilian space agency, an entity that had not previously existed in the Soviet Union. According to an unknown European manager familiar with the Soviet space program, "Until now, the Soviets have not a central civil national space organisation such as NASA in the USA, the French CNES or Japan's NASDA. The

[10] Interview with V. Nikolaev, June 30, 2024.

Fig. 9.1 Glavkosmos employees affix a sign in 1985. Source: V. Nikolaev

Academy of Sciences' two institutes that deal with space matters [IKI and the Vernadsky Institute of Geochemistry and Analytical Chemistry—O.D.] really did not have the power to make decisions. Glavkosmos was intended to hold decision-making responsibilities, serve as the interface between industry and Soviet laboratories, and participate in international cooperative programs".[11]

Dunaev deftly played along with his Western colleagues, both by comparing Glavkosmos to NASA and by deliberately exaggerating the number of his subordinates mentioning hundreds of employees in his interviews. The intentional exaggeration of the total workforce when discussing Glavkosmos' development in the 1980s with foreign journalists, is indicated by the figures provided in a much later interview in 2006. In that talk, Dunaev clarified that "Glavkosmos initially employed 20 people, later expanding its staff to a maximum of 30 employees".[12]

The unifying nature of the new organisation was accurately recognised by external observers. In the report on his 1987 trip to the USSR, the American scientist James W. Head from Brown University (Rhode Island)

[11] *Aviation week & Space Technology*, March 24, 1986, 77.
[12] Interview with Dunaev. (2006).

described his impressions during his journey to Moscow where he visited the major Soviet civilian space organisations. Regarding Glavkosmos, he emphasised its coordinating role in the Soviet civilian space program. "…the Soviets appear to be more flexible and resilient than we do, and this has fundamental implications for international cooperation and competition. There is reason to believe that the recent successes of the Soviet program are beginning to close the gap in the «coagulant»: a senior Soviet space official described the formation of Glavkosmos as bringing "viscosity" to the system".[13] Drawing upon his own observations, he foresaw a promising future for both Glavkosmos and the continued advancement of the commercial sector within the Soviet rocket and space industry:

> The formation of Glavkosmos and the nature of the Gorbachev regime strongly suggest that the Soviets are working toward concerted development and implementation of commercial activities in the next decade, particularly since recent events in the West have opened the door considerably to launch services, remote sensing data, etc. With their reliable, production-line approach to hardware (rockets and communications satellites, for example), and their desire for foreign currency and ability to undercut the market, they should be able to make major strides.[14]

From the official brochure distributed by representatives of Glavkosmos to potential foreign partners, the entity was described as the "Main Directorate for the creation and use of space technology for the national economy and scientific research, in cooperation with ministries and departments of the USSR and carried out work on international cooperation, including joint projects on a commercial basis".[15] Oleg Baklanov outlined the primary objectives of Glavkosmos, which focused on several key areas. First, in collaboration with relevant Soviet ministries and departments, it aimed to define the main directions for research and development in the creation and utilisation of space technology for international economic interests, scientific research, and international cooperation programs. Additionally, Glavkosmos was tasked with promoting domestic space equipment, technologies, and services to the global market on a commercial basis.[16]

[13] Head, J.W. (1987). *Field trip for the NASA advisory Council international relations task force*, Moscow, U.S.S.R., April 11–17. In NASA HQ Archives, 15,591, Part I.

[14] Ibid.

[15] The Main Tasks of the Glavkosmos of the U.S.S.R. In ESA-28130, HAEU.

[16] Baklanov, Rogozin. (2013), 456.

Thus, by the end of the Soviet period, an attempt was made in the USSR to concentrate the management of the peaceful exploration and exploitation of outer space within the framework of the single *civil* state institution. As Evgeny N. Bogomolov, Head of the Glavkosmos' department, said to the USA scientist James W. Head,

> The purpose of Glavkosmos is to coordinate industry and science. Glavkosmos is mostly related to industry, but science must consider industry as well. Glavkosmos plans, coordinates, and pays. It not only represents and coordinates scientific projects in industry, it makes the decision as to whether or not industry can do it. Government gives permission and money to Glavkosmos. Funding was allocated to Glavkosmos for the implementation of projects within the industrial sector and the Academy of Sciences (for instrument development). In addition to overseeing financial distribution, Glavkosmos was responsible for managing the projects.[17]

Officially, its main task was the practical implementation of the scientific knowledge gained over the previous decades in the interests of the national economy. Emphasis was placed on international cooperation, which was also to be supervised by Glavkosmos. However, in practice as it turned out later, its job was not so much the deepening of international cooperation but the commercial promotion of Soviet space technologies on the international stage, in order to acquire much-needed funds for the political leadership to replenish the national budget. This perspective is supported by the testimony of a direct participant from the scientific community. According to the memoirs of Genrikh Avanesov, "Glavkosmos was created to work with foreign customers of space technology. It had nothing to do with science".[18]

Notwithstanding the distribution of brochures and interviews conducted by Glavkosmos delegates, the primary function of Glavkosmos was not distinctly defined or comprehended, either internally or externally within the USSR. It was only in 1989 that the purpose of its establishment was unequivocally articulated: the commercialisation of the space sector. Doubts diminished following an interview that Dunaev gave to a *Sovetskaya Rossiya* correspondent. The question, posed in a distinctly provocative way, clearly hinted at the main purpose of Glavkosmos and Dunaev's answer, not only failed to refute this, but clearly confirmed it.

[17] Head. (1987). Part II.

[18] Interview with G. Avanesov, July 13, 2023.

> Journalist (Filippov): The USSR Glavkosmos has already been in existence as a commercial organisation for several years and, in spite of idle accusations about squandering national resources, brought in large sums in foreign currency for Soviet space exploration. What deals were concluded 'beyond the reach of" the USA Government and what did they give to the sides participating in them?
>
> Dunaev: We are currently collaborating with 35 countries. Under agreements for joint manned flights, each participating nation is expected to contribute approximately $10 million to the Soviet side.[19]

The data presented by Dunaev to demonstrate Glavkosmos' efficiency requires further clarification. Viktor Nikolaev highlighted the relative nature of the agency's financial contribution to the rocket and space industry during the *perestroika* period:

> In the late 1980s, Glavkosmos could not ensure the inflow of financial resources necessary to sustain the industry. We had to sell what the industry produced. If the industry did not produce anything and lost its qualifications, we had nothing to sell. The revenue generated from the sale of technology was insufficient to sustain the industry. These were called extra-budgetary funds, which were in addition to the state budget. Without government investment, it was akin to a dead poultice. At that time, commercial cooperation was just beginning to emerge. We had to start from scratch, needing time to enter the international market and adapt to its requirements.[20]

Glavkosmos: Navigating Internal Challenges

However, despite pursuing ostensibly innocuous and essential objectives aligned with Soviet political leadership, Glavkosmos represented a considerable challenge to the long-established scientific and industrial space sector of the USSR. Internal rivalries emerged, as the agency's functions and competencies overlapped with those of existing institutions. The primary conflict arose between Glavkosmos and the Academy of Sciences, particularly with its constituent bodies—the Intercosmos Council and IKI.

Internal disagreements undermined the work of Soviet space institutions to such an extent that Roald Sagdeev, the director of IKI, did not

[19] Soviet-U.S. space cooperation obstacles viewed. *Sovetskaya Rossia*, November 29, 1989.
[20] Interview with V. Nikolaev, June 30, 2024.

hide his hostility towards Glavkosmos by letting this leak into the national and foreign press. Sagdeev saw the creation of the new agency as an interlocutor between scientists and space equipment manufacturers without much enthusiasm. "When in the past we contracted with industry for spacecraft, systems of subsystems, we had to deal with individual companies. The role of Glavkosmos is to act as a top-level body. We would simply give an overall contract, and it would work with industry".[21] At the same time, Valery Barsukov, Director of the Vernadsky Institute from 1976 to 1992, outlined the international role of Glavkosmos, explaining: "Its important task should be to make the contacts between the Soviet space industry and foreign countries on cooperative programs instead of doing this at ministry level".[22] He compared this new space organisation with Glavatom, the atomic Soviet institution that oversaw international cooperation in the field of atomic energy.

Later, the contradictions between Glavkosmos and IKI became blatantly clear after the unsuccessful launch of two *Phobos* spacecraft in July 1988 (see Chap. 6). Sagdeev recalled that one of his colleagues, Konstantin Gringauz, published in *Pravda* a critical article "directed toward the space barons". The main backlash was directed at Albert Galeev, then director of IKI. He "was surrounded by angry directors from Glavkosmos, attacking him for the article in *Pravda*. They were bewildered by Gringauz' statements, comparing the mismanagement of *Phobos-1* with the mismanagement of the *Admiral Nakhimov*.[23] But even more, they were angered by the criticism of the Soviet shuttle Buran project. Gringauz had said he could not understand why now, when the country needed resources and money, we should copy the American Shuttle"[24] (see Chap. 1). The conflict between Glavkosmos officials and scientists from Intercosmos and IKI was also apparent on international forums. According to the USA correspondent, who participated in the International Astronautical Federation meeting held in India in 1988, "the Soviet Glavkosmos civilian

[21] *Aviation week & space technology*, March 24, 1986.

[22] Ibid.

[23] *Admiral Nakhimov*, built in March 1925, was a passenger liner of the German Weimar Republic later converted to a hospital ship, then a Soviet passenger ship. On August 31, 1986, *Admiral Nakhimov* collided with the large bulk carrier *Pyotr Vasev* in the Tsemes Bay, near the port of Novorossiysk and quickly sank. In total, 423 of the 1234 people on board died.

[24] Sagdeev. (1994), 318.

organisation headed by Aleksandr Dunaev was embroiled in a power struggle with the Intercosmos group of Soviet-bloc researchers".[25]

After Henry S.F. Cooper, the American writer and journalist, visited Moscow in 1990 and talked to Soviet scientists, including Sagdeev, he formed his own opinion about the complex relationship amongst the Soviet space institutions. He noted that there was a very important difference between Glavkosmos and NASA that influenced its relationship with scientists.

> ... Glavkosmos, a huge and unwieldy agency roughly comparable to NASA, but a NASA plus much of the rocket and space industry and minus the cosmonauts (who are under the Ministry of Defence) and the scientists (who are under the Academy of Sciences). In the United States, too, engineers usually have greater influence on a planetary mission than scientists do. But Glavkosmos and the Academy (of which IKI is a subsidiary) are highly competitive, and rivalry between government agencies, and between their dependent institutes, is far more intense in the Soviet Union than it is in USA. Moreover, *glasnost* and *perestroika* have penetrated much farther among Soviet scientists than among the engineers at Glavkosmos, many of whom are quite elderly; they are regarded by the scientist as being—in current Soviet parlance—"stagnant".[26]

Cooper was correct in his assumption as employees of IKI, accustomed to communicating with foreign colleagues through Intercosmos programs, hardly recognised the primacy of Glavkosmos. This was barely the picture of harmony and cohesion painted in the glossy Glavkosmos promotional brochure. The leading and organising role of Glavkosmos was challenged not only by astrophysicists from IKI, but also by space biologists and physicists from IBMP, whose distinctive activities were not subject to the competence of Glavkosmos. A Department for the Implementation and Promotion of Scientific Achievements had been founded in 1988 under the leadership of Mark Belakovsky to promote international assignments (see Chap. 2). One of its main competences was (and still is) international cooperation. For their part, Glavkosmos officials, and indeed representatives of the rocket and space industry, perceived scientists as freeloaders, who were trying to exploit the industry

[25] *Aviation week & space technology*, November 14, 1988.

[26] Cooper, H. S.F. (1990). Annals of space. The planetary community. I – Phobos. *The New Yorker*, June 11.

through the opportunity for free communication with foreigners. As Viktor Nikolaev recalled,

> Intercosmos was concerned that we would usurp their role in international cooperation. They participated in international flights, benefiting from free travel abroad, which was their bread and butter.
>
> Consider the negotiations with the French on a manned flight in 1982. When the French posed a question, the representatives of Intercosmos would leave the hall and go to the neighbouring room where the representatives of Energya were sitting, because specialists from manned space were strictly prohibited from interacting directly with foreigners. The Intercosmos representatives would relay the French question to the Energya specialists, then return to convey the answer to the French.[27]

He also cited the case of the second flight of the French cosmonaut Jean-Loup Chrétien in 1988, which Glavkosmos initially supervised in an attempt to get the French side to pay for the space flight.

> Glavkosmos had already negotiated that the French were to pay for the flight. However, before Gorbachev left for Paris in 1985, officials from Intercosmos advised him to invite the French for a free month-long flight with spacewalks. Naturally, the French gratefully accepted this entirely free offer.[28] Consequently, the internal conflict between Intercosmos and Glavkosmos—i.e., between scientists and industrialists—was politically won by Intercosmos. They simply granted the French this flight for free and claimed the victory.
>
> It was regrettable that this was done out of spite. Formally, the scientists from Intercosmos had harmed our field, to which they had no connection. They were not involved with the industry, serving merely as its superstructure. The industry did not claim dominance but sought to have its role clearly defined.[29]

While scientists tried to avoid Glavkosmos' interference in their internal affairs, space hardware manufacturers, on the contrary, saw it as an opportunity to enter the international arena. In 1987, the Foton Design Bureau at TsSKB Progress in Kuybyshev was established. This structural unit was

[27] Interview with V. Nikolaev, June 30, 2024.

[28] According to Yu. Baturin, in 1988 France reimbursed 10% of the cost of the second flight by French cosmonaut J.-L. Chrétien. See Baturin (Ed.), (2005), 682.

[29] Interview with V. Nikolaev, June 30, 2024.

responsible for the international cooperation of the still closed rocket and space plant, but, unlike IBMP, it worked in close cooperation with Glavkosmos, which acted as an interlocutor in relations with foreign partners. When, in 1989 Foton Design Bureau managers first took part in a business trip abroad, or, more exactly, a commercial tour of Western Germany, the entire campaign was organised by Glavkosmos. As part of the delegation, representatives of Yuzhmash, NPO Energya (successor to the famous Korolev's OKB-1) and TsSKB Progress advertised the products of the Soviet space industry. Specifically, the TsSKB Progress engineers presented the spacecraft *Foton* and *Bion* to their Western partners.

As Valery Abrashkin recalled, "In 1989, I participated in the inaugural Soviet delegation to Germany, collaborating with representatives from NPO Energya, Yuzhmash, and other prominent Soviet enterprises. Our delegation, comprising ten members, embarked on a comprehensive ten-day tour across Germany, visiting various firms from Hamburg to Friedrichshafen. During this visit, each member delivered presentations on their respective areas of expertise, ranging from the Mir space station to launch vehicles. I specifically addressed the delegation regarding the *Foton* spacecraft, providing insights into its applications and capabilities".[30]

Furthermore, Abrashkin asserted that the initial contacts between him and Wim Jansen, an ESA employee, were established in Hanover during the same trip. The primary objective of these contacts was to explore the potential utilisation of Soviet automatic spacecraft for the advancement of ESA scientific programs. "Accompanied by a small team of five individuals, Jansen presented documentation generated through their collaborative efforts with American counterparts. Reciprocally, we shared insights into our own work. Ultimately, we reached a consensus to sustain ongoing communication and collaboration".[31] Such an ambitious start encountered a major challenge from CoCom, which proved to be one of the hardest obstacles to the Soviet space industry's envisioned "triumphal march" onto the global stage.

[30] Interview with V. Abrashkin, March 4, 2023.

[31] Ibid.

Technology Transfer Hurdles: The CoCom Challenge

According to American space experts, the Soviet Union was aiming mostly at the European satellite market and at international organisations. Due to the number of launch vehicles, the variety of their technical characteristics, and low prices, the USSR emerged as a tempting alternative to the USA for those countries that did not have their own launch vehicles but needed to launch their own satellites. However, despite a promising "take-off", the Soviet leadership encountered several significant obstacles in their pursuit of dominance in the international space market.

One of the main hurdles for Glavkosmos came from export-control laws under an agreement by the Paris-based CoCom, prohibiting satellites made with USA components from being transferred to socialist bloc countries. Given that most satellites contained some American components, commercial cooperation between Western European firms and private USA companies with the Soviet civilian space program was significantly hindered. Domestic communications, land remote sensing satellites, similar payloads, and other incorporated technology were considered "Defence Items" by the USA Department of State and were listed on the Department's "Munitions Control List". These payloads could not be exported without an appropriate licence being issued by the Department. Of significance was that in 1986, the Department responded negatively to two inquiries from USA companies proposing, on a hypothetical basis, a launch of communications satellites aboard Soviet vehicles.[32]

During the first half of the 1980s, the Reagan administration undertook an ambitious defence build-up that called for the development and application of a wide range of advanced technologies to military systems and purposes. Administration officials regarded the tightening of national security export controls as essential to safeguarding these new investments. Supported by intelligence data, they concluded that the Soviet Union had systematically sought to acquire military technology and had done so consistently and successfully throughout the 1970s. In April

[32] The United States Government's position, in Glavkosmos/ Licensintorg briefing. In ESA 29182, HAEU.

1982, the CIA released a redacted version of a classified report that provided information on Soviet methods, targets, and acquisitions.[33]

According to that report, the Soviet Union had devoted since at least the 1930s, vast amounts of financial and manpower resources to the appropriation of Western technology that enhanced its military industry and improved the efficiency of its armed forces. In the 1980s, as CIA officials sustained, this Soviet effort was massive, well planned, well managed, and approved at the highest party and governmental levels. Both legal and illicit acquisitions of USA and Western technology and equipment were coordinated with intelligence gathered through the Soviet Union's extensive network of international governmental, scientific and technical agreements, and exchanges with advanced industrial nations. These efforts encompassed the collection of know-how, equipment, and computer database information by Soviet scientists and engineers participating in academic, commercial, and official S&T exchanges.

Analysing the current situation and trying to predict the future, CIA experts highlighted that the massive military development and plant expansion activity were taking place in the early 1980s when the Soviet economy had reached its lowest level of growth since World War II. Under these conditions, they reasoned that it would be impossible for the Soviets to maintain the current growth in military production without using Western technologies acquired illegally. A new generation of Warsaw Pact weapons would require selected critical components and modern manufacturing technologies and it was in these areas that unlawful Soviet acquisitions would most likely be over the next five to ten years. So, "the task of stopping Soviet bloc intelligence operations aimed at Western military and industrial technologies posed a formidable counterintelligence problem, both in the United States and abroad".[34]

The Soviet invasion of Afghanistan provided a convenient pretext for further tightening controls over the transfer of technology to the Warsaw Pact countries. As an expert on USA foreign policy, Michael Mastanduno argued that the United States adopted its position for both foreign policy reasons—to penalise the Soviets for the invasion—and strategic

[33] Mastanduno, M. (1992). *Economic containment: CoCom and the politics of East-West trade.* Cornell University Press, 278.

[34] CIA, *Soviet Acquisition of Western Technology*, Washington DC., April 1982. A copy of the report can be found at https://www.cia.gov/readingroom/docs/CIA-RDP83M00914R001200050005-3.pdf

reasons—to weaken Soviet military capabilities. Consequently, the Reagan administration prioritised the revitalisation of CoCom, elevating its political significance in 1982 and establishing the foundation for three related initiatives: the expansion of the control list (which now included spacecraft and space launch vehicles), enhancements in enforcement, and the strengthening of CoCom's institutional framework. In fact, the multilateral mechanism was boosted during the 1981–1985 period and attention from Western governments and tech manufacturers to export control issues increased. The control list was further expanded to include previously underemphasised areas of military significance. Enforcement became a higher priority, particularly in the United States, Great Britain, and France, and a mechanism was established to promote a higher level of defence within the control process. Additionally, the transfer of items listed under the Munitions Control List was regulated by the Arms Export Control Act of 1976, which required all foreign governments receiving weapons from the United States to use them exclusively for legitimate self-defence purposes.

The impending détente in US-Soviet relations in 1985 did not lead to a large-scale relaxation of export controls. However, the improved diplomatic climate made it politically less contentious to support the deregulation of less sensitive items. In 1987, USA officials lifted restrictions on energy technology and equipment, which had been imposed on the Soviet Union nearly a decade earlier for foreign policy reasons.[35] Nonetheless, as of the spring of 1992, nearly all space-related hardware was still included in the USA Munition List and regulated by the Arms Export Control Act and its International Traffic in Arms Regulations (ITAR).[36] These protectionist measures were intended to benefit both USA national security and the nascent private launch industry.[37]

The control of technology transfer remained a significant obstacle, not only to commercial relations but also to purely scientific cooperation. During the negotiations preceding the signing of the US-Soviet Treaty on Scientific Cooperation in 1987, one of the most challenging issues for negotiators was technology transfer, driven by concerns that advanced

[35] Mastanduno. (1992), 277–301.

[36] See more in Daniels, M., Krige, J. (2022). *Knowledge regulation and national security in postwar America*. The University of Chicago press.

[37] Long Callahan A. (2013). Russian-American cooperation in space: privatization, remuneration, and collective security. In Krige, Long Callahan, Maharaj. (Eds.), 156.

sensors, computers, and other scientific technologies would inevitably be diverted to the Soviet military. As a result, the agreement signed on April 15, 1987, was notably vague in its provisions, primarily calling for the exchange of data between various missions and the coordination of mission schedules. It is enough to analyse the Agreed List of Cooperative Projects[38] at the end of the Agreement. Half of the envisaged projects were related to the exchange of scientific data, while the others were related to the coordination of activities in areas of interest to both sides. It carefully avoided proposing that researchers from one side install instruments and hardware on the other side's spacecraft and even with restrictions, each mission proposed for cooperation would first have to pass through a painstaking process of interagency review.[39]

One notable example is the case of Frederick L. Scarf, a space plasma scientist, and his plasma wave instrument, which he intended to place on the 1988 Soviet *Phobos* spacecraft. The USA equipment, developed between 1974 and 1976 by Scarf, a scientist at TRW[40] and a member of UCLA's Institute of Geophysics and Planetary Physics, was initially designated as a backup for NASA's Voyager interplanetary mission. Regrettably for Scarf, the authorisation for his instrument's involvement in the Soviet *Phobos* project was not promptly granted by the American authorities and the Soviets turned to other international sources and arranged for new, more sophisticated instruments to be built in an effort involving ESA, Poland, Czechoslovakia, and the Soviets themselves. Scarf subsequently applied to be a co-investigator on this instrument without transfer of hardware, and he received USA approval as well as NASA funding for this activity.[41]

Laws restricting the transfer of technology to the USSR sometimes extended beyond reasonable limits. A notable example is the case of the Topaz II nuclear reactor, which the Soviets showcased at the Symposium on Space Nuclear Power Systems in January 1991 in Albuquerque, New Mexico. The Topaz II, brought to the USA by a Soviet delegation, was a non-operational model of a reactor that Moscow sought to sell to an

[38] Retrieved March 4, 2025 from https://aerospace.org/sites/default/files/policy_archives/Cooperation%20in%20Peaceful%20Use%20of%20Space%20-%20USSR%20Apr87.pdf

[39] Technology transfer is the issue in space cooperation. *Science*, 236, June 12, 1987.

[40] Thompson Ramo Wooldridge Inc. (1901–2002), U.S. Corporation operating in aerospace, electronics, automotive and credit reporting.

[41] *Aviation week*, October 12, 1987.

American scientific consortium, which was partially funded by the USA military. Following a successful promotional tour in the US, complications arose concerning the return of Topaz II to the USSR. The American Nuclear Regulatory Commission (NRC) declared that the compact Soviet reactor, designed for use in space, could not be returned until the Soviets obtained an official export licence. The NRC further said that it was not permitted by law to grant such a license, so this meant that the nuclear reactor was trapped in a legal limbo, as property of the Soviet Union that must remain in the United States.[42] It took five months of wrangling and pressure from both sides, including the intervention of New Mexico Senator Pete Domenici, to define the Topaz II exhibit as a non-"nuclear utilisation facility" so it could be sent home.[43]

Soviet Strategies in Overcoming Export Controls

The Soviet leadership recognised the urgent need to guarantee the "technological security" of Western equipment and implemented extensive measures to reassure their Western partners in this regard.

During a discussion on the obstacles to the commercialisation of space products, Boris Maiorsky, head of the Department of Peaceful Uses of Nuclear Energy and Outer Space at the Ministry of Foreign Affairs, expressed his frustration to USA scientist James Head over restrictions imposed by the United States. He criticised the ban on shipping spacecraft or instruments containing USA components to the Soviet Union, calling it unfair—especially given the "state-approved guarantee" and the personal assurance of Nikolay Ryzhkov that the equipment's integrity would remain intact. Maiorsky denounced these restrictions as "handcuffing Glavkosmos" and amounting to blackmail.[44]

The primary challenge confronting Glavkosmos centred on addressing the underlying issue, namely the pervasive risk of industrial espionage. At this point, it is intriguing to reflect on the attendance of Aleksandr Dunaev at the Space Future Forum held in Moscow in 1987 and his replies to key questions. Most of his speeches addressed various areas of cooperation on

[42] Smith, R.J. (1991). U.S. won't let Soviets take reactor back home. *The Washington post*, April 20.

[43] Dabrowski, R. (2013). U.S.–Russian Cooperation in Science and Technology: A Case Study of the Topaz Space-Based Nuclear Reactor International Program. *Connections The Quarterly Journal* 13(1), 71–87.

[44] Head. (1987). Part II.

mutually beneficial terms. First, he spoke about the leasing of communications satellites, the offer of remote sensing capabilities, and the launch of foreign satellites on a variety of rockets. Moreover, he tried to promote the Soviet Proton on the launch vehicle market. To sponsor his "product", Dunaev emphasised that "the Soviet boosters were quite reliable and would be able to put a wide range of payloads into any orbit". It is noticeable that he also added that the Soviets would "guarantee total protection of technical secrets" that may be on the satellites of other countries. In the same talk, Dunaev divulged a proposal to create an "international launch centre in order to exercise the maximum potential of Soviet (rocket) technology".[45]

Confirming their vision of commercial cooperation, during the promotion campaign organised in the Western countries in search of future partners, the Soviets also insisted on giving the guarantee that there would be no technology transfers when shipping a foreign satellite to the USSR. The governance of Space Commerce Corporation (SCC), which served as the American intermediary and consultants (private American firms had less doubt about cooperation with Soviet space enterprises than federal authorities, see below), also stressed this point. They assured that foreign satellites would not be inspected by customs or any other authorities and that they could be continuously escorted by Western technicians, who would independently handle their integration onto the launcher without Soviet intervention. The Soviets were also ready to provide all relevant information concerning their launchers but did not go so far as to granting visits to manufacturing plants that were military establishments.[46] To allay the fears of Western governments, the Soviets promised that foreign satellites would be left sealed until launch, thus ensuring that no secrets of advanced technology would be stolen.

Ryzhkov also underscored the significance of "technology protection" during the mentioned discussions at the Forum in October 1987. Remarkably, it was the sole aspect that elicited a positive response from the European partners—an assurance to uphold the security of advanced technologies and knowledge owned by Western partners. As Jean Arets affirmed, it was "the important step in the new Russian policy for

[45] Dunaev stresses cost effectiveness of Soviet space technology. *Aerospace Daily*, October 20, 1987.

[46] Langereux, P. (1987). Glavcosmos propose sept lanceurs aux utilisateurs étrangers. *Air&Cosmos*, 1148, Juin 20.

providing guarantees for unwanted transfer of technology".[47] This admission was an exception, as the Soviet leadership's increasing focus on the commercialisation of the space industry generally provoked serious concern among European and American counterparts.

Despite these definitive strides towards transparent commercial collaboration under the auspices of *perestroika*, some Western scientists and space managers persisted in voicing their frustration with what they perceived as Soviet managerial clumsiness, citing the persistence of various "vestiges" of the still unsuccessful approaches. Wolfgang Pietsch of West Germany's Max Planck Institute for Extra-Terrestrial Physics, whose team had an X-ray detector aboard the Kvant module, first learned that the experiment had finally been launched when he heard the announcement on an East German radio broadcast. Similarly, Andrea Caruso, head of Europe's Eutelsat cartel, expressed frustration with Soviet business practices, stating: "From a business point of view, they still have a lot to learn. I keep corresponding with them, and they keep sending me back telexes in Russian. It is a disaster".[48]

The assurance of confidentiality for Western technology was just one aspect of the Soviet Union's new approach to the international space market. Another key element was the diverse range of products and services provided by the Soviet rocket and space industry. In this regard, Soviet leadership, scientists, and technical developers took pride in showcasing both the quantity and quality of their offerings.

Conclusion

The establishment of Glavkosmos coincided with the onset of *perestroika*, marking a new phase in the development of the Soviet space sector. Previously, the space field had been entirely dependent on substantial state subsidies. With economic reforms underway, it was then expected to transition towards self-sufficiency, in line with other sectors of the economy. Glavkosmos was envisioned as the driving force behind this transformation, gradually enabling the space sector to sustain itself and contributing to the national budget.

[47] Arets, J. *Mission to U.S.S.R.*, January 16, 1987. In ESA-28023, HAEU.

[48] Lemonick, M. (1987). Surging Ahead, the Soviets overtake the U.S. as the No. 1 spacefaring nation. *Time*, October 5.

Such a revolutionary shift within the framework of a planned economy inevitably provoked objections, criticism, and internal contradictions within the sector. The mere fact that Glavkosmos was established within MOM and staffed primarily by personnel from the VPK led to scepticism and distrust within the scientific space community. Many viewed Glavkosmos as a potential competitor rather than a collaborative entity. At the same time, international observers highlighted the positive aspects of the new institution, seeing it as a means to unify various branches of the Soviet space field and foster closer cooperation with foreign partners in the commercial utilisation of space infrastructures.

Glavkosmos also faced external challenges, particularly the restrictions imposed by the Coordinating Committee for Multilateral Export Controls. The Soviet space and political leadership had to employ extensive diplomatic efforts—at times even appealing to wounded national pride—to reassure potential partners that the USSR had no intention of engaging in technological or scientific espionage under the guise of commercial cooperation. The extent to which these efforts succeeded in convincing Western clients of the Soviet Union's good intentions will be explored in the following chapters, where we will analyse the outcomes of Glavkosmos' activities during its years of operation in the USSR.

CHAPTER 10

The Soviet Space Sector's New Offerings

The shift from a planned economy to a gradual market-based transition had a profound impact on the space sector. The *perestroika* era was not only a time of significant scientific advancements and military achievements, but also marked the Soviet Union's first concerted effort to enter the global space market. This chapter explores the early initiatives of the Soviet military-industrial leadership as they worked to establish an international presence by offering space-related goods and services. It examines the diverse range of Soviet contributions and the strategies employed by these "pioneers" to promote their capabilities in the international space market. Operating in an unfamiliar and rapidly evolving environment, they had to adapt in real time, acquiring essential knowledge and experience to achieve their objectives.

The principal Soviet commodity on offer was rockets. Their quantity, variety, and reliability promised a rosy future for the Soviet rocket and space leadership. Foreign partners were invited to use Soviet rockets to launch their payloads (satellites of various purposes) into Earth's orbit. Rockets were not the only commodities available for sale; Soviet satellites were also offered, along with the accommodation of scientific equipment on board. The Mir orbital station was also at the centre of attention for Soviet officials selling off space infrastructure. Available for purchase were "tickets to Mir", services for conducting various experiments on the

O. Dubrovina, *The Soviet Space Program and International Collaboration during Perestroika, 1985–1991*, Palgrave Studies in the History of Science and Technology,
https://doi.org/10.1007/978-3-032-03157-0_10

scientific Kvant orbital module, and the opportunity to launch scientific instruments from other countries into outer space. Secret military hardware was gradually transitioned to civilian use and employed for the interests of the national economy and commercial profit acquisition. This primarily involved the transformation of former reconnaissance satellites into remote sensing instruments capable of capturing high-quality images of the Earth's surface. The Energya rocket also served as an example: originally designed for military purposes, it underwent "conversion" and emerged as one of the most prominently advertised products in the Soviet market. Moreover, in the domains of space medicine and materials science, the USSR held an indisputable advantage.[1]

As the country's economic situation worsened and state funding for the space sector plummeted, the Soviet Union's space-related goods and services took on increasingly unconventional forms. These ranged from a contract to conduct zero-gravity experiments with Coca-Cola to rumours—circulated in the USA press—about the potential sale of the Mir orbital station to American clients. Alongside these extreme measures, the Soviet Union expanded its presence in the international market by diversifying its offerings and strategically targeting customers across different regions. Notably, successful efforts were made to attract buyers from developing countries, further broadening the reach of Soviet space commerce.

Rockets for Sale

In the 1980s, the Soviet Union paved the way for commercial space ventures. With valuable contributions to the international space market, Glavkosmos aimed to promote Soviet manned and unmanned launch capabilities at competitive prices. In the mid-1980s, the USSR could offer no fewer than seven conventional rockets capable of launching Earth-orbiting payloads ranging from 450 kg to 21 tons. In addition to the renowned Proton and Soyuz launchers, Glavkosmos also offered the Vertikal geophysical sounding rocket, Cosmos, Vostok, and Molnya rockets, as well as the three-stage Tsyklon vehicle. All these launchers were well-tested and had been in use for 20–30 years. By the late 1980s, the most recent rockets available for foreign satellite launches included the

[1] Postyshev, V. (1989). Mezhdunarodny kosmicheskiy rynok. *Zemlya i Vselennaya*, 4, 21–27.

Zenit (commissioned in 1985) and the new giant Energya, which had its maiden flight on May 15, 1987.[2]

The reliability of the Soviet launch vehicles was discussed at length in the Western press. The most reliable rockets were respectively Energya and Zenit (100%), but they were still not widely used, so the statistic was not yet significant. Next were ranged Vostok (98.8%), Soyuz (97.9%), Tsyklon (97.3%), Cosmos (95.8%), Molnya (94.8%), and finally Proton (92.4%), which was the least reliable of the Soviet rockets available.[3]

In the 1980s, the Soviet Union surpassed the USA both in the number of objects launched into space and in overall launches.[4] The Soviets certainly had a broad range of launch opportunities and operational experience. Moreover, thanks to the system of planned economy and direct state intervention in the pricing process, they were able to strategically manipulate launch costs, posing a significant threat to American and European competitors. Experts attributed the frequent launches to the need to renew the Soviet satellite fleet, as their operational lifespan was shorter than that of their American counterparts. The financial aspect was also a factor; the overall cost of launches in the Soviet Union was lower than in the US, and state financing of the space sector was not fully accounted for until nearly the very end of the Soviet era.

Indeed, the Soviet planned economy pulled the strings on launch prices, leaving Western competitors far behind. Western newspapers gave various examples of Soviet dumping prices. For example, in 1986 a satellite could be put into orbit for as little as $25 million—a quarter of Western rates. According to *Aviation Week & Space Technology*'s information, the Soviets were quoting launch prices of about $30 million for a 4000-lb spacecraft to geosynchronous orbit—a rate more than half of what USA commercial booster companies or Europe's Arianespace could offer.[5] The French *Air & Cosmos* quoting Aleksandr Dunaev cited the cost of launching a payload of 7 tonnes on the Vostok and Soyuz vehicles equal to $10–14 million. By comparison, according to the NSDD-181 (National

[2] Langereux. (1987).

[3] Langereux, P. (1989). Statistiques inédites sur les fusées russes. *Air&Cosmos* 1240, Juin 3.

[4] Graphs of Annual number of objects launched into space from United Nations Office for Outer Space Affairs are available https://ourworldindata.org/grapher/yearly-number-of-objects-launched-into-outer-space?time=earliest 1997 and of total launches by country https://aerospace.csis.org/data/space-environment-total-launches-by-country/.

[5] Proton Marketing team finds U.S. interest, opposition. *Aviation week and Space technology*. May 25, 1987.

Security Decision Directive) issued on July 30, 1985, titled "Shuttle Pricing for Foreign and Commercial Users", the minimum acceptable bid for a Shuttle launch was set at $74 million.[6] Western experts agreed that Soviet pricing undermined the international rocket market; indeed, the WEU (Western European Union) recommended that Western users boycott Soviet rockets.

Market forces also exerted their influence on Soviet entrepreneurs, who, by the end of that period, were compelled to raise prices for their services. In 1989 at the Paris Air Show, Glavkosmos unveiled its new commercial rates for the rental and sale of Soviet space equipment and services. Even with the rise in prices compared to previous years, Soviet launchers remained considerably economical in absolute terms, nearly half the cost of Western launchers. However, the seemingly low price of the service did not consistently align with high quality. According to the American press, Glavkosmos charged the annual rental of a *Gorizont* telecommunications satellite at just $1.5 million. But the cost of buying the satellite, including launch (by Proton) was between $35 and $40 million. USA experts pointed out that it was too expensive, given the satellite's only modest performance. Weighing four times as much (2150 kg), with four times fewer receivers (eight) and a shorter lifespan (three years), *Gorizont* was deemed unattractive by Western specialists compared to the American *HS 376*, for example.[7]

The USA private space launch industry experienced significant growth in 1986 when President Ronald Reagan issued an order prohibiting the Space Shuttle from carrying commercial payloads.[8] Additionally, USA analysts predicted that the Soviet Union would be unsuccessful in its attempts to market rockets in the United States. Two primary reasons were typically cited. Firstly, the concern revolved around the threat of technology leakage. As argued by Peter Zimmerman, former senior analyst at the Carnegie Endowment for International Peace in Washington, "space is not an arena where free trade can happen". Secondly, engaging in business negotiations

[6] Logsdon, J. (2019). *Ronald Reagan and the space frontier*. Palgrave, 264.

[7] On the NASA official website, the Gorizont's technical specifications look more optimistic. The Gorizont spacecraft possessed an initial mass in excess of 2.1 metric tons and have demonstrated a lifetime of nearly 10 years, although a 5-year service life was more common. Seven separate transmission antennas allowed a variety of reception patterns for both broad and localised terrestrial regions. Retrieved March 4, 2025 from https://nssdc.gsfc.nasa.gov/nmc/spacecraft/display.action?id=1978-118A.

[8] Logsdon. (2019), 310–311.

with the Soviet space industry was viewed as sending a "bad signal to the fledgling commercial space industry in the US", according to John Logsdon, a Soviet space analyst at George Washington University at that time.[9]

While Dunaev asserted that Western organisations and firms had shown interest in the Soviet space industry, he lamented that "it often amounted to nothing more than a mere expression of interest on their part". He attributed this limitation to certain factors that, in his view, were beyond the control of the Soviet side. As the head of Glavkosmos complained,

> according to a discriminatory law passed in the United States, the import of space technology components manufactured in the United States was prohibited into the territory of the USSR. These components were necessarily present in almost all objects intended for launching of Soviet potential partners into space, although the Soviet side gave full guarantees of their safety during transportation on the territory of the USSR. But obstacles remained, despite the appeal to the USA State Department of our potential customers.[10]

Soviet representatives of Glavkosmos and other government agencies made no secret of their outrage at the implied blocking of their actions by the USA administration. According to the USA press, even Mikhail Gorbachev had raised such issues at summit meetings, but there was no serious relaxation of USA policy in that direction before the collapse of the USSR. Until 1991, no Western satellite had been launched by a Soviet rocket. Nevertheless, during the same period, the Soviet fleet remained valuable for commercial transactions, as exemplified in the case of Indian satellites.

The USSR sought to expand into the international market not only by diversifying its goods and services but also by targeting a geographically diverse customer base. As a non-member of CoCom, India experienced much fewer impacts from these USA restrictive measures against commercial cooperation in space technology. Notably, the carrier rocket Vostok marked a shift for the commercial use of Soviet rockets, launching foreign satellites not in Europe but in Asia. In 1988, Vostok successfully delivered

[9] McClintock, P. (1989). NASA says nyet to Soviet offer to launch the U.S. space station. *Washington times*, May 12.

[10] Interview with Dunaev, A. (1987). Kosmos dlya narodnogo khozyaystva otkryl novyye gorizonty, kotorykh ran'she u nas ne bylo. Orbity "Glavkosmosa". *Argumenty i fakty*, April 4.

the Indian satellite IRS into a sun-synchronous orbit for the study of natural resources. India paid $7.5 million for this service.[11] In August 1991, the satellite IRS-1B was also successfully launched from Baikonur. These were the fifth and sixth launches of Indian satellites by Soviet rockets, but with one very important difference compared with the previous four launches: for the first time, the USSR launched Indian rockets on a commercial basis. As succinctly reported in the newspaper *Novosti kosmonavtiki*, founded shortly before the collapse of the USSR: "The work to prepare for the launch of the satellite was carried out by the NPO Lavochkin in cooperation with other enterprises of the MOM in accordance with a commercial agreement concluded in 1988 by the USSR Glavkosmos and ISRO through the Licensintorg. The Indian scientists-constructors of the IRS-1B satellite conducted a final test of the satellite at the cosmodrome".[12] However, Soviet officials did not abandon their hopes of penetrating the Western market. Another endeavour involved the commercial promotion of Energya, which was the most powerful launch vehicle at that time.

Energya: Seeking New Horizons in the Space Market

In 1987, the year of Energya's successful test flight, Dunaev presented a new "space product" to the international community, advertising its capabilities for multiple purposes and validated the use of Energya for the construction of the future Mir-2 orbital station: "Space stations weigh up to several dozen tons. What is needed are stations that weigh several hundred tons. We should soon learn to build big structures out there, not tens of metres but kilometres across multifunctional platforms". Clearly, the military purposes of the super missile were not disclosed by the Soviet leadership. In spite of public statements by the Soviets regarding the already planned applications of Energya, a frenetic search for users had already commenced to offset the substantial costs associated with its production and operation. With a two-year hiatus in Shuttle launches since 1986, uncertainties surrounding private launches in the US, and the introduction of a powerful rocket like Energya into the international market, American companies were inevitably intrigued, especially in light of the orbital station project initiated by President Reagan.

[11] Evich. (1991).

[12] *Novosti kosmonavtiki*, 3, 1991.

The initial enthusiasm from potential customers swiftly waned due to the challenge of competition. The arrival of Soviet launchers into the rockets market for delivering international commercial satellites into orbit seriously alarmed both European and American manufacturers. The competition throughout the 1980s had gradually advanced the interests of many: the American private launcher manufacturers; NASA as a main Shuttle launches provider; Arianespace with the European Ariane rocket (launch prices varied only slightly per kilogram compared to American competitors), but also Chinese and Soviet launcher constructors. The intricacy of the situation for the European side was underscored by Guy Collins, the author of "Europe in Space", in his groundbreaking study on the history of ESA. He specified that the primary challenge for Arianespace was competition from the Soviet Union and China, which could offer launch services at significantly lower commercial prices because, as he argued, they operated on a non-profit basis.[13] Indeed, this assertion holds merit, especially during the *pre-perestroika* period. A notable instance occurred during negotiations with ESA in the late 1970s for the launch of a European satellite, Marecs-C, using a Soviet rocket carrier. At that time, the Soviets offered the Proton launcher at a significantly lower cost compared to their American counterparts. Though, it must be noted that for Soviet space authorities this was more a matter of international prestige—a first attempt to break the American monopoly and gain a Western business partner. During *perestroika* the economic crisis in the rocket and space sector became more pronounced, compelling the leadership to seek an escape route. Primarily, the focus shifted towards transitioning to market relations to salvage the economy and, more straightforwardly, to supplement the national budget.

A question emerged regarding the commercial utilisation of Energya, not solely to benefit domestic industry (which lacked the financial resources to bear the cost of constructing and operating the world's heaviest rocket) but also to cater to international demand. In fact, after a successful test flight in 1987 of the Energya super-class launch vehicle, its engineers faced the question of its utility and the payback of its production. As Boris Chertok stated: "Intensely, we sought payloads for it. We developed the most intriguing and promising projects, the realisation of which held the potential for groundbreaking achievements in fundamental astrophysical

[13] Collins. (1990), 68.

research, global communication systems, informatisation, and monitoring in the pursuit of national economic and security interests".[14]

In a document detailing the technical characteristics of Energya, Boris Gubanov also gave a list of possible applications for the vehicle, not only for the needs of individual countries, but also for international cooperation.

> The Energya launch vehicle capabilities to put heavy spacecraft into geostationary orbit made it possible to construct heavy platforms for placing global communication systems, Earth ecology and international agreement control systems. The first estimates showed that heavy communication platforms can provide general telephone communication, not only for our country but for other countries concerned as well. Powerful large antennas of these platforms would have substantially simplified ground communication means and make them available for the majority of users including distant regions of the planet.[15]

Gubanov also underscored the potential utilisation of Energya as an auxiliary tool for establishing a mechanism for international collaboration in outer space within the framework of global institutions. As he articulated,

> ...heavy information platforms make it possible to create a world ecological monitoring system for large-scale and long-term observation of the Earth climate, nature processes on the continents and in the oceans, the state of forest, pasture and crop areas, solid erosion, air contamination and Earth resources. Such ecological platform could become "common" platform for wide cooperation of all the countries which are concerned about ecology of their region and planet as a whole.[16]

Thereafter, some of the projects proposed by Gubanov received initial development, but for various reasons they did not come to fruition. Such an example would be the Universal Space Platform which was designed by an NPO Energya headed by its General Designer Yuri Semenov. In 1989, Boris Chertok personally endeavoured to convince French specialists and officials to participate in the development of a global communication system, utilising a heavy Universal Space Platform weighing 18 tons, a payload that only the Energya rocket could launch into geostationary orbit.

[14] Chertok, B. (1999). *Rakety i lyudi. Lunnaya gonka*. Mashonostroyenie. Vol. 4., 368.
[15] Gubanov, B. (1989). *U.S.S.R. space transport systems*. In ESA-28130, HAEU.
[16] Ibid.

However, the French ultimately abandoned this costly venture due to a lack of necessary funds. At that time, the "prevailing interests of France were effectively focused on their own rocket, Ariane".[17] The German corporation Bosch had also shown interest in the Universal Space Platform but even they, as Chertok reported, did not dare to invest in the project which, according to a very optimistic business plan, would begin to make a profit after only 5 years.

During the same period, under the direction of Soviet astronomer Nikolay Kardashev, NPO Energya developed a project for a space radio interferometer to study the universe, in collaboration with the Academy of Sciences. Only the Energya rocket would have been capable of launching this spacecraft, equipped with a parabolic antenna measuring 25 metres in diameter, into orbit. This mission was supposed to be implemented in cooperation with ESA but neither the Soviets nor the Europeans were able to find funding for such an expensive project.[18]

The erstwhile political and ideological adversary was also engaged in exploring practical applications of Energya. Amidst the transformations brought about by *perestroika*, few in the USSR were astonished by the involvement of USA experts from NASA who were invited to discuss potential options. In this regard, from 6–14 July 1988, IKI organised an international seminar on long-term studies for the prospects of Mars exploration, and delegates from Western countries were invited to participate. During the seminar, Roald Sagdeev briefed the participants on the long-term Soviet program for Mars studies in the 1990s and 2000s. Scientific objectives were discussed as well as possible options of the *Mars-94* projects, depending on whether a Proton or Energya carrier-rockets would be used.[19] The new and candid climate surprised the American scientists whose opinions were welcomed by their Soviet colleagues. *Aviation Week & Space Technology* quoted "one veteran USA planetary scientist": "It was an amazing example of the Soviet's new *glasnost* policy. We were urged to come to the seminar's working group sessions and toss out our ideas—which were listened to. Can you imagine

[17] Chertok. (1999), 368.

[18] Ibid., 369.

[19] International Seminar on the Long-term studies of Mars. Final document, July 6–14, 1988. In ESA-28051 (3), HAEU.

NASA inviting the Soviets to do the same during a planning meeting in Washington for one of our proposed new missions starts?"[20]

Foreign scientists were asked to share their ideas about the possible use of Energya. They suggested using the Energya heavy-lift vehicle to place a multipurpose satellite network around Mars in support of Martian rovers, atmospheric balloons, and other systems operating at the planet. As Thomas Heinsheimer, vice president of California's Titan System, commented: "The satellite network launched by Energya could be used as a science support system for all missions to Mars and its moons. Such a network could serve as a high-bit-rate relay for transmitting scientific data to Earth, providing navigation and positioning information on Mars, and updating data compression algorithms for surface rovers and landers".[21]

Another notable attempt to secure a buyer for Energya was a proposal to provide launch services for the construction of NASA's Freedom International Orbital Station. This event was particularly revealing, as it highlighted the Soviet Union's efforts to integrate its space capabilities into Western-led projects amid ongoing economic difficulties. The Soviet Union proposed that NASA would pay Soviet rockets to launch Freedom through its intermediator, the Space Commerce Corp, mentioned below. Soviet intentions were also evidenced by statements made in May 1989 at the American Institute of Aeronautics and Astronautics (AIAA) in Washington D.C. Among several Soviet space officials present at that meeting was Boris Gubanov, who declared that "the Soviet Union is ready to engage in direct negotiations in the commercial sphere",[22] clearly hinting that his brainchild was also available for commercial purposes to both private and public USA clients. Gubanov even added that the space shuttle Buran had a larger cargo bay than the USA Space Shuttle for carrying commercial payloads. NASA, however, was not eager to rely on the Soviet space program for launching services. Don Miller, senior Soviet Desk officer at NASA declared, "we are going to wish them well, but we are not in the same business. We are aware of their interest to launch the space station".[23]

[20] Energya-Based Mars mission plan proposed by Western specialists. *Aviation week & space technology*, August 29, 1988.

[21] Ibid.

[22] Press excerpts. NASA HQ Archives. 15,583.

[23] Ibid.

Hence, notwithstanding earnest endeavours, Soviet producers of the Energya system failed to identify a viable application for it. Yuri Mozzhorin, a participant in the Energya-Buran project, highlighted two primary factors that contributed to the lack of demand for the Energya-Buran system: the dissolution of the primary customer, the Soviet VPK, and the absence of adequate funding to promote the system in the international market. "With the end of the Cold War, the disappearance of a primary adversary, and the waning competitive fervour that characterised the space race, the driving force behind scientific and technological advancement dissipated. Consequently, the reusable space system Energya-Buran found itself bereft of its *raison d'être*—the primary payloads for which it was originally conceived. Faced with this dilemma, the prospective customer balked at shouldering the financial burden for a product lacking immediate utility. Despite efforts to identify scientific or national economic applications for this remarkable yet costly system, success proved elusive as its original purpose remained unfulfilled".[24]

Despite setbacks in the commercial deployment of rockets, the Soviet leadership remained determined to attract foreign partners to its space initiatives. While rockets proved to be a challenging product to market, the Mir orbital station possessed a crucial advantage: its uniqueness. No other space endeavour could match Mir in terms of originality, as well as its profound political and social significance.

Mir Space Station: Was It Really for Sale?

Western journalists sustained that it was in the desperate rush to find the means for the survival of the space program, that the Soviet space sector leadership went ahead with the sale of the Mir station. French weekly, *L'Express*, announced that representatives of the Soviet "space leadership" unofficially approached Americans with an offer to sell or lease several key elements of the Soviet space program. Powerful engines, a satellite control centre, the Soyuz spacecraft, and even the Mir-2 orbital station, which was planned to replace the Mir-1 station, were included in the list of items for sale.

Indeed, in early September 1991, the news leaked to the American press that the Soviets were ready to sell Mir. "...last week the Soviets announced the unthinkable. They are considering selling the station

[24] Mozzhorin. (2000), 344.

altogether. For an estimated $600 million to $700 million, some lucky buyer could have a slightly used 100-foot-long space station 240 miles out in space. Considering NASA's plan to spend $30 billion to build a similar model, the Mir offer looks like a bargain".[25] The alleged Soviet proposal stirred up the USA space community and sparked a debate about the need to buy a Soviet orbital station. The main argument against the Mir purchase was the need to develop America's own technology and preserve the jobs created by the Freedom Project. The many arguments in its favour were economic (saving the national budget by using technologies already created in the USSR), practical (using the Soviet experience in long-term stay in space), and political (supporting democratic processes in the USSR itself).

Another issue of particular concern to USA space experts was the issue of ownership. Indeed, on the eve of the collapse of the USSR, in addition to the problems of financing and maintaining production ties in the crumbling union, issues of ownership of rocket and space technology were naturally aggravated. The most high-profile example was the declaration by Kazakh President Nursultan Nazarbayev, who proclaimed the Baikonur cosmodrome as the property of the Republic of Kazakhstan. Similarly, Yuri Semenov asserted that the Mir orbital complex belonged to the NPO Energya work collective,[26] as it designed and produced the station. In fact, an important question for possible buyers of the entire Mir station, its individual modules or on-board services, was the question of its ownership and who had the authority to sell it.

In her written report to the USA Congress, Marcia Smith, a specialist in rocket and space policy from the Science Policy Research Division, drew attention to the distinction between the existing Mir and the proposed new Mir. According to her, the ownership of the existing Mir complex was unclear. Glavkosmos was reportedly responsible for the overall operations of Mir, but the station was owned by the MOM, whose future remained uncertain. Smith expressed her doubts for the validity of an agreement signed with Glavkosmos in a month from then, or even a year. In fact, Yuri Semenov denied that Mir was for sale, and asserted that Glavkosmos had no right to sell it. He argued that Mir should be an international laboratory and any agreements to that effect must be signed with his organisation. Then came the question of ownership by the residual Soviet central

[25] A space station of one's own. *New York times*, 8 September 1991.

[26] Tarasenko, M. (1992). *Voennye aspecty sovetskoy kosmonavtiki*. TOO "Nikol'".

government (or by the Russian republic), where Glavkosmos, NPO Energya, and Khrunichev (modules' manufacturer) were located.[27]

During that period in the United States, the individual most closely connected to the Soviet space industry was Arthur (Art) Dula, a prominent figure in the USA business sector who actively sought to establish commercial relations with the Soviet space infrastructure. A Texas-based lawyer, Dula founded SCC in Houston to promote and market Soviet rocket services. Dula harboured scepticism regarding the likelihood of Soviet authorities taking such an unprecedented step as selling the Mir space station. He held the belief that Yeltsin, at that time, would not endeavour to dismantle the Soviet space program: "If it offers a way to make hard currency, I suspect it will be supported".[28]

Finally, Reuters reported on October 30, 1991, that NASA had refused to buy Mir, preferring to develop its own Freedom system. However, the news of the Mir sale reverberated throughout the Soviet Union. According to the announcement released by the Moscow Associated Press, one of two cosmonauts aboard Mir "plaintively asked in early December 1991 on a live TV broadcast whether the dramatic changes in his homeland while he drifts weightless in space meant that the Mir space is to be sold together with them".[29] The rumours were definitively denied by Semenov. He declared that the space station would remain the property of his collective (Fig. 10.1).

Journalists from *Komsomolskaya Pravda* sought answers by putting their questions about the deal to Glavkosmos and to NPO Energya. Glavkosmos refused to comment on the situation, only stating that if it was a question of deals with Americans, it was for a joint use of Soviet complex. Officials from NPO Energya announced that nothing had been sold to the Americans and indicated that Glavkosmos had leaked incorrect information. The article concluded with journalists expressing doubts about the statements made by both organisations. Considered that Glavkosmos and NPO Energya were "the most secretive organisations in the country, their belated denials seemed rather strange".[30] This

[27] Smith M.S. (1991). *Soviet space commercialization: selling the Mir space station*. CRS Report for Congress, 30 September 30.

[28] Broad, W. (1991). Almost broke, Soviet Union's space efforts go on sale. *New-York Times*, September 3.

[29] AP Moscow, 9 December, 1991.

[30] *Space authorities deny rumoured U.S. deal*, in Komsomolskaya pravda, 9 November 1991. NASA Archives.

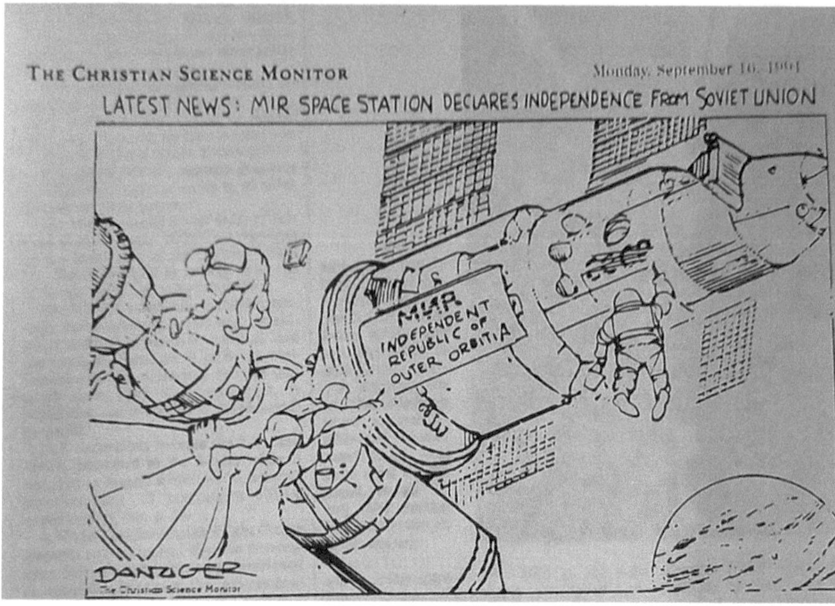

Fig. 10.1 Cartoon from *The Christian Science Monitor*, September 16, 1991. Source: NASA HQ Archives

observation came against the backdrop of heated deliberations in the USA Congress and within NASA regarding the potential acquisition of the Mir space station. The authors implied that the delayed and seemingly reluctant denials may evoke suspicion, echoing the journalistic maxim that where there is smoke, there may be fire.

Yuri Semenov contributed a personal statement on the subject, as documented in the *Novosti Kosmonavtiki* article titled "Selling the Mir Orbital Station: Myth or Reality?" In response to the question posed in the article, Semenov conveyed that he became aware of the matter through radio reports while commuting to his workplace. "There is no real reason for this message. The information about the cancellation of another expedition and closing of Mir, as well as about its sale or lease to foreign entrepreneurs or organisations does not correspond to reality. The Chief

Designer's Council decided on September 5, 1991, that the Mir complex is to be fully exploited".[31]

Contentious debates surrounding the sale of Mir within both the Soviet Union and the USA underscored the gravity of the matter. However, the sale of "tickets to the orbital station" represented in a sense a viable solution. This approach allowed the Soviet leadership to maintain its political prestige while simultaneously generating substantial revenue from the venture.

Tickets to Mir: A Breakthrough Into the International Space Market

Amid unsuccessful attempts to launch Western satellites with domestic launch vehicles, the Soviet leadership sought to compensate by offering "space tickets" to the Mir orbital station. This unique worldwide service at the time positioned the Soviet space infrastructure beyond competition. Long-duration flights attracted foreign partners interested in conducting various experiments in weightlessness. Additionally, the matter of national prestige played a significant role, particularly when a representative of a specific country first opened the door to space for his compatriots. Even if it was on a Soviet rocket and aboard a Soviet station.

On December 2, 1990, the Soyuz TM-11 spaceship lifted off from Baikonur, carrying to the outer space an unusual passenger—Japanese citizen Toyohira Akiyama, who later went down in history as the first Japanese citizen, the first journalist to have been in space and also the first "private" paying guest to Mir. The deal to send a Japanese TV reporter into space on board a Soviet rocket netted Glavkosmos (according to evidence from the time) GBP £6.7 million ($11.9 million dollars[32]) paid by the Japanese news company TBS.[33] More plausible data is given by Yuri Baturin in the encyclopaedic edition *World Manned Spaceflight*. As he states, "the most ambitious commercial campaign of Japanese television cost $37 million. Half of the sum was provided by the sponsors, Sony and Minolta corporations, and the rest was recouped through advertising".[34] In any case, it was far less than the costs incurred by NASDA to fly Japanese astronauts on

[31] Novosti kosmonavtiki, 3, 1991, 8.
[32] At 1989 exchange rates.
[33] Postyshev. (1989), 21–27.
[34] Baturin (Ed.), (2005), 682.

the USA Shuttle.[35] Toyohira Akiyama returned from his 8-day visit to Mir, thus demonstrating that the Soviets were indeed taking paying customers into space.[36] However, the accomplishment, while garnering considerable attention in the international arena, generated much less enthusiasm domestically.

The approval of Glavkosmos on March 29, 1989, for a Japanese journalist to go on board a Soviet spacecraft to Mir struck a chord with the Soviet journalists' ambition as it would be (and was) the first spaceflight by a professional journalist. Curiously, Soviet journalists missed the first ever opportunity to fly in space back in 1965, when Sergei Korolev was still alive, and at a time when he supported the idea of Soviet journalists recruiting the next cohort of cosmonauts from members of their profession. In January 1966, Korolev died, and the journalists' flight was forgotten. The subject of sending a reporter into space was not revisited by the Soviet space leadership until the second half of the 1980s, but clearly did not resonate with national journalists despite Gorbachev's personal commitment. During his meeting with *Pravda* editors in October 1989, Gorbachev said that the decision to launch a Soviet journalist into space was made and that the Soviet journalist was supposed to blast off before the Japanese reporter, whose flight was scheduled for late 1990. However, this promise of the General Secretary was never implemented. NPO Energya officials maintained that it was impossible to prepare a Soviet journalist for the mission in the time remaining before the Japanese flight. However, as space historian Yuri Karash pointed out, the real reason was NPO Energya's lack of interest in a non-commercial flight. As a result, a Soviet journalist never flew in space, despite six journalists being admitted to the General Cosmonaut Training program in Star City.[37] Karash interprets this episode within the broader context of the diminishing influence of political authorities over decision-making in the rocket and space sector. This shift took place alongside the growing consolidation of power within the management of major industry enterprises, such as NPO Energya under the leadership of Yuri Semenov.

It was a decade that prioritised commercial advantages over national prestige, shifting away from the politically driven pursuit of space records

[35] Harvey, B. *Japan In Space: Past, Present and Future*, Springer, 2023, 185.

[36] Dubbs, Ch., Paat-Dahlstrom, E. (2011). *Realizing Tomorrow: The Path to Private Spaceflight*. University of Nebraska Press, 92.

[37] Karash. (1999), 169–170.

that had been a hallmark of Khrushchev's and Korolev's era. Naturally, the decision by NPO Energya and Glavkosmos to select Akiyama was met with strong criticism from those who felt disadvantaged by the outcome. This decision had a significant negative impact on the professional pride of Soviet journalists who, empowered by *glasnost*, openly expressed their frustrations in the free press. One such reaction came from Yaroslav Golovanov, a well-known journalist and historian of Soviet cosmonautics, whose essay vividly captured the sentiment of the time:

> ...we sell the right to the first flight of a journalist (!) to a Japanese broadcasting corporation that wants to celebrate its 40th anniversary in this way! Why? And how could this happen in the country that gave the world Gagarin and Tereshkova! Is it not clear that such a deal is an affront to our civic dignity? Don't our high-ranking "space merchants" understand that? We journalists have welcomed and encouraged the first and rather timid attempts of Glavkosmos to make space flights profitable. But have they not learned to distinguish between profit and loss and do not they understand that the agreement with the Japanese, however good it may be economically, is in the end inevitably loss-making, because it hurts the national pride and this is exactly the case when any trade is unacceptable![38]

Another Soviet journalist, Andrey Tarasov, who—along with 25 other colleagues, including Yaroslav Golovanov—underwent medical selection for journalist-cosmonauts at IBMP in 1989–1990, raised the issue even more forcefully. He emphasised not only the practical concerns but also the moral implications of the decision.

> The Glavkosmos believes that the time has come. That is fine. But his gaze slid indifferently past the Soviet press, radio, and television to his colleagues at TBS. No need to wonder why. Alas, we all get ordinary Soviet rubles for our work and cannot pay for the flight in convertible currency that Glavkosmos so badly needs. Alas, even our biggest national newspaper, *Pravda*, cannot pay for us in dollars, pounds, and yen. So once again the matter is being postponed until an unknown time. Glavkosmos is obviously not interested in our commercial abilities.[39]

[38] Golovanov, Ya. (1989). Prestizh na vynos. Pochemu yaponets na sovetskoy orbital'noy stantsii dolzhen stat' pervym kosmicheskim zhurnalistom?! *Komsomolskaya pravda*, March 30.

[39] Tarasov, A. (1989). Ot bednykh rodstvennikov. *Pravda*, March 30.

It was the flight of Akiyama to Mir that publicly revealed a new attitude for space research and exploitation of outer space that did not yet exist in the Soviet Union: its commercial aspect. Both expert opinion and the broader public needed time to adjust to the new image of space, shifting from the heroic to the pragmatic. The gap in opinion had been breached and from that moment on the road to private flights by civilian passengers aboard the Soviet Soyuz had been cleared. It was just a matter of finding new private partners. The next client was a British chemist, Helen Sharman, who took off from Baikonur on May 18, 1991. According to the British *The Independent* a "further 10 million to GBP £15 million was earned by Glavkosmos thanks to a privately financed Briton-in-space deal".[40] As Brian Harvey reported "a consortium of British companies undertook to raise GB £16 m to fly the first Briton into space".[41] However, preparations for the flight of the British cosmonaut did not go according to the envisaged scenario. Of the required 16 million GBP by 1989, only half could be raised which, as one can assume, greatly worried Glavkosmos. Lack of government support, gradual withdrawal of private investors from the project, and problems with advertising[42] led to the cancellation of the British scientific programme of the mission which was rescued by the Moscow Narodny Bank (based in London), NPO Energya, and Licensintorg.[43]

As Viktor Nikolaev recalled from his time at Glavkosmos, "The British flight in 1990 was funded with a minimal amount. Only the initial payment was made, and subsequent payments ceased. We sought to persuade Yuri Semenov, head of NPO Energya, to cancel the flight, as it would be futile. The British government showed no interest. However, for some reason, Semenov believed that the British would take offense. The flight was funded by the state budget. Semenov acted independently and disregarded Dunaev's opinion. Semenov was nicknamed 'the tree with roots at the top' due to his close ties with the Politburo".[44]

An Austrian citizen became the last foreign cosmonaut to fly aboard the Soyuz spacecraft and the Mir orbital station, although the Austrians were ahead of other foreign claimants in finalising the agreement. The so-called

[40] *The Independent*. May 22, 1989.

[41] Harvey. (2021), 192.

[42] See more about Juno mission in Harvey (2021), 192–195.

[43] Baturin (Ed.), (2005), 683.

[44] Interview with V. Nikolaev, June 30, 2024.

Austromir-91[45] project was set in motion in 1987, when the Austrian government accepted an invitation from Nikolay Ryzhkov to carry out experiments on board the space station. A mixed commission of experts was assembled to deliberate on all conditions of the flight, with the inaugural meeting taking place in Moscow. On April 5, 1988, the Austrian Federal Government officially approved the spaceflight and confirmed its willingness to fund the mission at an agreed price of $6.46 million.[46] As Johannes Ortner, Director of the Austrian Space Agency, commented: "It was a unique chance. Without this, our scientists would never have been able to get into space".[47]

Viktor Nikolaev explained the lengthy interval between the contract's conclusion and its execution. He recalled,

> As they were the first to sign a treaty to send their cosmonaut into space, we offered to launch them first. However, they declined, which surprised us greatly. They stated that they would require three years to prepare the hardware and scientific instruments, and an additional two years to analyse the results. This highlighted to us that they were securing guaranteed employment for five years in one go, hence the salaries. They just did not need to fly earlier… It underscored a completely different mindset, and we realised there was much for us to learn.[48]

Soviet-Austrian Austromir-91, the first and only flight by an Austrian cosmonaut, took place in October 1991 as the Soyuz TM-13 spacecraft, commanded by Aleksandr Volkov, carried Franz Artur Viehböck and Kazakhstan's first cosmonaut-researcher Toktar Aubakirov, to the Mir orbital complex. Over the course of the eight-day expedition, Viehböck conducted several scientific experiments on the station, the results of which were later utilised in medical research and development.

In the USA, there was also interest in purchasing a ticket to Mir. However, given the historical rivalry in the space race, flying in a Soviet spacecraft to the Soviet orbital station was more of a political (or even image) challenge than a financial one. The Houston-based Space Travel

[45] For further details, see Besser B. Ph. (2004). *HSR-34 Austria's History in Space.* ESA Publications Division ESTEC, 39–42.

[46] Baturin (Ed.), (2005), 684.

[47] Lon Rains. (1989). U.S. Experiments ready for Dec.20 launch to Mir. *Space News*, December 11.

[48] Interview with V. Nikolaev, June 30, 2024.

Services Corp., in collaboration with Art Dula's Space Commerce Corp.—already established as an exclusive partner of the Soviet rocket and space sector—approached Soviet officials to explore the possibility of flying an American astronaut to the Mir space station. Representatives of Glavkosmos duly confirmed that the Soviets would allow a "sweepstakes" winner to go aboard the Mir. Upon hearing this news, a group of American entrepreneurs raised funds to pay for "an" American's flight, and a sweepstake was organised whereby anyone could enter by calling a special telephone number and paying a $2.99 phone charge. A winner would be selected at random in a draw in December 1991 for a space trip in 1992 or 1993 plus $500,000.[49] According to the Houston company, the price for the flight was established at $12 million.[50] Political authorities took the necessary steps to stop the fundraising, and the first space crowdfunding initiative did not yield the expected results.[51]

The approach to addressing the challenge of earning hard currency received mixed reactions from Soviet space observers. Some expressed disappointment, noting that "aside from relatively modest hard currency receipts, which essentially only covered costs for joint flights, the Soviet side gained little from such collaborative expeditions. In this view, the existing model did not fully capitalise on the potential for more productive cooperation within the market, including the space market". Critics argued that "given the agreements with Western partners allowing them to conduct experiments on Soviet spacecraft, these collaborations could provide an opportunity for the Soviet side not only to receive currency but also to acquire a share of the advanced technology—essential 'know-how' in modern times". Such an exchange, they contended, would be a more equitable arrangement, recognising the significant value of technological expertise that developed countries hold in high regard.[52]

In addition to offering costly tickets for missions to the Mir station, the Soviets also sold opportunities to conduct microgravity experiments aboard Mir using Soviet equipment. Through these two avenues, Western countries gradually began to engage with the USSR's commercial space offerings. Mir became a gateway to the international space market,

[49] Space odyssey moves ahead one, back one. *Washington times*, December 20, 1990.
[50] Soviet space program to fly first American. *Washington post*, December 1, 1990.
[51] For further details, see Dubbs, Paat-Dahlstrom. (2011), 94–95.
[52] Avduevsky, Leskov. (1990), 22.

enabling the Soviet space program to secure much-needed resources for its survival, by leveraging its own technological capabilities.

Dual-Use Satellites: From Military Surveillance to Commercial High-Resolution Imagery

Rockets and the orbital station were not the only products for sale despite being the most spectacular. High-resolution satellite images of the Earth's surface could also be purchased, with the exception of restricted areas, which were excluded for national security reasons. More than thirty American firms expressed interest in signing contracts for the acquisition of satellite images from the Soviet Soyuzkarta foreign trade association.

In 1987, the French *International Herald Tribune* announced the Soviet Union's plans to enter the international market for satellite images as part of the Earth Observation (EO) program. As Karin Barbance argued in March 1988: "until now the USSR had been very reluctant to open up to international exchanges in the field of EO due to the fact that on their activities the civil and military components were not neatly separated".[53] So, what explains such a dramatic change in the Soviet approach?

On July 25, 1987, the largest EO orbital station, *Cosmos-1870*, was launched using a Proton booster. *Cosmos-1870* was a 20-ton-class remote sensing vehicle which featured a side looking Synthetic Aperture Radar (SAR)—Mech[54]-K—that, according to the Soviet news agency TASS, was "for the remote sensing of the surface of the Earth and the world ocean, regardless of weather conditions and the time of the day or night".[55] The spacecraft was originally created under the *Almaz-T* program of 1976 to carry out military, scientific, and economic tasks within the USSR Ministry of Defence.[56] The new automatic orbital station (*Almaz-T*) was equipped with a unique onboard high-resolution radar, transmitting images via radio channel to a ground receiving station. The automatic SAR Mech-K, part of the first generation of radars (1978–1992), was initially designed for military reconnaissance. However, it was also employed for a variety of civilian applications, including the study of natural resources, environmental

[53] Barbance, K. *U.S.S.R. Meeting with Klemanov/Intercosmos.* 10 March, 1988. In ESA-28037, HAEU.

[54] In Russian "Mech" means sword.

[55] Soviets launch polar orbiting satellite. *Aerospace Daily*, August 24, 1987.

[56] Polyachenko, V. (2008). *Na more i v kosmose: Vospominaniya.* MORSAR AV.

monitoring of the ocean surface, ice cover analysis, agriculture and forestry, geology, and cartography.[57] Under this program, three spacecraft were built: *Cosmos-1870*, *Almaz-1A*, *Almaz-1B*.[58] The latter was never launched, while the first two stations will be discussed below.

The best publicity for the radar onboard *Cosmos-1870* on the international stage was provided by none other than Mikhail Gorbachev during a summit meeting with USA President Ronald Reagan in Washington in December 1987. In a hugely exciting announcement, Gorbachev claimed that the Soviet Union had "a space means of detecting submarines while diving". This statement was received in two ways by the USA. On one hand, it sounded like an ill-concealed warning, while on the other, it hinted at the potential for cooperating with private clients to earn valuable foreign exchange. Certainly, *Cosmos-1870* left a significant impression on American experts, due to its radar capable of detecting the topography of the ocean floor. They were particularly astounded by a remarkable image that unveiled the underwater topography of Puget Sound Strait (State of Washington) at a depth of 300 m. Traditional microwaves typically penetrated the water surface only superficially and did not reach significant depths. However, this new type of radar, under specific conditions, had the ability to "see" and capture an image of the seafloor. Western militaries immediately wondered if it would be possible to detect submarines whilst submerged because, until the end of the 1980s, missile-launching submarines were considered undetectable and therefore invulnerable. According to information from the French magazine *Air & Cosmos*, the Soviet military apparently used the *Cosmos-1870* to obtain information on the state of the sea and American nuclear submarine formations.[59]

Gorbachev's statement also sounded like an invitation to bilateral cooperation in the EO field with the Soviet Union and a wish to cooperate closely with American partners in the field of space photography. Considering the broader context of the statement and Gorbachev's overarching foreign space policy, rooted in the principle of mutual disarmament and collaboration, it can be inferred that the primary objective of his announcement was to showcase Soviet technological achievements and

[57] Verba, V., Neronskiy, L., Osipov, I., Turuk, V. (2010). *Radiolokatsionnyye sistemy zamleobzora kosmicheskogo bazirovaniya*. Radiotekhnika, 577.

[58] Retrieved March 6, 2025 from official website of NPO Machinostroenya. http://npomash.ru/history/ru/history.htm.

[59] Langereux, P. (1991). Almaz film la terre au radar. Air&Cosmos, Avril, 15–21.

attract foreign customers. The very content of the negotiations in Washington, which resulted in the signing of the Treaty on the elimination of Intermediate-Range and Short-Range Missiles (INF Treaty), testifies to this fact. Furthermore, the gradual access to Soviet technologies and the readiness to convert military space into civilian space rather indicated a new comprehension of the leadership of the late USSR. Finally, the Soviet military was not interested in using a new radar that did not meet their needs.

As defence experts pointed out, the resulting images with a resolution of 25–30 m did not particularly suit Soviet military customers. According to unofficial data, the resolution of a similar American Lacrosse reconnaissance satellite system ranged from 1 to 3 m, and this limit was fundamental since Lacrosse was created primarily to track Soviet armoured vehicles and mobile launchers. In addition, the problem of real-time data relaying had not yet been solved on *Cosmos-1870*, and the locator worked on average 10 min per orbit, with only 20% of the information being recorded.[60] Thus, although the spacecraft was used by the military, it appears that it was not of primary importance to them.

Although the *Cosmos-1870* orbital station was also employed for military-strategic purposes, the data received from the satellite was mainly used in the strategic and economic interests of the Soviet Union, while the Priroda Centre[61] used the radar satellite for civilian purposes in agriculture, hydrology, geology, cartography, and environmental monitoring. Given the potential for a wide range of applications of the automatic *Cosmos-1870*, the Soviet leadership decided to expand its scope and make attempts to enter the international market.

The collaboration in the EO field and especially the business prospect of images taken from space was facilitated by the formal resumption of space cooperation between the Soviet Union and the USA and in April 1987, the Agreement on Soviet-American collaboration in Space was signed in Moscow, by the USA Secretary of State George Schultz and Soviet foreign Minister Eduard Shevardnadze. The Agreement concerned cooperation in the exploration and use of outer space for peaceful

[60] Tarasenko. (1992).

[61] The State Centre "Nature" was established on December 15, 1973, by order of the Main Directorate of Geodesy and Cartography under the Council of Ministers of the USSR, dated December 12, 1973, to conduct studies on the Earth's natural resources and environmental monitoring using space-based tools.

purposes and established five joint working groups (solar system exploration, space astronomy and astrophysics, earth sciences, solar-terrestrial physics, and space biology and medicine[62]) that would have begun their work over next year.[63] As mentioned earlier, the agreement was kept in a very general form to prevent overly close cooperation and the risk of unauthorised technology transfer. Nevertheless, the mere act of its conclusion established a formal basis for broader international activities of Soviet space enterprises, concurrently fostering closer Soviet-ESA collaboration in space exploration.

Shortly after the launch of the *Cosmos-1870*, the Soviets announced the creation of the Soyuzkarta company which was responsible for marketing the images from space. For a more effective cooperation, Western space officials and intelligence services tried to ascertain the organisational scheme of the Soviet institutions and to define the responsibilities of the various scientific and governing bodies. It was Barbance who worked out that inside the Soviet Union the EO activities had no common governing umbrella. According to her information, the Academy of Sciences managed purely scientific aspects of EO, Glavkosmos was responsible for technology aspects of EO and for collocation of foreign remote sensing instruments on the Mir station and that Soyuzkarta oversaw the commercialisation of Soviet remote sensing data.[64]

Quoting the CIA research paper "Soviet Commercial Space Photography: Offering Resolution as the Solution", the USSR began "aggressively marketing satellite imagery of the Earth" in 1987. National budget constraints made Soyuzkarta primarily to market film-based products while Glavkosmos entered in the digital imagery market in 1988.[65] The authors of the report pointed out that the financial restraints within the Soviet space program had forced the Soviet leadership to emphasise space missions with the potential to generate hard currency and improve the national economy. According to CIA opinion, the remote-sensing orbital station could accomplish both objectives. By marketing imagery,

[62] Karash. (1999), 162.

[63] Logsdon. (2019), 354.

[64] Barbance, K. *U.S.S.R. Meeting with Klemanov/Interkosmos*, 10 March, 1988. In ESA-28037, HAEU.

[65] Directorate of Intelligence, Central Intelligence Agency. Soviet Commercial Space Photography: Offering Resolution as the Solution, June 1990. National Security Archive. Retrieved March 6, 2025 from https://nsarchive2.gwu.edu/NSAEBB/NSAEBB501/docs/EBB-47.pdf.

much-needed hard currency could be generated, while imagery also held the promise of improving Soviet agriculture.[66]

The new Soviet entity, Soyuzkarta, came into direct competition with the Landsat program, managed in the late 1980s by a private USA Earth Observation Satellite Company (EOSAT), which sold images from USA satellites and with Spot Image, which marketed images from the French Spot satellites. According to the French press, the creation of Soyuzkarta was proof of the Soviets embarking on the commercialisation of space services and competing with the Americans and Europeans on the world market.[67] Judging by assessments from experts published in the Western press, the Soviet station was quite competitive on the world stage and to many, it offered better resolution than commercially available options in the West. Here, however, it is worth noting that while the Soviet authorities permitted the sale of such photos abroad, the USA government restricted these types of civilian operations "for fear that high-resolution space photographs would reveal military secrets".[68]

However, even Soviet distributors of space images had to contend with significant obstacles imposed by military departments, as Viktor Nikolaev recalled.

> Glavkosmos initiated efforts to negotiate with the Ministry of Defence for authorisation to acquire high-resolution images from EO satellites, specifically the *Almaz* and *Resurs* satellites. The military-imposed restrictions on the resolution level of ground imagery available for sale, as it was in the USA. The department within Glavkosmos responsible for earth sensing from space endeavoured to secure permission to sell imagery with resolutions of one metre or better. However, these efforts were protracted and largely unsuccessful due to significant opposition from the military, who felt that there were too many issues related to concealing certain sensitive facilities.[69]

However, quoting Viktor Yashchenko, Head of the Main Directorate of Geodesy and Cartography under the Council of Ministers of the USSR, the *International Herald Tribune* mentioned that "Soviet trade organisation Soyuzkarta had already sold photographs taken by Soviet spacecraft

[66] Ibid.

[67] E.G. (1987). *Les Soviétiques vont vendre les images prises par leurs satellites. Le Monde*, 13 Aout.

[68] Russians selling space photographs. *International Herald Tribune*, July 10, 1987.

[69] Interview with V. Nikolaev, June 30, 2024.

to Syria, while Australia, Kuwait, Angola, North Korea, Vietnam, and East Germany expressed their interest".[70] On another occasion, Yashchenko indicated that the prices for Soviet space photos were lower than those of the French, but higher than those charged by the Americans. A colour photo by Soyuzkarta, along with its negative film, cost an average of $1000.[71] The disadvantage of Soyuzkarta images was their limited commercial value due to the restricted number of available zones.[72]

In 1987, Soviet executives from Soyuzkarta approached the USA company GeoSpectra Corporation of Ann Arbor, Michigan. They processed and marketed pictures taken by civilian cameras in space, with the proposal to become an American distributor for its satellite photographs of the Earth. Even if the American company was eager to work with the Soviets, GeoSpectra officials feared that forming a Soviet link could jeopardise the company's relationship with the Pentagon who had earlier proposed certain secret work for the Defence Department.[73] Thus, the initial attempt by entrepreneurs from Soyuzkarta to expand their business into the USA proved unsuccessful. Apprehension about potential sanctions dissuaded private businesses from collaborating with what was perceived as a risky partner.

In October 1988, reports appeared in the Soviet press that a deal had been struck between USA Contitrade Services Corporation and Soyuzkarta on the purchase of Soyuzkarta photographs taken in orbit by Soviet satellites and the sale of space information to practically the entire Western world. As Miron Lazerson, chief of the USA firm declared, "over a period of 5 years we will be able to buy very high-resolution space photographs from the Soviet Union and sell them at home and also in Canada, South America, and the Caribbean countries… It is a business with a future. There is a lot room for using the information. The space probe materials help to identify land suitable for cultivation and irrigation and natural feed and water resources. Finally, given the right methods, the space

[70] Russians selling space photographs. *International Herald Tribune*, July 10, 1987.

[71] Berg, S. (1988). Premier contrat étranger pour Soyouzkarta. *Air & Cosmos*, 26–30 Janvier (1175).

[72] Wirin, W. B. (1989). Commercial Space Opportunities from the Soviet Union. *Small Satellite Conference*. Retrieved January 13, 2025, from https://digitalcommons.usu.edu/smallsat/1989/all1989/39/.

[73] Broad, W. (1987). Soviet space photos stir U.S. debate. *The New York times*, October 13.

photographs will make it possible to discover geological structures with oil, coal, or gold potential…".[74]

As far as Europe was concerned, CNES had the privilege of being the first Western space agency to see the radar images from *Cosmos-1870*.[75] In November 1987, ESRIN (ESA) took its first steps, expressing its intention to "place an order with Soyuzkarta for the supply of basic products in order to better evaluate the type of images and services they provided".[76] In January 1988, Soyuzkarta signed new contracts with the Finnish firm FM-Projects, giving it sales exclusivity in Western Europe and Australia of the photos taken by the Soviet satellites. Several orders, according to Yashchenko, had since been placed also by Italy and the Netherlands.[77]

However, according to the aforementioned report delivered by American intelligence services, Soyuzkarta's film-based products had several disadvantages. The Soviet competence of supplying the digital imagery required by Western customers was impaired by limited data handling and computer processing capabilities, together with software and computer incompatibility problems. In addition, SAR data from *Cosmos-1870* was of average quality, with evidence of hand splicing that degraded the images. Despite claims to the contrary from the Soviets, digital imagery from the satellite was unavailable, further reducing its benefit to Western consumers. In addition to these intrinsic disadvantages were other negative factors, such as dissemination policy, limited photographic archives, a poor distribution system, the failure to offer related goods and services, and a rigid pricing structure. To offset the weakness and to improve the quality of the offer, it was circulated that the USSR intended to introduce three new Earth resources space platforms to be launched in the early 1990s. These systems would enhance the Soviets' product line by providing digital data, better spatial and/or spectral resolution, and all-weather imaging capability.[78]

To improve the quality of radar images, the next orbital station was powerfully upgraded and launched into orbit on 31 March 1991, under its own name, *Almaz-1A*. At the time, *Almaz-1A* featured the world's highest resolution, a digital system for data transmission via the Geyser

[74] U.S. Grain magnate on satellite photography deal. *Sovetskaya Rossia*, 26 October 1988.

[75] P.L. (1989). Satellite-Radar militaire. *Air & Cosmos*, 2 décembre. (1261).

[76] Calabresi, G. *Contacts with the Soviet Union*, 16 November 1987. In ESA-28037, HAEU.

[77] Berg. (1988).

[78] Directorate of Intelligence, Central Intelligence Agency, *Soviet Commercial Space Photography*.

relay satellite, and a ground facility for digital radar information processing.[79]

Almaz-1A and Domestic Competition for International Markets

The ground resolution of *Almaz-1A* was increased to 15 m, and information was relayed to ground, both directly and through geostationary satellites. The system was capable of penetrating overcast weather conditions as well. From the outset, this satellite was announced as a multi-purpose spacecraft for remote sensing of the Earth, with a focus on the wide commercial distribution of its radar images abroad, moving beyond its original use for reconnaissance purposes.[80] As Soviet Army Maj. Gen. Alexey Shumilin, deputy Baikonur director, recounted in an interview to *Aviation Week& Space Technology*, "it was the most significant unmanned Soviet space mission in 1991 [...] and a major test of Soviet plans to convert massive segments of its rocket and space industry to revenue-generating enterprises".[81]

The promotion of unique radar images on the international market, initiated in 1987, received a significant boost with the launch of *Almaz-1A*. This success was largely attributed to NPO Mashinostroeniya,[82] the manufacturer of the automatic spacecraft, which also established a dedicated foreign trade organisation for the commercial distribution of images obtained from the *Almaz-1A* station. The organisation marketed its data in many countries, including Australia, China, France, Germany, India, Indonesia, Italy, and the Scandinavian nations. The cost of one image of a 40 × 40 km area of the Earth's surface, recorded on magnetic tape or floppy disk, was estimated at $1600.[83] At the end of 1989 and for the first time the Soviets invited a Western country, France, to provide equipment and explore data from the future *Almaz-1A* radar observation

[79] Polyachenko. (2008).

[80] Tarasenko. (1992).

[81] Covault, C. (1991). Soviets Launch Largest Earth Resources Satellite on Modified Salyut Platform. *Aviation week & Space technology*. April 8.

[82] NPO Mashinostroeniya (Research centre of Machine manufacturing) is a rocket design bureau based in Reutov (Moscow region) and headed until 1984 by Vladimir Chelomey. During the Cold War it was responsible for the military Almaz space station program.

[83] Soviet Radar Earth observation satellite gets U.S. attention. *European space report*. October 22, 1990.

station.[84] Gerbert Efremov, chief-constructor of Almaz orbital stations at the NPO Mashinostroeniya, proposed in 1989 that France participated in the exploitation and even in the distribution of the radar images already obtained by the precursor *Cosmos-1870*, and provide observation instruments for the next *Almaz* satellites.[85]

Negotiations with the French side were unsuccessful, so in 1990, the Soviet *Almaz* manufacturer turned to American entrepreneurs, specifically the previously mentioned USA firm, SCC of Houston. This marked the first large-scale commercial Soviet-American joint venture in space. According to the European Space Report, the American SCC secured exclusive marketing rights for *Almaz* data for one year, with the condition that the imagery would be processed not in Moscow, but in the USA. The data would be received in Moscow at the primary Soviet ground station and then transmitted to the USA through conventional satellite links.[86] Although the enterprise of Almaz Corp was indeed created through mutual Soviet-American effort, it appears that the reaped benefits did not meet the expectations of the American partner: the Houston SCC's total revenue from the commercial sale of the images was about $250,000, although the original planned sales value in 1992 alone was estimated at $2 million.[87]

In summer 1991, EOSAT owned by General Electric and Hughes which operated Landsat satellites planned to sign an agreement with the Almaz Corp. for marketing imagery data from the Soviet *Almaz* spacecraft. However, the plans for the deal were very soon dropped, owing to the Soviet economic crisis and instability sparked by the attempted August coup d'état. Almaz Corp. signed instead a deal with Spot Image Corp. (SICORP) that marketed and distributed images from the French Spot spacecraft.[88] Thus, in August 1991, SICORP was granted the rights to distribute *Almaz* data throughout the United States and Canada.[89]

[84] L'URSS prépare une série de satellites-radar. *Air & Cosmos*, 2 décembre 1989. (1261)

[85] Ibid.

[86] Soviet Radar Earth observation satellite gets U.S. attention. *European space report*, October 22, 1990.

[87] Garbuk, S., Gershenson, V. (1997). *Kosmicheskie sistemy distantsionnogo zondirovaniya*. Izdatelstvo A i B, 156.

[88] Kremlin coup kills deal; Spot Image to sell data. *Aviation week & space technology*, August 26, 1991.

[89] SPOT Image Corporation European Space Agency funding proposal, December 24, 1991. In ESA-28037, HAEU.

Almaz manufacturers extended their search for potential customers by advertising their products and services within the framework of the UN Assembly. In February 1991, Gerbert Efremov introduced to the Committee on the peaceful uses of outer space (COPUOS) a working paper on the implementation of the *Almaz* program in the Soviet Union. At the end of the presentation, Efremov emphasised that "more than twenty Soviet and foreign organisations would participate in these programs". He then invited the international scientific community to participate on board the spacecraft and foreign firms to purchase *Almaz* products: "All interested countries, their research organisations, and firms, as well as scientists and specialists, are invited to take part in the research and experiments conducted on board the *Almaz-1A* space platform. Moreover, foreign partners will have ample opportunity to use and purchase photographs of the Earth's surface taken by equipment on board the *Almaz* spacecraft". Going further Efremov specified that the "work would be carried out on a commercial basis at reasonable rates for scientific research".[90] As a result of the efforts made by the management of NPO Mashinostroeniya to promote its services on the world market, cooperation was established with more than forty firms and organisations in the US, France, the UK, Italy, Germany, and other countries.[91]

At that stage, a conflict emerged between different divisions of the VPK over primary access to foreign sources of funding. In our case, it manifested between Soyuzkarta and NPO Mashinostroeniya over the right to establish direct commercial contacts with international clients for the distribution of radar images. Therefore, the Soviet space technology manufacturer began direct negotiations with foreign companies bypassing Glavkosmos and Soyuzkarta. If in the case of *Cosmos-1870*, Soyuzkarta was responsible for the international market after the launch of the *Almaz-1A*, the spacecraft manufacturer itself assumed responsibility for the commercial distribution of photographs. The NPO Mashinostroeniya entered into an agreement with the American SCC, which handled marketing in the USA and South America. It was no coincidence that Soyuzkarta officials believed that NPO Mashinostroeniya was making a

[90] Union of Soviet Socialist Republics: working paper, February 19, 1991. In ESA-28037, HAEU.

[91] Postnikov, I. *31 marta – ispolnilos' 20 let so dnya zapuska avtomaticheskoy kosmicheskoy stantsii «Almaz-1»*. Retrieved January 10, 2023, from http://npomash.ru/press/ru/podrobnee310311.htm?l=0.

mistake by trying to market such important data through an "untested" firm like Space Commerce. It is likely that poorly concealed competitive dissatisfaction was evident in this criticism. Efremov's active position was supported by his far-reaching plans. The fact is that Efremov intended to use *Almaz* station as a springboard to expand NPO Mashinostroeniya's commercial contacts in the West. Moreover, according to information released by *Aviation Week& Space Technology*, the NPO Mashinostroeniya also had intentions to take part in an effort with *First Boston, Inc.*, a New York-based investment bank, to create an "international conversion fund" to help the Soviet defence industry convert to joint commercial ventures with USA companies. The management of NPO Mashinostroeniya was compelled to take all these actions to save the enterprise from imminent collapse in the conditions of an economic crisis and to prevent the re-profiling of production activities. Otherwise, as Efremov claimed, "NPO Mashinostroeniya risked switching to the development of machinery to cut and package chickens".[92]

Amidst these business developments, the dissolution of the USSR and the economic downturn in post-Soviet Russia pushed the space industry to the brink of collapse. This, in turn, significantly affected the continuation of the *Almaz-T* remote sensing program. The *Almaz-1A* radar station, launched in early 1991, was scheduled to re-enter the atmosphere around Autumn 1992, and there was limited funding available for the assembly and launch of the future *Almaz-1B*. However, the launch of the third station was critical because data continuity was vital to the international marketing efforts of *Almaz* data. According to information from *Aviation Week & Space Technology*, the program's officers were searching for funding from the Commonwealth of Independent States and international investors, to not only assemble a third *Almaz* for launch in 1994 but also to build an advanced *Almaz-2* follow-on radar satellite for launch around 1996.[93] Despite all the efforts made, the *Almaz-T* remote monitoring program was shut down in the 1990s.

[92] Covault. (1991).

[93] Collapse of Russian Economy Threatens Almaz Radar Spacecraft Program. *Aviation week & Space technology*, February 3, 1992.

The New Frontier: Advertising Placement

Taking its first steps in the world of capitalism, Soviet cosmonautics gradually began to discover its secrets and driving forces. Advertising was no exception. With the catastrophic fall in state funding, the management of the space sector had increasingly to resort to making its sites and products available to foreigners for their own advertising campaigns.

Soviet space hardware quickly became an ideal platform for advertisers. In true capitalist fashion, even the shell of the rockets was available for commercial branding. In 1988, during the launch of a Proton rocket carrying the second *Phobos* spacecraft, foreign guests at Baikonur witnessed an unusual sight—markings painted on the vehicle, including the names and logos of two Western European steel industry companies. Glavkosmos officials announced that the Western firms—Italian Danieli and Austrian Voest-Alpine—were "sponsors" of the specific booster being used for the launch of *Phobos-2*, and therefore, their names appeared on the lower portion of Proton's second stage (Fig. 10.2).

Cecilia Danieli, Managing Director of the northern Italian-based Danieli firm, said her company, working together with Voest-Alpine, had been supplying steel factories and related equipment to industries in the Soviet Union for some time. She told journalists that her firm did not even pay an astronomical price to be a sponsor. She believed that it was reasonable that their names appeared side by side even though they did not have any direct contact with the Soviet rocket and space industry or space program; they took the opportunity to send their names and the Danieli star logo into space.[94] Helmut Richter of Voest-Alpine's Moscow office would not say what his company and that of Danieli had paid the Soviets for the ads, but recognised that "because we were the first, we got a special discount".[95] Next year another Italian company decided to place its advertisement on the outer shell of Proton. In September 1989, the launch of Soyuz TM-8 bore the logo of Generali, an Italian insurance company.[96]

It was in August 1991 that after months of preparation, two Soviet cosmonauts on board the Mir station conducted an experiment for the

[94] Lenorovitz, J. (1988). Launch of two Phobos spacecraft begins ambitious mission to Mars. *Aviation week & Space technology*, July 18.

[95] Soviets Launch Mars Craft on Rocket with Foreign Ads. *The Washington Post*, July 12, 1988.

[96] Chu, J. (1990). Building the Future on the Past: The Expansion of Soviet Space. *Harvard International Review*, 13 (1).

Fig. 10.2 Western European companies' advertisements on the Soviet Proton. Source: *Aviation Week & Space Technology*, July 18, 1988. NASA HQ Archives, 15301

Coca-Cola Company. According to the official statement "the Coca-Cola company, together with NPO Energya, a major Soviet space company, successfully tested a Coca-Cola 'space can' on board the Soviet Mir orbital station this week. The test showed the reliability of the can, which was specially designed for use in zero-gravity conditions".[97]

Coca-Cola had been sold in the Soviet Union since 1985, but by the early 1990s, it aimed to surpass its rival, Pepsi. Before 1985, Pepsi held an exclusive license for sales in the USSR and dominated the Soviet market. As part of this "soda race", Coca-Cola sought to secure a larger share of the Soviet market. As Polly Howes, the company's Moscow representative, pointed out, "as part of its cooperation with Soviet partners, Coca-Cola planned to build 12 plants to produce and distribute Coca-Cola in the USSR". She also added that the company had "far-reaching plans" in the Soviet Union, but that concrete steps would depend on

[97] Interfax. August 29, 1991.

"developments in the USSR".[98] Coca-Cola was quick to publicise the test but did not reveal how much it had paid the Soviets. The company was not permitted to run commercials about the experiment, but the event did generate some publicity in the Soviet Union. It is no coincidence that in a movie produced by Coca-Cola Russia[99] the voiceover, paraphrasing Neil Armstrong's famous expression, proclaimed: "It was a small sip for man and a huge leap for the company".

Selling State-Funded Innovations: The Paradox of Transitional Economy in the Space Sector

As part of its international commercial activities, the Soviet Union offered simulators for those wishing to fly into space with independent means. For example, at the end of 1991, the USSR Ministry of Aviation Industry (MAP) and the French CNES signed an agreement to conduct ten test flights for French astronauts on the Tu-154 simulator, which was created for practicing manual and automatic landings under the Buran program and had previously been used by Soviet cosmonauts. It was further expected that CNES would consider a long-term contract for the use of the Soviet simulator to practice the landing of the ESA reusable Hermes spacecraft. As reported in the Soviet press, if the test flight was successful, "the contract to train French astronauts would bring MAP a multi-million-dollar profit".[100]

By 1991, commercial euphoria had reached its peak, with seemingly unbelievable proposals reflecting the desperate state of the entire Soviet rocket and space sector. Former cosmonaut Alexey Leonov, Deputy Director of the Gagarin Cosmonaut Training Centre, even proposed renting out Baikonur to European nations and transforming it into an international spaceport. "There should be a German zone, a French zone, a British zone, then the entire geographical region can be developed as in all civilised countries". Vladimir Shatalov, former chief of the Gagarin Cosmonaut Training Centre, advocated renting modules of the Mir orbital station to other countries. "It would be very good if the Europeans paid one-half or even one-third and continuously carry out experiments on

[98] Ibid.

[99] Soviet cosmonaut takes Coca-Cola to orbit in 1991. Retrieved January 13, 2025, from https://www.youtube.com/watch?v=7rioo4m-v1i.

[100] *Novosti kosmonavtiki*. (1991), 4.

board and install their equipment on one of our modules. When I was in Germany, I said we would offer them a module for their equipment so that they could have permanent European representatives on board".[101]

The assets of the Mission Control Centre (famously known as TsUP) in Kaliningrad were also incorporated into efforts to integrate them into commercial circulation. Confronted with significant financial challenges, the TsUP management made its scientific databases available for purchase. It was Vladimir Lobachev, TsUP General Director, who initially promoted this idea. During an interview he announced, "We have a widespread telecommunications and computer network usually used for spaceflight. It is possible to use it for other purposes. Following *perestroika* and the other changes in our country we have begun to find some other outlets for our databases, not only space but meteorology, biology, metallurgy, etc. We have information on the effects of radiation after Chernobyl, for example". Efforts to develop commercial databases at TsUP began as early as 1989.

The K-36RB safety ejection system, originally manufactured by NPP Zvezda for use on the Buran space vehicle, was proposed for ESA's Hermes manned spaceplane.[102] Zvezda, the Soviet Union's primary designer of flight crew protection and ejection systems for both military and civilian aircraft, won the bid in collaboration with Italian Fiat Spazio. They defeated the proposal from Aermacchi and Martin-Baker, who had been offering a seat based on the British manufacturer's design.[103]

The sale of the Topaz II reactor warrants a more in-depth analysis, as it represented a case widely discussed in the Soviet press of that time and became famous due to the extreme CoCom restrictions (see Chap. 9). In 1991, journalists uncovered the USA government's intention to acquire a compact Soviet space thermo-emission nuclear power plant, specifically the Topaz II, with the purpose of providing energy for artificial satellites. The idea of creating a space power system with a nuclear reactor and thermo-emission energy conversion originated in the USSR in 1958 at the Obninsk Institute of Physics and Power Engineering. After successful tests of two ground samples of Topaz-I in the 1980s, it became possible to

[101] Nadler, G. (1991). Cosmonauts try capitalism. *The Washington Times*, October 29.

[102] See more about ESA-Soviet (Russian) collaboration on Buran ejection seats and spacesuits in L. van den Abeelen. (2017), 494–502.

[103] Lenorovitz, J.M. (1991). Soviet ejection seat for Buran Shuttle qualified for deployment at up to Mach 4. *Aviation week & Space technology*, June 10.

proceed to in-flight tests. The world's first space launches of thermoelectric units took place in February 1987 (*Cosmos-1818*) and July 1987 (*Cosmos-1867*).[104]

The USA interest in this system stemmed from the development of the SDI program which, in the early 1980s, reignited USA investments in thermionics. The device, intended as an alternative to solar panels, embodied technologies that the Strategic Defence Initiative Organisation (SDIO) had not yet tested through its own research program.[105] In France, the Netherlands, Sweden, Germany, and the Soviet Union, extensive evaluations were undertaken on thermionic research. Among these nations, solely the Soviet Union had executed power tests on the Topaz thermionic reactor-converter during the early 1970s. Additionally, the USA was aware that in 1987 the Soviets had deployed a Topaz reactor into space on two separate occasions.[106]

The story of the Soviet-USA commercial collaboration on Topaz-II began in January 1989, when Soviet scientists, led by Academician Nikolay Ponomarev-Stepnoy of the Kurchatov Institute, first described the Topaz-II reactor at a scientific symposium held in Albuquerque. In May 1990, USA scientists visited the laboratories of the Obninsk Institute, where Topaz was developed. According to their conclusion cited in the British magazine *New Scientist*, "the USSR was far ahead of us in the field of energy for space purposes".[107] A number of USA engineers, perceiving the commercial viability of the technology, effectively secured funding from the SDIO. This financial support facilitated the arrangement for the delivery and testing of six operational models in Albuquerque.[108] The principal advancements in the project transpired during the initial half of the 1990s, a period during which approximately 250 Soviet scientists were actively engaged in the United States to further develop the reactors.

Space News published data regarding the cost of the deal. The newspaper quoted Richard Verga, SDIO director of key technologies, that SDIO would spend $12 million on the Topaz-II purchase, but less than half of

[104] Gryaznov, G., Pupko, V. (1991). "Topaz-I". Sovetskaya kosmicheskaya yaderno-energeticheskaya ustanovka. *Priroda*, 10.

[105] Washington news initiative. Special report N. 2. Soviet space assets for the American space program, October 30, 1991. Prepared for the New Initiative Office Johnson space center. In NASA HQ Archives, 15,583.

[106] Dabrowski. (2013).

[107] *New Scientist*, 1752, 1991.

[108] Dabrowski. (2013).

that money would go to the Soviets. More than $6 million was to be spent on the facility that housed the reactor, and on tests that had to be conducted on the hardware.[109]

The news about the sale of the ground-based prototype of the Topaz-II space thermo-emission nuclear power generator to the United States had provoked lively discussions both among Soviet experts and in the mass media. The Soviet and USA press expressed different opinions on the feasibility of such a commercial operation, the possibility of using this facility in the United States for military purposes and more generally, on the prospects and safety of using nuclear energy sources in outer space. The developers of the Topaz project, Georgy Gryaznov (director of NPO Krasnaya Zvezda) and Viktor Pupko (director of the Obninsk Institute of Physics and Power Engineering), had to give a detailed account of the history of Topaz, explain its exclusivity and importance for the USA market. Furthermore, they insisted on the "harmlessness" of both the use of nuclear energy in space and the sale of the reactor to the United States where, according to them, it was to be used exclusively for peaceful space exploration purposes.

Concluding the analysis of the Topaz development and its applications, the authors highlighted the uniqueness of this invention, situating it within the context of Soviet realities and international research in the field of nuclear energy use in space. "Our country is currently at the forefront of the world's development and use of the most advanced space nuclear power sources - thermionic reactor nuclear power generators. As noted in our press, this is the 'accident where the USSR retains a priority uncontested by anyone'. Circumstances had an impact: the Americans, who started the development of space nuclear installations at the same time with us and achieved significant initial successes in this field, in the 1970s, having failed to find markets for these developments, unexpectedly and significantly reduced government funding in this area and now, after the resumption of work cannot make up for lost opportunities. In our country, the work on thermoelectric nuclear power generators continued almost continuously, systematically and on a broad front using budgetary funding. The concept of 'self-financing' in science had not yet been implemented. As a result, the Soviet Union was a decade ahead of the United States in some of the most promising technologies in space and energy".

[109] Lawler, A. (1991). Defence teams leading way to Soviet store. *Space news*, September, 23–29.

The authors drew a significant conclusion about the essential role of state funding in fostering scientific and technological advancements. They subtly warned against the potentially devastating consequences of transitioning to a market economy in the field of advanced technologies, emphasising the need to be "very cautious about transferring market-based principles from external contexts to domestic ones".[110]

The paradox of cases like this was that technologies developed with state funding generated income for their developers precisely because they were able to enter the international market, offering their products on a commercial basis to interested buyers—something that was not possible within their own country due to the prevailing conditions at the time. In a broader context, the sell-off of state-funded projects from the Soviet era during the transition period that began with *perestroika* has led to complex and contradictory socio-economic consequences in Russian society.

Conclusion

One of the defining features of the *perestroika* period that characterised the space sector should be highlighted: the beginning of the USSR's new course to establish a presence in the space market. At the time, such assertive actions in the international space market sparked concern, if not panic, among USA analysts. In fact, Donald E. Fink, an observer for *Aviation Week & Space Technology*, expressed concerns about the future of USA leadership in space. He emphasised the USSR's serious intentions to challenge the United States' dominance in international markets and to attract European partners to its side. As he argued, "the all-too-familiar alarms about the aggressive space program under way in the Soviet Union" were raised by American experts. They were concerned about the fact that the Soviet Union was "fulfilling its space objectives and actively sought to draw Western partner nations into cooperative projects". According to some USA experts, the Soviets "were in position to offer exciting cooperative possibilities to nations that once had little choice but deal with the US, too often on whatever terms the USA chose to dictate".[111]

[110] Gryaznov, Pupko. (1991), 36.

[111] Fink, D. E. (1991). Picking up the space banner. *Aviation week & space technology*, October 19.

Indeed, the Soviet Union owned extensive experience and reliability, a wide range of goods and services, and a competitive pricing advantage. However, it faced two major challenges: identifying and attracting potential customers for these offerings and fostering commercial activities amid the country's inevitable collapse.

CHAPTER 11

The Soviet Space Industry's Marketing Strategies

Confronted on the one hand with the challenge of transitioning to self-financing and compelled to operate under new socio-economic conditions on the other, the leaders of the rocket and space sector had to devise new strategies and approaches for engaging in the international market—essentially making a leap into the unknown. They envisioned a transition similar to other industries, towards self-financing, thereby achieving greater autonomy in their commercial endeavours, including the sale of products and services to Western countries. They had to operate under the harsh reality of a "fly-or-fall" world, paraphrasing the famous proverb. This chapter will examine the efforts of the Soviet space industry elite to enter the global market for space goods and services.

The accelerating decline in economic performance, the failure of reforms, and the need to reallocate the national budget to more urgent needs led to serious cuts in funding for the VPK and the directly related rocket and space sector. By the conclusion of *perestroika*, even the most senior rocket and space officials were no longer concealing this reality. At the celebration of the 30th anniversary of Gagarin's flight in April 1991, Oleg Shishkin, the last Soviet Minister of MOM, stated at a press conference that government subsidies for the Soviet space exploration program were steadily declining. In 1989, the government allocated 6.9 billion

O. Dubrovina, *The Soviet Space Program and International Collaboration during Perestroika, 1985–1991*, Palgrave Studies in the History of Science and Technology,
https://doi.org/10.1007/978-3-032-03157-0_11

rubles for this purpose, but, in 1991, the sum shrank to 6.3 billion rubles.[1] When referring to the "space exploration program", Shishkin was not only addressing its civilian and scientific aspects but also its military component. According to experts' calculations, between 1988 and 1991 the financing of the space program in the USSR was cut 1.5 times.[2]

This critical situation compelled the sector's leadership to seek new ways to ensure its survival. The key approach was the active development of international commercial relations, which primarily involved identifying, establishing, and maintaining connections with foreign partners. By the end of the Soviet era, after decades of operating within a planned economy, the outcome of these fundamental shifts in the sector's functioning remained uncertain and unpredictable. In this new context, strategic measures were crucial for effectively promoting space products in the global market.

International Conferences

Despite a near-total lack of experience of advertising in the capitalist world, Soviet hardware manufactures had resolutely embarked on a new path to spread the word about its activities and the wealth of resources within the domestic rocket and space industry. The most familiar, tried and tested means of communication over the years have been international conferences. The more prestigious the venue, the more thorough and representative was the selection of the Soviet delegation. For example, at the 37th Congress of the IAF in Innsbruck in October 1986, Soviet space engineers and scientists, as an American journalist put it, "abruptly and radically changed their style". Led by Aleksandr Dunaev, they presented "some of the most open, albeit limited, expositions of Soviet space technology ever showcased in such a public forum". At the centre of the Soviet exposition was Mir station recently launched in space. It was described in detail and in "good English" by designer Viktor Legostaev. Alongside Legostaev was his boss Yuri Semenov, Director of NPO Energya, a veteran cosmonaut Valery Ryumin, Director of Manned Flight Programs, and Aleksandr Dunaev who presented Soviet achievements in space and explained to the international audience the goal of his new department, Glavkosmos: "We do everything NASA does except military missions:

[1] Moscow TASS, April 9, 1991. *FBIS*, April 11, 1991.

[2] Baturin (Ed.). (2005), 477.

launch vehicles, launch facilities, spacecraft development, but Intercosmos does the science".[3]

The landmark 1987 Moscow Space Future Forum, held from October 2–4, served as both a showcase of Soviet cosmonautical achievements and a strategic advertisement aimed at promoting these successes to the international market. With over 900 participants from the Soviet Union and abroad, it was one of the largest international space meetings ever held. The USA delegation, led by Thomas Rona, Deputy Director of the White House Office of Science and Technology Policy, included six NASA officials and astronaut Kathryn D. Sullivan. In total, approximately 160 Americans attended.[4] Beyond representatives from various space agencies, the Soviet organisers also extended invitations to industry delegates from Japan, Europe, and the United States. This deliberate move underscored a key objective behind hosting such a large-scale event: positioning the USSR as a contender in the global space market. Considering all these factors, the Forum primarily played a commercial role, serving as a breakthrough moment in making the international community aware of the Soviet Union's new strategy to establish itself in the space industry.

Trips Abroad

Officials of Glavkosmos refined their advertising through trips abroad and meetings with potential foreign partners. The already mentioned German visit, which took place in 1989, was far from the first "turn-out" for Glavkosmos. According to Viktor Nikolaev's testimony, "In autumn 1985, the first representatives of Glavkosmos began travelling abroad with the purpose of establishing contacts. Events were attended where it was possible to foster relations in industrial cooperation, promote technologies, and sell space images".[5]

Two years later, in May 1987, Glavkosmos conducted a tour of the USA to promote launch services using Soviet boosters. The Soviet delegation met with numerous American satellite manufacturers, operators, and federal officials in Washington, Houston, and New York. Additionally, a Glavkosmos team visited Los Angeles in an effort to captivate an international group of scientists with the prospect of launching research payloads

[3] Harford, J. (1986). Mir today, Mars tomorrow. *Aerospace America,* December.
[4] Forum Participants, October 2–4, 1987. In ESA-28130, HAEU.
[5] Interview with V. Nikolaev, June 30, 2024.

on a Soviet booster. They also visited the Commercial Space Group of Heron, Burchette, and Ruckert&Rothwell, with the aim of assisting American companies in the evaluation of Soviet launch capacities. The international tour finished in Geneva where the director of Glavkosmos, Aleksandr Dunaev, presented new proposals for the use of Soviet launchers for scientific and commercial missions. A dozen Western firms and organisations including Eutelsat, Martin Marietta and British Rocket and Space attended that presentation which further expanded the panoply of launch services offered by the Soviets.[6]

That same year, a Soviet delegation comprising representatives from Glavkosmos, Licensintorg and the Soviet Ministry of Foreign Trade embarked on marketing trips to France, Japan, and Australia. For the first time, Yuri Makarov, Vice President of Glavkosmos, publicly raised the possibility of utilising the three Soviet cosmodromes—Baikonur, Kapustin Yar, and the highly secretive Plesetsk, which was primarily reserved for military launches—for commercial missions.[7]

At the invitation of NASA Administrator Richard H. Truly on July 9–16, 1991, Soviet space officials made a week-long visit to the United States. A fifteen member Soviet delegation headed by Oleg Shishkin, a MOM minister, and Yuri Semenov, Chief Designer of the NPO Energya, included officials from MOM and the State Military and Industrial commission.[8] It was briefed on NASA's programs at NASA Headquarters and toured Space Shuttle and Space Station facilities at several flight and space centres.[9] It was the first official visit to the United States of top officials from these Soviet organisations since the time of the Apollo-Soyuz project.[10]

The high ranking of the delegation's participants, the presence of Dunaev and the length of the stay in the USA clearly demonstrated the importance that the late Soviet Union attached to international cooperation, particularly with the US, not only in the political field but also commercially. The participation of Oleg Shishkin, the highest representative of the space industry, indicated the seriousness of the Soviet rocket and space

[6] Langereux, P. (1987).

[7] Langereux, P. (1987). L'URSS ouvre ses trois cosmodromes aux lancements commerciaux. *Air&Cosmos*, 1138, Mars, 28.

[8] NASA HQ Bulletin, August 5, 1991. In NASA HQ Archives, 15583.

[9] Soviet space officials to visit NASA *facilities*. NASA News, July 5, 1991. In NASA HQ Archives, 15583.

[10] Eisenhower. (2004), 23.

Fig. 11.1 Soviet delegation in USA July 1991. Source: NASA HQ Archives, 15583

sector in establishing and developing business contacts (Fig. 11.1). As Yuri Karash pointed out, both sides agreed that 50% of the Soviet's space technology was 5–6 years ahead of the United States, whereas the United States was ahead of the Soviet Union on the other 50%. They also agreed that the two countries could complement each other in space ventures.[11]

Invitation to Foreign Guests

Another way of publicising the valuable resources accumulated by the Soviet space program was by inviting foreign guests to attend rocket launches. Previously classified Baikonur cosmodrome had been transformed into a venue for a breathtaking show of launching scientific missions. The launch of *Phobos-1* (see Chap. 6) was particularly illustrative in this regard. The Soviet space authorities opted to showcase their offerings

[11] Karash. (1999), 165.

on July 7, 1988, coinciding with the launch of the initial Proton rocket. This marked the first of two launches intended to dispatch robotic missions to Mars' moon Phobos. Invited were Western journalists, several USA Air Force generals, NASA officials, and other foreign dignitaries who had received personal invitations from the Soviet government.

The group was flown on special Aeroflot charters from Moscow on a two-day expedition to witness the launch. Visiting Americans were impressed by the flexibility of the Soviet system, which contrasted with its image of hard-line and narrow-minded bureaucracy. The atmosphere described by an American journalist was much closer to an elite glamour occasion than to an ordinary space program event. "A tentative sort of *glasnost*, or openness—featuring warm Pepsi, American-style French fries, and press conferences—had reached the long-secret cradle of the Soviet space program, about 1600 miles southeast of Moscow. [Soviets] seemed unconcerned about having their visitors swam through hangars filled with craft being readied for launching, poking the rockets' 'skin' and sticking their hands up the nozzles. Tables of soft drinks were set up in the same pink pastel brick hangar with two Proton rockets".[12]

International Exhibitions

Another promotional manoeuvre involved organising exhibitions abroad to showcase Soviet space technology—highlighting both its historic breakthroughs (the glorious past) and its latest scientific and technological advancements (a no less glorious present)—with the goal of attracting potential foreign customers and securing a radiant future. An exhibition of this kind was organised in the summer of 1990 at the Boston Museum of Science, marking the first display in four decades of superpower rivalry in space. The "Soviet Space" exhibit, expected to travel to other USA museums that autumn, opened to the public featuring a fleet of spacecraft and models never seen outside the Soviet Union. Among the artefacts were a replica of *Sputnik-1*, a lunar rover, and components of the last unmanned Moon probe. The exhibit was prepared by the Museum of Science in cooperation with Glavkosmos and accompanied by technical experts, translators, and two heroes of the Soviet space program—cosmonauts Aleksey Leonov and Valentin Lebedev. An important viewpoint was made

[12] Sawyer, K. (1988). Alines invade Soviet space center. *Washington post staff writer*, July 17.

by Kelly Beatty, editor of *Sky and Telescope* magazine and a technical adviser to the Boston Museum. She pointed out that the Americans who travelled to the Soviet Union in the previous December to arrange the exhibit were astounded at the access they enjoyed. She added that they saw everything and that the Soviet side did not hide anything from them.[13]

Another striking international exhibition event, in which Soviet space hardware and equipment manufacturers took part was *Moscow Rocket and Space 90 Show*, held in September 21–26, 1990. Here, 300 companies from 28 countries gathered in the VDNKh exhibition hall. For the Soviet industries, this marked a significant breakthrough on the international stage, as attested by the participation of major aircraft and helicopter manufacturers, as well as equipment suppliers from the USSR. Numerous research institutes, consortiums, production facilities, and NPOs were also present, accounting for half of the exhibition stands. They showcased hundreds of engines, aircraft and rocket components, models, and various other samples. As a journalist from *Air&Cosmos* reported: "Western specialists were able to examine and photograph them at their leisure from every angle, with the added bonus of superb examples of turboprop engines, cryogenic engines from the Energya rocket, and so on".[14]

France was the most prominently represented foreign country, alongside Germany, which organised the exhibition and was eager to assume a leading role. In fact, the event was largely dominated by announcements of contracts, projects, and cooperation agreements with Western organisations and industries, particularly from Germany. Additionally, some unexpected statements emerged from representatives of Soviet firms, who revealed insights into their upcoming projects. Officials from NPO Molnya announced their intention to collaborate with Western partners on the mini-airborne shuttle project. Additionally, Soviet and German firms unveiled their joint project for *Romantis*[15] telecommunication satellites to serve the USSR.[16]

In March 1991, another major event was organised, combining an exhibition on Soviet Union scientific achievements in space and a conference on various aspects of scientific and commercial cooperation. The

[13] Daly, Ch. (1990). "Soviet space" exhibit opens in Boston. *The Washington post*, June 2.

[14] Le choc de la perestroika. *Air&Cosmos*, 1299, septembre 29, 1990.

[15] See more: Friederichs L., Kantor L. (1991). ROMANTIS: A communication satellite system for the U.S.S.R *ESA, Second European Conference on Satellite Communications.*

[16] Menaces sur l'espace soviétique. *Air&Cosmos*, 1299. Septembre 29, 1990.

US-USSR-East European Scientific and Technological Exhibit and Symposium was held at the Centre of Adult Education, University of Maryland at College Park, and was hosted by The East-West Science and Technology Centre at the Department of Physics and Astronomy and the Soviet American Ventures Initiative (SAVI) and organised by Roald Sagdeev, by then former director of IKI. He had married Susan Eisenhower, moved to the USA, and become Distinguished Professor of Physics and Director of The East-West Science and Technology Centre at the University of Maryland.

The content of the presentations suggested an intention to provide a comprehensive overview of advancements in space science within the USSR while simultaneously highlighting their potential for commercial application in the USA market. The primary objective of the event, namely to secure foreign customers for Soviet space technology, was further underscored by the keynote speeches made by invited guests. These speakers offered analyses and perspectives on the political and economic landscape, business conditions, and forecasts for future development in the USSR, aligning with the overarching goal of promoting Soviet space technology internationally. Among Soviet participants there were about twenty leading research and development centres, space hardware and equipment manufacturers such as NPO Energya, NPO Krasnaya Zvezda, Kurchatov Institute of Atomic Energy, NPO ELAS (communication satellites, system of data acquisition, transmission and communication, etc.), Soyuzkarta, VNIITransmash[17] (with a small Martian Rover), IKI (with Regatta project, Ecos-A project, etc.), and Institute of Applied Physics, etc. Around thirty Soviet and American newspapers were invited to cover the event, as well as electronic media such as CNN, Voice of America, Radio Free Europe, etc.[18]

Conclusion

For Western stakeholders in the space sector, all of the aforementioned marketing efforts were not new. Exhibitions, official visits, conferences, business trips, and invitations to potential business partners had long been part of the capitalist economy, with an established market and well-defined behavioural patterns among its participants. However, it is important to

[17] Mobile Vehicle Engineering Institute

[18] *Proceedings from the U.S.-U.S.S.R. - East European scientific and technological Exhibit and Symposium*. In NASA HQ Archives, 15591.

understand that to Soviet officials of the highest rank, whose ideological thinking was deeply rooted in Marxist-Leninist principles, such practices appeared completely foreign. They were now required to operate in entirely new conditions, developing their activities according to rules that, not so long ago, they had not only rejected but also strongly criticised. As a result, they had to exert significantly greater effort to align with the behavioural patterns of the Western capitalist business world. The scale of the promotion efforts—along with the substantial investments made to market Soviet space technology and services internationally—demonstrated the seriousness of the leadership's intentions to open the sector to foreign partners, and the continued dominance of state structures in shaping market relations.

This was further evidenced by the establishment of new organisations designed to collaborate directly with foreign firms, founded on the basis of existing industrial enterprises and closely linked to state management structures, such as ministries. However, this process also had another dimension. It marked the early stages of the transfer of state-owned property into private hands—particularly within the industrial nomenclature—a practice that became widespread in the 1990s following the collapse of the USSR.

CHAPTER 12

To Buy or Not to Buy, That Is the Question

The response to the Soviet proposals from American and European partners was, on the one hand, predictable as the emergence of competition was seen as potentially detrimental to their domestic producers. On the other hand, different professional categories perceived the opportunity to cooperate with this new promising player on the international market in distinct ways. This chapter will analyse the various reactions of Western partners to Soviet proposals, as well as the efforts of private entrepreneurs to establish commercial, mutually beneficial cooperation with Glavkosmos in promoting Soviet goods and services in the international market. We will assess the effectiveness of these initiatives and identify the areas in which they have yielded tangible results.

America's Reluctance: Understanding the Hesitation

The reactions of American officials and private business to the Soviet Union's increased openness in the space sector and its wide range of offerings varied significantly. USA administrators categorically rejected the idea of issuing an export license to any USA entity seeking to launch American satellites using a Soviet booster, believing that such a move would harm the USA commercial launch vehicle sector. The transfer of any USA spacecraft was deemed illegal under existing export regulations. They

O. Dubrovina, *The Soviet Space Program and International Collaboration during Perestroika, 1985–1991*, Palgrave Studies in the History of Science and Technology,
https://doi.org/10.1007/978-3-032-03157-0_12

interpreted the goal of protecting foreign customers' "know-how", as articulated through Glavkosmos officials' declarations, as a strategic tactic to undermine USA influence over its allies. Furthermore, they believed that the Soviet policy, which aimed to protect Western technologies from the risk of industrial espionage, deliberately complicated any potential objections from the United States regarding future Western European firms seeking to utilise the Proton launcher. In turn, the Soviets accused the USA of "doing all it could to exclude the Soviet Proton from the list of competitors" in the market of launchers.[1]

In contrast, commercial satellites manufacturers and communications service companies like Hughes Communications and Space Commerce Corp. were concerned that foreign satellite manufacturers such as those in Europe would strike package deals with the Soviets, enabling them to offer craft cut-rate launch prices to the international communications market. Consequently, they collected information on the Soviet technical, cost and insurance terms without losing commercial space opportunities with the USSR.[2]

Some USA firms seized new opportunities. For example, Glavkosmos could proudly highlight its agreements with Payload Systems Incorporated,[3] a subsidiary of Space Industries International. This Massachusetts-based consulting firm assisted companies in realising private space projects by leveraging the advantages of long-duration spaceflight and the ability to launch on short notice.[4] In 1989, Payload Systems signed a contract with the Soviet Union for experiments to be conducted by Soviet cosmonauts aboard the Mir space station. These experiments, sponsored by clients including pharmaceutical companies, aimed to utilise the near-weightlessness of space to grow protein crystals for the development of new drugs.[5] For the first time, an American company signed a contract with the Soviet Union to transport Western commercial payloads in orbit.[6]

[1] D'Anastasio, M. (1987). Soviets step up efforts to win share in launching of commercial satellites. *The Wall Street Journal*, January 20.

[2] Proton Marketing team finds U.S. interest, opposition. *Aviation week and Space technology*. May 25, 1987.

[3] Bond, P.R. (1990). Watch this space for a Russian. *The Independent*, April 23.

[4] For further details on the American position in this venture, as well as the history of obtaining a license to cooperate with the USSR, see Manber, J. (2009). *Selling Peace. Inside the Soviet conspiracy that transformed the U.S. space program*. An Apogee Books Publication.

[5] Broad, W. (1988). House Panel Chief Questions Pact with Russians on Space Payload. *The New York Times*, February 26.

[6] Broad, W. (1988). Soviet agree on space venture. *The New York Times*, February, 21.

Granted an export licence by the USA Commerce Department in February 1988 Payload Systems, resulted as the only USA company authorised to fly on Mir and reserved six 3-month experiments over a four-year period. Nevertheless, this manoeuvre faced resistance from influential members of the Reagan administration and high-ranking officials, who were mostly opposed to any technology exchange with the Soviet Union. According to Bill Nelson, then member of the USA House of Representatives from Florida and Head of the House Space Subcommittee of the Committee on Science, Space and Technology: "the decision-making process was questionable and set a bad precedent. [...] It is not in the best interests of this nation to become dependent on foreign entities for our space research. [...] We would do everything we can to prevent the experiment from flying". On the contrary, a Commerce Department spokesman declared that the Commerce Department's actions "were totally appropriate within the regulations and it did not have to go through the State Department" as "there was no military sensitivity. Even the Defence Department had no objection".[7]

Supporters of cooperation with the USSR had another compelling argument in their favour. The agreement with Payload Systems underscored the growing commercial interest in the utilisation of orbital stations. It is worth noting that this episode highlighted the ongoing construction of the USA space station Freedom, whose investors were actively seeking potential customers at the time. The experiments that USA companies intended to conduct aboard the Soviet Mir station provided clear evidence that the international space station would eventually attract buyers.

However, within the American establishment, there was another perspective regarding the use of Mir as a testing platform for the future Freedom station. As Jeffrey Manber recalls "If the pharmaceutical crystals grown on board Soviet space station were inconsequential when compared to crystals grown in normal gravity, it could very well dampen Congressional enthusiasm for the NASA multi-billion dollar orbiting laboratory. It was surprising to me how much of the emotional debate about reliance on the Soviet Union and concerns over technology transfer were nothing but a fig leafs for protecting the eight-billion-dollar space station Freedom project".[8]

[7] Broad. (1988). House Panel Chief Questions Pact with Russians on Space Payload.
[8] Manber. (2009).

Nearly two decades later, Manber reflected on this initial step in Soviet-American commercial cooperation, noting that although it did not result in the establishment of stable relations at the time, it held considerable long-term significance. "Payload systems demonstrated that Russia could serve as a supplier of hardware on a commercial basis. Equally illustrative, it was also clear that Russians had no desire or capability to undertake their own market-driven projects, but were ready to support those of other customers".[9]

Another initiative by the American private sector to establish a business leveraging Soviet infrastructure was exemplified by the activities of the aforementioned Artur Dula. In November 1987, Dula and two consultants spent two days at Baikonur inspecting facilities and rockets for launching satellites to space. As he pointed out during an interview, "This is the first time in history that a Soviet booster to launch a satellite had been checked out by a foreign technical team. The Soviets are obviously trying to open up the system".[10] The team was allowed to extensively review several facilities related to the assembly and launching of Proton rockets. Dula emphasised: "I touched the rocket on the pad. We saw all the fuelling apparatus, the gantry system, the clean rooms, the rocket assembly areas, and the areas where they fuel satellites. They answered our questions very freely".[11]

Art Dula, a shrewd and experienced space industry executive, recognised the commercial potential of Soviet rockets and sought to facilitate their entry into the USA market. He also identified the growing demand for launchers in the United States, particularly following the limited availability of USA Space Shuttles after 1986. Many American companies were eager to explore the possibility of launching their satellites using Soviet rockets, and Dula aimed to capitalise on this opportunity.

Not surprisingly, the interests of the Soviet space industry aligned with Dula's efforts. To facilitate its entry into the American market, Glavkosmos established commercial ties with SCC. The joint venture commenced official operations in the summer of 1989. Its primary mission was to support advertising and marketing activities, aiming to foster business connections between American enterprises and the Soviet rocket and space industry. As William Wirin, Executive Vice President and Chief Operating Officer of

[9] Ibid.
[10] Broad, W. (1987). Americans See Rocket in Soviet. *The New York Times*, December, 1.
[11] Ibid.

the SCC concurred, the agreement made available "to the entire world the vast array of goods and services in the Soviet space program". He emphasised that "this opportunity was truly unique because prior to now, every space program in the world has been instigated, managed, and controlled by a government. The restrictions and labyrinth of procedures necessary to obtain space services have had a chilling effect on the businessman. Therefore, it is truly remarkable that it is through a socialist society that the free enterprise businessman will have the first legitimate opportunity to utilise space".[12]

One of the first attempts to concretely realise these ambitious plans was the project to introduce Soviet rockets to the American satellite market. The partnership began in September 1988 when the USA granted export licences for the launch of three American Hughes communications satellites from the Peoples Republic of China. Congress, however, blocked these exports after the Tiananmen Square events of June 1989. As a result, the SCC signed a Memorandum of Understanding with the Soviets to launch a Hughes communications satellite aboard the Proton rocket. However, the corporation still required a USA government license to build and operate these satellites atop a Soviet rocket. This arrangement ultimately served as SCC's test case for a potential shift in USA policy, which had previously prohibited the export of communications satellites to the Soviet Union.[13]

Their calculations for obtaining a licence for the launching of Hughes communications satellites turned out to be wrong as the Department of State imposed a ban on the granting of the licence. After some time, Aleksandr Dunaev recounted the outcome of the first cooperation in launching USA satellites with Soviet launchers: "The USA Congress did not give its consent to the launch from the Baikonur cosmodrome of USA satellites made by the very large and well-known rocket and space firm 'Hughes' for international communications. If we had placed them in geostationary orbit, we would have had about $35 million for each launch. Alas...".[14] He expressed great disappointment, noting, "A ban was also placed on the launch of the 'Inmarsat' satellite, built by the British rocket and space firm. Now it turns out that the satellites, which cost over $100

[12] Chu. (1990). 58.
[13] Wirin. (1989).
[14] Soviet-U.S. space cooperation obstacles viewed. *Sovetskaya Rossia*, November 29, 1989.

million, are still sitting in warehouses, waiting for their turn to be launched by Ariane or another launcher".[15]

In 1989, another attempt was made to launch a USA satellite with a Soviet rocket. The contract was concluded between Soviet-USA joint enterprise SCC and the private firm Energetic Satellite Corporation to launch USA satellites atop Soviet rockets from the Baikonur Cosmodrome. In his interview released to *Izvestia*, Dmitry Poletaev, a high-ranking official of Glavkosmos who led the Soviet side in the negotiations, gave a detailed description of the contract:

> ...the satellites themselves are designed and manufactured by Energetic Satellite Corp. Their functioning is ensured via a ground communications centre in Denver, Colorado. [...] It is planned to use Soviet Proton rockets to place the satellites on geostationary orbit as associated freight. With the aid of special adapter, the USA craft will be installed between the main satellite and the rocket's final stage. The technical realisation of the project in the Soviet side has been entrusted to our well-known space firm—the NPO Lavochkin, which created the legendary lunokhod (Moon rover) and is responsible for the vast majority of Soviet automated space stations. A group of specialists from the NPO Lavochkin spent 2 months working on the technical aspects of organising the launch.[16]

The cost of launching the first two satellites was approximately $13 million, and the total value of the contract amounted to $54 million. Both partners were fully aware of the "notorious CoCom restrictions" and aimed to overcome them by publishing all satellite data to demonstrate that the satellites were solely peaceful and commercial in nature, without employing prohibited technology. However, even in this instance of cooperation, the entrepreneurs encountered failure, and their space satellites ultimately met with an unfavourable outcome.

As Ch. Dubbs and E. Paat-Dahlstrom, authors of a volume about space commercialisation, ironically pointed out, the job would have been a lot easier had it not been for restrictions on sending USA technology to the Soviet Union.[17] Despite the ban, SCC persisted with its sales campaign, as its prospects for success were strongest in areas such as in-space experiments conducted aboard the Mir space station, where USA regulations

[15] Ibid.
[16] Konovalov, B. (1989). Kommercheskiye starty v kosmos. *Izvestiya*, November 27.
[17] Dubbs, Paat-Dahlstrom (2011), 90.

were more flexible. Space Commerce expanded its portfolio by promoting additional services and developing new applications. One notable example was the introduction of advertisements aboard the orbital station. As articulated in their company brochure, the following statement encapsulates this endeavour verbatim: "Imagine a trio of cosmonauts floating about their space cabin sipping from cans of your company's beverage while a Russian-accented version of your product jingle plays in the background... The possibilities are endless!"[18]

In 1990 the company also proposed using an Energya rocket to launch components of the USA space station into orbit. It predicted a cost saving of several billion dollars over the Shuttle-C, an unmanned cargo pod version of the American Space Shuttle system being considered for development at that time. SCC upheld the proven reliability of the Soviet product given that the USSR space industry had the potential to rise above a growing crowd of competitors that included Europe's Arianespace, China's Great Wall Launching Company, Japan's NASDA, and six private USA companies.[19]

Despite energetic efforts to promote Soviet space technology in the USA amidst promising signs of commercial cooperation, Space Commerce did not achieve significant results. On the one hand, the USA administration's fear of exporting sensitive national technology to the USSR posed a major obstacle. On the other hand, the reluctance of large monopolies within the USA space industry to share market niches with Soviet companies further hindered progress. This inevitably led to a gradual fading of SCC's dynamism, and it is almost a gesture of desperation to see an article published in 1991 by Art Dula in which he translated the commercial cooperation between American companies and NASA with the Soviet rocket and space sector into a political issue. He sought to convince the USA government not only of the economic and technological advantages of cooperation with the USSR, but also of the need to help build democracy by involving the Soviet space complex in international market relations.

> The Soviet people do not want a handout. They are proud people. They have a rocket and space industry that can earn the money they need to make democracy a success. [...] The Soviets need to earn hard currency and they

[18] Burgess, J. (1989). U.S. Company trying to market Soviet satellite launches. *The Washington Post*, September 6.

[19] Chu. (1990), 58.

> recognise their rocket and space industry in one area where Soviet industry can compete on a basis of equality or even superiority with industries in the West. [...] Political reality requires that the United States and other Western countries do everything possible to ensure that democracy and capitalism succeed in the Soviet Union. The Soviet space industry is the finest (industry) they have. They want to work with the West to sell in the world market. If these industries cannot sell their goods, then what other industry in the Soviet Union will be able to succeed? If the Western countries refuse to share their markets, is there any chance that capitalism in the Soviet Union can succeed?[20]

As Viktor Nikolaev recalled, "Obviously, Art wanted to make money. However, he also sincerely aimed to promote Proton in the American market. He was not naive; he understood that this could capture a significant share of the market, and being the first to introduce Soviet technology to the USA meant substantial profit. He made efforts to achieve this, but ultimately, it did not succeed".[21] For his part, Manber explains Dula's failure: "The Dula agreements with Glavkosmos were not possible in the political climate of the time. In other words, Dula was too far in front of American policy to reap the practical benefits".[22]

European Perspectives: Navigating the Emergence of a New Player

The Europeans' major interest in the field of commercial cooperation with the Soviet Union was for various reasons, the possibility of conducting scientific microgravity experiments on board the *Foton* automatic spacecraft (see Chap. 7). The *Foton*'s first commercial customer was the French company Matra and CNES, with a 20 kg Sepha (Separation de Phases) experiment, which flew on *Foton-5* in 1989. According to Glavkosmos this maiden commercial mission cost 6.5 million rubles. The second French payload flew on *Foton-6* in 1990 with the CNES Crocodile experiment aboard containing materials to produce high-quality organic crystals for new generation electronics. The third commercial flight of *Foton-7* was scheduled for obtaining protein crystals and semiconductor materials

[20] Dula, A. (1991). Help Soviets launch democracy. *Space news*, September.

[21] Interview with V. Nikolaev, June 30, 2024.

[22] Manber. (2009).

under microgravity conditions. This time the spacecraft also carried scientific apparatus devised by German and French specialists.[23]

Another customer on board the automatic Soviet spacecraft was the German firm Kayser-Threde which, in turn, served as a contractor for the West German company Intospace. Valery Abrashkin emphasised the pivotal significance of this initial step in establishing collaboration with the FRG. "Through our partnership with Intospace, we successfully launched German equipment aboard an Earth-sensing satellite. This endeavour involved identifying a suitable place for equipment installation and executing the launch. The process provided our Western partners with tangible proof of equipment delivery to space, followed by the acquisition of results".[24]

The agreement to conduct experiments in space materials science with Kayser-Threde was one of the largest commercial transactions made by the Soviet space industry with Western commercial organisations during *perestroika*. As Antonio Verga recalls, "Munich based Kayser-Threde also had branches in East Germany, in Leipzig and Dresden where their two respective universities were very advanced in materials research, as well as in Stuttgart (FRG). So, it was easy for Kayser-Threde to find connections in the GDR to get in touch with Glavkosmos".[25] In 1989, working through Kayser-Threde, several European sponsors had taken advantage of new flight opportunities offered by the Soviet Union on board of *Resurs*, an Earth Observation vessel carrying payloads of 15–30 kg, in addition to its main optical equipment (Fig. 12.1). Thus, four flights took place before the collapse of the Soviet Union, wherein Western European experiments found their seats on board: Cosima 2 (Crystallisation of Organic Substances in Microgravity for Application) on *Resurs-F#5* (September 1989), Cosima 3 on *Resurs -F#6* (May 1990), and Casimir (Catalyst Studies for Industry through Microgravity Research) on *Resurs -F#9* (September 1990).[26] In October 1991, Cosima 4, crystallisation techniques were successfully performed also on a Soviet re-entry *Foton-7* capsule.[27]

Antonio Verga explained the reason for the German company's interest in the *Foton* in particular. As he puts it,

[23] Langereux, P. (1989). NIKA, nouveau satellite technologique. *Air&Cosmos*, 1244.

[24] Interview with V. Abrashkin, March 4, 2023.

[25] Interview with A. Verga, March 23, 2023.

[26] Retrieved March 10, 2025, from http://www.astronautix.com/r/resursf1-14f43.html.

[27] Henning, M., et al. (1994). Cosima—protein crystal growth facility for automatic processing on unmanned satellites. *Journal of crystal growth*, 135 (3–4), 513–522.

Fig. 12.1 Resurs at the Space Expo, Noorwijk (NL). Photo by the autor

Foton originated as a surveillance satellite, primarily serving remote-imaging purposes. Given the absence of digital cameras during that era, the films containing images captured by onboard cameras were stored and automatically transferred to a sub-capsule, fashioned in the shape of a pot. This "pot" was ejected from the *Foton*, following ballistic calculations that ensured its return to Soviet territory. Subsequently, the films were recovered, developed, and the images were gathered by the Soviet authorities.

Kaiser-Threde expressed interest in these technologies not solely for aerial imaging applications, but rather due to its keen interest in developing materials suitable for atmospheric re-entry. Specifically, materials capable of withstanding high heat fluxes and pressures during the transition from space to the atmosphere were of particular interest. At that time, they had already developed a capsule Mirka featuring a lining composed of part-ceramic and part-carbon wire materials. Well before approaching ESA, Kaiser-Threde

> had established contact with Glavkosmos to explore the feasibility of testing this material. This initiative was driven by Germany's historical advancement in such technology, and NATO's collective interest in materials capable of enduring elevated temperatures, particularly for applications such as jet nozzles and launcher nozzles.[28]

In addition to Kayser-Threde, the Finnish company EKE Group was promoting sales of Glavkosmos space technology services. EKE offered potential clients "better and more competitive terms, e.g. following satellite services: satellite launching to orbit with reliable rockets such as Proton, Vostok, Soyuz, Tsyklon etc.; hiring of Soviet satellites for weather, communication, TV broadcasting and other purposes, such as commercial satellite Gorizont; manufacturing of space technology including cooperation with Western manufactures; manufacturing of extreme pure material in microgravity conditions; special metal product manufacturing according to customer's drawings and wishes".[29]

Conclusion

Summarising the outcomes of promoting Soviet space technology and services in USA and Western Europe, some analysts reached unfavourable conclusions for the USSR. Edward Crawley, Director of the Space Engineering Research Centre at MIT, assessed the Soviet's first steps into the international space market arguing that "an early commercial casualty for the Soviets, both in terms of policy and marketing, was their attempt to enter the commercial satellite launch business. It is an area where the Soviets had no clear commercial advantage, and no unique technical capability". Nonetheless, according to Crawley's acknowledgment, the Soviet space industry boasted several products that captured the interest of the American consumer: "Soviet hardware that was ripe for USA exploitation included rocket engines, space nuclear power technology, space suits and other life support systems. They also had unique ground test facilities".[30] In addition, as an example, he cited Soviet high-speed wind tunnels that could prove valuable to USA work on the National Rocket and Space Plane. At the same time Crawley added that Soviet marketing skills left

[28] Interview with A. Verga, March 23, 2023.

[29] *EKE Engineers. Glavkosmos services available from the U.S.S.R.* In ESA-28182, HAEU.

[30] David, L. (1991). Failed coup muddles future of U.S.-Soviet joint venture. *Space news*, September, 16–22.

much to be desired. "They have yet to learn how to identify a need in a market, then how to effectively propose something they have to fill the need".[31]

In 1989, drawing conclusions from the results of Glavkosmos activity in Europe, the French space expert, editor-in-chief and deputy director of *Air & Cosmos*, Pierre Langereux stated that the USSR had only sold a few *Foton* microgravity experiments to Germany and France up to that point. Indeed, *Foton*'s transformation into a "sales outlet" was yet to come. Its active use for commercial projects occurred in the 1990s and 2000s, while in the late 1980s, there were only initial attempts at cooperation in microgravity on Soviet spacecraft. A much more serious failure was the promotion of launch services on rockets. Despite the attractive prices, Glavkosmos had not managed to sell a single rocket launch. As Langereux put it, this was mainly due to the USA embargo on sending Western satellites to the Soviet Union.[32]

There were several Glavkosmos "outlets" in both the United States and Europe. Concrete evidence, particularly from the USA, France, and West Germany, attests to the appeal of Soviet technology to Western partners. Based on the collaborative efforts of that period, cooperation primarily focused on providing spacecraft designed for European scientific experiments. Like the situation with the United States, Glavkosmos faced difficulties in offering Soviet rocket launch services for European satellites across various missions. Despite its efforts and the interest shown by potential buyers in its goods and services, the overall results of its activities were unimpressive. What were the reasons behind Glavkosmos's failure and, consequently, the Soviet state's unsuccessful attempt to enter the international market?

[31] Ibid.

[32] Langereux, P. (1989). Nouveaux tarifs des lanceurs soviétiques. *Air & Cosmos*, 1243.

CHAPTER 13

Evaluating the Impact of Glavkosmos: Successes and Setbacks

Established from the ground up in 1985, Glavkosmos was introduced as a new division within the Ministry of General Machine Building to align with the evolving demands of the era and the overarching political direction of the Kremlin's new leadership. The organisation was tasked with addressing emerging challenges, formulating tactics for an entirely new operational environment, and developing the strategic vision necessary to integrate effectively into the domestic space industry while also achieving success in the international market. This chapter examines and evaluates both the accomplishments and shortcomings of Glavkosmos.

According to Dunaev's memoirs, Glavkosmos reached its peak performance at the very end of *perestroika*. Reflecting on the institution's history, he argued that its gradual decline was driven more by shifting public opinion regarding the space program than by financial difficulties in the sector. "We were entrusted with a colossal task, and a huge number of foreign guests came to us. In 1988–1989, we collaborated with over 800 people. In 1989, we also carried out a significant number of launches. At that time, we were still thriving. Public criticism of the space sector began in 1989, intensified in 1990, and by 1991–1992, it had reached extreme levels".[1] Indeed, during this period, criticisms of the Soviet space program

[1] Interview with Dunaev. (2006).

O. Dubrovina, *The Soviet Space Program and International Collaboration during Perestroika, 1985–1991*, Palgrave Studies in the History of Science and Technology,
https://doi.org/10.1007/978-3-032-03157-0_13

began to emerge in the Soviet press. A notable example was a commentary published in *Izvestia* and later transmitted by the American press: "The monstrous hybrid of departmentalism and monopolism in the management of Soviet space research has led to economic, political, and moral losses, transforming it from a source of well-deserved pride for the Soviet people... into a soup kitchen".[2] However, while Dunaev attributed the failure of Glavkosmos and its dissolution to the prevailing negative public opinion towards the space sector, this sentiment appears to have been more a consequence than a primary cause of the challenges faced by Soviet cosmonautics at the end of the Soviet era.

In reality, Glavkosmos employees had to struggle not only with growing public dissatisfaction over excessive spending on space activities but also with a range of other obstacles. These challenges included severe economic difficulties, exacerbated by the ongoing dissolution of the entire socialist bloc and the USSR itself, as well as geopolitical constraints, such as bloc confrontations, which led to various restrictions on commercial transactions.

Difficulties Along the Way

To understand the obstruction of Soviet space production in the international market, it is essential to recognise that Soviet space manufacturers were compelled to compete with well-established entities from the USA rocket and space sector on the global stage. The initiative to integrate the Soviet Union into the international space market, including the establishment of Glavkosmos and the provision of relevant opportunities to selected Soviet organisations, was initiated only during the era of *perestroika*. Unlike the Soviet Union, private rocket and space companies have existed in the USA since the beginning of the space age.[3] While in the Soviet Union all enterprises starting with scientific research centres, design institutes, and manufacturers were state-owned, USA space program participants were private organisations that took part in the creation and development of the national space program on a competitive basis. In the early 1980s, the process of commercialisation and privatisation of the space sector in the USA gained new momentum. When Republican

[2] Aggressive Soviet space program threatened by budget, policy changes. *Aviation week & Space technology*, March 18, 1991.

[3] Dubbs, Paat-Dahlstrom (2011), 32–60.

Ronald Reagan assumed power, the issue of private launches and the transfer of certain areas of space exploration, such as meteorological observations and remote sensing of the Earth, into the hands of private entrepreneurs was brought to the forefront.[4]

Consequently, the primary sectors of international space commerce yielding the highest revenue became dominated by American rocket and space corporations in partnership with prominent space civil institutions, most notably NASA. Specifically, these sectors encompassed the manufacturing and provision of equipment for satellite ground stations, the reception and processing of data derived from Earth's remote sensing, consumables, associated services, apparatus for space navigation systems to be installed on various vehicles, and consumer goods developed based on space technology.

Having saturated the international space market with its products and services, the United States effectively precluded its primary competitor from displacing them from their monopolistic position. The Soviet space program, attempting to navigate the already divided landscape, confronted the formidable challenge of engaging in a multifaceted "trade war" on economic, political, and cultural fronts simultaneously. Operating within a capitalist market characterised by an established system of financing, investment, insurance, business negotiations, and other components inherent in a capitalist economy posed significant challenges for Soviet experts accustomed to the workings of a planned economy. Thus, in addition to intense competition and the prevalence of American firms in the launch vehicle and space services market, manufacturers of Soviet space technology encountered a lack of pertinent practical experience. The intricacies of commerce remained unfamiliar to Soviet entrepreneurs.

Cultural difficulties, misunderstandings in project management approaches, language barriers, and ingrained mistrust and suspicion on both sides complicated the already thorny path of the Soviet space program to international markets. In fact, despite declarations of the Soviet space authority, much of the Soviet space bureaucracy remained hidden. Jacques Blamont stated, "They are trying hard not to let us understand how it works. But if Glavkosmos wants to sell Soviet products, they must become civilised".[5]

[4] Logsdon. (2019), 171–207.

[5] Sawyer. (1988). Soviets move aggressively in marketing space services.

The orientation towards different standards, variations in working culture, and divergent levels of achievement across different fields further complicated the integration of the Soviet Union into the international market. Consequently, potential customers in the satellite industry harboured reservations regarding Soviet launch services. Criticisms were directed towards Proton for vibrations that could potentially impact the delicate USA satellite technology. Additionally, concerns were raised regarding the adherence to Western standards in "clean rooms" (satellite processing rooms), questioning whether they met stringent cleanliness criteria or might permit moisture or dust to compromise the sensitive parts of satellites. Soviet satellites were perceived as having simpler designs that did not necessitate the meticulous conditions characteristic of Western satellite manufacturing practices.[6] While it is conceivable that such statements were made for competitive reasons, the opinions voiced in the American press inevitably cast a shadow over the reliability of the Soviet space industry.

The problem of insufficient infrastructure was also emphasised by Evgeny Velikhov, Vice President of the Soviet Academy of Sciences. He pointed out that the Soviets had trouble using some advanced technology, such as main-frame computer technology. Nevertheless, according to Velikhov, "Our problem is not technology, it is infrastructure… We have not spent enough money on factories, clean rooms, and machines—what it takes to make a big industry". Velikhov envisioned international cooperation as a solution that could yield substantial benefits.[7]

The dearth of legal culture in the USSR also adversely affected the environment in which Soviet lawyers had to work. The deficiency of legal support for commercial space activities negatively impacted the Soviet opportunities of the space sector. As lawyer Vladimir Postyshev noted, there was not a single legislative act related to the exploration and use of outer space, and therefore it was impossible to enter into trade and commercial relations. Any commercial contract or agreement was in effect a legal document that had to meet all the requirements of the laws of the relevant states. Already in 1989, Postyshev called for the creation of a dedicated centre providing trade, intermediary and legal consulting services for interested scientific and industrial organisations.[8]

[6] Ibid.
[7] Ibid.
[8] Postyshev. (1989), 21–27.

In his memoirs, Wim Jansen provided a concrete example highlighting the difference in legal cultures: "When a Western businessman says 'There you have a good project,' such a statement reflects nothing more and nothing less than an abstract conclusion the businessman has arrived at. It does not necessarily imply that he intends to actively participate in the project. Once a written agreement is concluded, all verbal agreements become void if they were not included in the contract. In contrast, a Russian signs a contract only for his superiors. What was discussed, what was said—that is what he continues to regard as the only real agreement... The Western businessman does not comprehend this. Once he has signed the contract, he promptly forgets everything that was said".[9]

Divergent Trends

Amid the development of the space sector, characterised by the expansion of product offerings, the establishment of international cooperation on a commercial basis, and the potential for a systematic and controlled transformation of the state sector towards market principles, certain phenomena began to emerge. Notably, there was a growing desire for greater autonomy from state oversight.

Prior to the foundation of Glavkosmos in 1985, scientific entities like IKI and IBMP had proven their ability to establish and sustain contacts with foreign partners. Therefore, efforts to centralise and shift all international contacts and projects to a commercial basis through Glavkosmos generated significant discontent within these institutions. The very presence of Glavkosmos, a state structure designed to oversee the international activities of participants in the rocket and space sector, appeared to be a hindrance to an autonomous development. The transition from a planned to a market economy compelled several prestigious academic institutes to seek foreign sources of funds, marking a path towards further liberalisation from the control and coordination exerted by Intercosmos and Glavkosmos. An indication of this came in April 1989 when, according to information from Karin Barbance, a Soviet/German space association with commercial goals would be established. "It will comprise industry from the FRG and on the Soviet side IKI, IBMP, industries responsible for Mir development, Mars missions (*Phobos* and *Mars-96*) and materials

[9] Jansen. (2002).

processing".[10] In the list of Soviet scientific institutes and production centres that cooperated with German partners, she did not mention either of the two state coordinating bodies.

Other academic research centres were also not spared from the active cooperation and commercialisation sweeping across the space industry as it offered its technologies and devices for sale in a bid to earn hard currency. For instance, as Valery Barsukov stated to American journalists, their "catalogue of potential commercial products included sensors, biosensors, and advanced technologies ... The government is ready to provide the resources for the fundamental scientific research, but since the Ruble is devaluated and we are suffering from the shortages of novel scientific technologies and modern computing technology, many of the scientific institutes are thinking of organising enterprises to earn money".[11] To achieve this objective, the institute's leadership engaged with American counterparts from NASA's Solar System Exploration Division, bypassing Glavkosmos. This collaboration unfolded within the framework of the joint US-Soviet Solar System Exploration Working Group.

Although during the *perestroika* period, design and industrial centres operated strictly under the umbrella of Glavkosmos, by the final days of the Soviet Union, they had gained major autonomy to engage directly with Western partners, eliminating the need for Glavkosmos' mediation. *Aviation Week & Space Technology* quoted the words of an anonymous manager at one Soviet manufacturer about the choice to use Glavkosmos or to act in autonomy. He pointed out that it depended upon the type of product offered and the management personalities involved. "Some of us prefer the approach of going through Glavkosmos, while those with a more progressive sense of marketing are trying to go it alone. There is a considerable chaos these days within the Soviet industry, the result of things moving too fast after a history of being too tightly controlled".[12]

In this context, Manber's observation is also noteworthy, as it highlights the gradual separation of individual production centres—still under the control of communist hardliners—from a centralised leadership pursuing liberal reforms. As he argues, "the reform-minded Russian space

[10] Barbance, K. U.*S.S.R./IKI*, March 23, 1989. In ESA-28023, HAEU.

[11] Soviet Labs seek commercial ventures to raise research funds. In *Aerospace Daily*, October 22, 1990.

[12] Soviets exploring commercial space flight opportunities. *Aviation week & space technology*, March 19, 1990.

engineers and officials had to confront the fact that the factories, most of whom still supported the communists, were gaining month by month as the power of Gorbachev and the Kremlin ebbed away".[13]

These statements clearly revealed the significant steps towards greater independence for manufacturers and traders. Further deregulation of both domestic and foreign commercial activity meant that single institutions became less and less dependent on the political and financial influence of Intercosmos and Glavkosmos.

In addition to the above-mentioned challenges, another crucial factor that significantly influenced Glavkosmos' operations and its relationships with other state-owned entities must be considered: its leadership. The primary goal of Glavkosmos, established within the MOM, was to facilitate the penetration of the Soviet space industry into the global market. However, according to some scientists, Dunaev was deemed unsuitable for his position, harbouring perspectives that were considered out of touch with contemporary realities. One of his most vehement critics was Roald Sagdeev, who expressed scathing but cryptic criticism of the head of Glavkosmos. "Today international *glasnost* was operating to the point where the whole world knew that at the head of this organisation stands an odious figure. But here at home, nobody knows about this. It has not reached us yet".[14] In his memoirs Sagdeev recalls Aleksandr Dunaev as "a genuine personification of an apparatchik"[15] and a hawk who was one of the "influential members of the military and industrial sectors [...] who were involving us in a nonstop escalation of an SDI budget".[16]

Furthermore, Dunaev, an engineer by profession, exhibited a mindset that leaned towards advocating for the interests of the professional category he represented: design engineers for space technology. The *Phobos* case starkly exposed the accumulated dissatisfaction among those engaged in the implementation of the Soviet space program, particularly scientists directly dependent on the production of spacecraft such as probes, satellites, robots, etc. (see Chap. 6).

In summarising the challenges encountered by Glavkosmos within the Soviet Union and on the global stage, a crucial point emerges. Dunaev

[13] Manber. (2009).

[14] Bohlen, C. (1988). Soviet Scientist faults space bureaucrats. *New York Times*, November 5.

[15] Sagdeev. (1994), 324.

[16] Ibid., 273.

and the institute under his leadership found themselves caught between two conflicting trends. On the one hand, the push towards an intensified adoption of market relations and the inclusion of Soviet space technologies and services within world trade necessitated the concentration of efforts within a single department (Glavkosmos), under the framework of the existing socialist planned economy. On the other hand, these efforts led to discontent among disadvantaged institutions, where centrifugal tendencies were already gaining momentum, set against the backdrop of the ongoing crisis in the country. As a result, by the end of the *perestroika* period, Dunaev and the institution he led found themselves in a deadlock, due to the aforementioned reasons.

Matter of Proportions

The analysis of the civilian space sector in the Soviet Union, encompassing both scientific and commercial aspects, raises the question of its position within the overall space program during the *perestroika* period. Having had no access to relevant Soviet sources of information at the time of writing, we can only attempt to reconstruct the overall ratio of space launches for civilian and military purposes based on secret reports from USA intelligence agencies and open publications, from which information about the Soviet space program was made available. The depiction presented below does not claim to be comprehensive or exhaustive. However, it provides a general overview of Soviet space priorities in the 1980s and offers insight into the information available to the United States regarding the Soviet space program.

In the report "Soviet Military Power" produced in 1984 by the American Defence Intelligence Agency it was specified that "military systems accounted for more than 70% of Soviet space launches. Another 20% have combined military-civil application, with less than 10% devoted to purely civil/scientific activities. Very little of their effort was devoted to programs that have economic benefit".[17] In 1985, the situation had changed in favour of military use of space efforts: "In the past year, some 80% of Soviet space launches have been purely military in nature, with

[17] Defence Intelligence Agency, Soviet military power 1984. Chapter II: Forces for global warfare. Retrieved January 13, 2025, from https://irp.fas.org/dia/product/smp_index.htm.

much of the remainder serving both military and civil functions".[18] Approximately the same ratio remained in 1986: at least 90% of the launches and satellites were military related as support to both offensive and defensive operations. Furthermore, American experts expected the "military emphasis to continue in the years ahead. Of the approximately 200 operational Soviet satellites projected to be in orbit by the mid-1990s, about 150 will most likely have purely military missions, such as ocean reconnaissance, electronic intelligence, imagery reconnaissance, and special communications. Another 40 could support joint military-civilian functions, such as providing communications, navigation, and weather data. The manned program will fulfil both military and civilian missions. The approximately 10 remaining satellites could include interplanetary probes and other scientific missions".[19]

According to USA intelligence analysts, quoted by *Aviation Week* in a 1987 issue, "for hours each day, Soviet Cosmos military satellites drift over the USA photographing missiles silos and naval deployments. Other Soviet spacecraft lurk with sensitive electronic ears that can pick up telephone conversations in Washington while Meteor weather satellites monitor conditions over key USA targets. Soviet infrared satellites watch for the telltale head signalling a launch of USA ICBMs. At the military launch site in Plesetsk 500 miles northeast of Moscow, crews stand ready to launch additional intelligence satellites at the moment's notice".[20]

By 1988, though American analysts maintained that research and development trends suggested that the Soviets still fully intended to use space for both offensive and defensive purposes, they asserted that space would become the "fastest growing industry in both the military and civil sectors, although all available evidence suggested that military requirements drive the Soviet space program".[21]

By 1989, American intelligence services still expressed doubts about the sincere intentions of the Soviets to transform space from a military training ground into an economical and mutually beneficial international market. According to American experts, *glasnost* led to various public

[18] Defence Intelligence Agency, Soviet military power 1985. Chapter III: Strategic Defence and Space Programs.

[19] Defence Intelligence Agency, Soviet military power 1987. Chapter III: Strategic Defence and Space Programs.

[20] Lemonick, M. D. (1987). Surging ahead. *Time*, October 5.

[21] Defence Intelligence Agency, Soviet military power: an assessment of the threat 1988. Chapter IV - Soviet Strategic Programs and Space Forces.

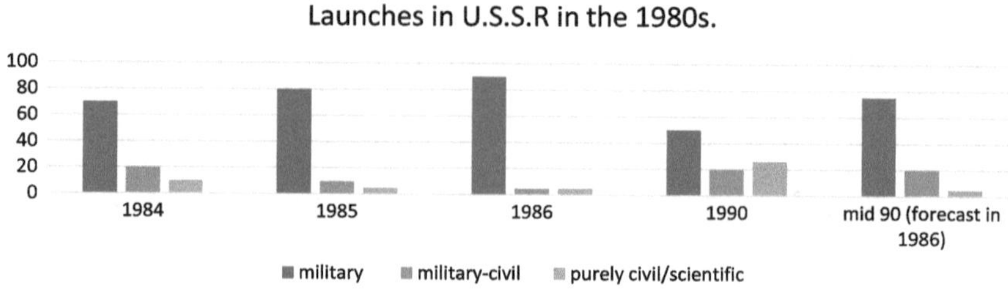

Fig. 13.1 Launches in the Soviet Union in the 1980s based on Defence Intelligence Agency data

reports on the Soviet space program. In their assessment, these reports exhibited a highly selective approach to information provision, often downplaying the military orientation of Soviet space systems and operations. Information released by the Soviet officials in those years continued to portray the Soviet space program as nonmilitary, whilst extolling advances in technology and promoting efforts to penetrate the market for commercial services. According to USA analysts the "Soviets apparently had made no effort to change their military space objectives as they maintained they were doing with some of their other military forces (Fig. 13.1). The Soviets sought sufficient supremacy to provide optimal space-based support for their terrestrial military forces and the capability to deny the use of space to other states".[22]

In the final year of the Soviet Union's existence, American analysts continued to emphasise that, even if the Soviets pursued commercial space initiatives to offset budget reductions and generate hard Western currency, these efforts remained minimal compared to the military and manned operations, which continued to dominate spaceflight activities.[23]

The data of the USA intelligence agencies is confirmed by independent Geoffery Perry's observations. Perry was a founder of the so-called Kettering Group based in England—a band of amateur radio enthusiasts that had sprung up in the early 1960s around the globe tuning into "secret" Soviet satellite transmissions.[24] In his interview to *Rocket and*

[22] Defence intelligence agency, soviet military power: prospects for change 1989. Chapter IV - Nuclear, Strategic Defence, and Space Programs.

[23] Aggressive Soviet space program threatened by budget, policy changes. *Aviation week & Space technology*, March 18, 1991.

[24] See Phelan (Ed.). (2013).

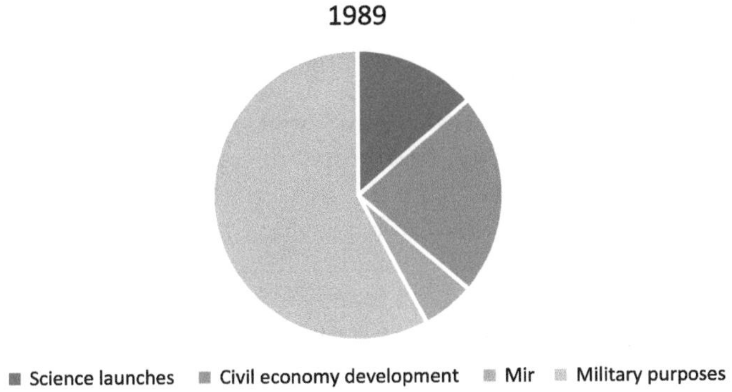

Fig. 13.2 Launch numbers in 1989 by target based on Geoffery Perry's calculations

space Daily,[25] he underlined that in 1989 the Soviet Union led the world with 74 out of the 101 space launches into orbit, compared to 18 for the USA and seven for Europe (Fig. 13.2).

In contrast to the data available in the West, the Soviets publicly insisted on the disarmament process, which was actively unfolding in the country, as a reflection of the political intentions of their leadership. Notably, in December 1988, during the UN General Assembly, Mikhail Gorbachev declared the implementation of a comprehensive program involving unilateral reductions in military initiatives by the USSR. The envisaged plan for 1989–1990 encompassed a 12% reduction in the size of the armed forces, a 14% decrease in the military budget, and a 20% cut in military production. The primary objective of the government was to ensure the provision of essential resources to the Armed Forces, guided by the principle of reasonable sufficiency.[26]

Oleg Baklanov, arguably the most well-informed—though perhaps somewhat biased—observer of the state of the VPK, noted a rapid shift in government spending priorities towards civilian initiatives. In his memoirs, he emphasised a significant and quick transformation in the distribution of financial resources. According to Baklanov, "in 1988–1989, the country's industrial potential was allocated as follows: 40% was directed toward the national economy, while 60% was devoted to the production of

[25] Soviets had 74 launches in 1989. *Aerospace Daily*, January 23, 1990.
[26] Bystrova. (2006), 298

weapons. In 1990, this ratio was equalised to 50/50. By 1991–1992, there was a significant opportunity to achieve a 60–40 ratio, prioritising peaceful production". Baklanov underlined the strategic emphasis on transitioning resources towards civilian initiatives, signalling a transformative phase in the Soviet Union's economic priorities during this period.[27]

The sole available Soviet data pertains to the ratio of military to space spending in 1989. Some evidence is provided by the data from the brochure "Where is Soviet Cosmonautics Going?" published in 1990. According to the information presented in 1989, a total of 6.9 billion Rubles were allocated for space programs in the country. This allocation comprised 1.7 and 3.9 billion Rubles designated for national economic/scientific and defence purposes, respectively. The expenditure on the development of the Soviet Shuttle Buran amounted to 1.3 billion Rubles. In the country's national income, total expenditure on the space sector amounted to only 0.26%. For comparison, analysts noted that this figure was ten times lower than the expenditure of the Ministry of Land Reclamation and Water Management in the same year and five times less than the financial aid provided by the Soviet Union to other countries.[28]

Another Soviet specialist, Anatoly Evich, commented on this data. According to him, total expenditure on the space sector in 1989 constituted 1.4% of the state budget. He also added that over a span of 14 years, the most expensive space programs in the Soviet Union saw a total expenditure of 14 billion Rubles on the Energya-Buran program alone, while 1.7 billion Rubles were allocated to the Mir orbital station during the same period. Nonetheless, according to the analogy Evich offered, "if we arrange all industries in descending order based on the allocated funds for their development, the space sector would not occupy the top position. It is worth noting that in the same year, the USA allocated $29.6 billion for space, which accounted for 2.6% of the federal budget. Clearly, the results of this comparison do not favour the domestic space program".[29]

As for the civilian space sector, the gradual reduction of state budget allocations for its development led to its steady decline, set against the backdrop of publicly announced prevailing political priorities. On April 12, 1990, Dunaev announced that the 1990 space budget would be 20%

[27] Baklanov. (2012). Vol. 2, 661.
[28] Avduevsky, Leskov. (1990), 3.
[29] Evich. (1991).

smaller than in 1989,[30] yet in September, the West learned from Soviet sources that the 1990 budget had actually been cut by 37%, compared to the previous year.

There is a clear challenge in drawing definitive conclusions due to the contradictory data provided by American and Soviet sources. However, the following can be asserted. First, the military space sector clearly held dominance over the civilian sector in the late Soviet Union. Second, there was a discernible trend towards narrowing this gap, though a significant disparity persisted. The scientific and commercial branches represented only the visible tip of the huge iceberg that was Soviet cosmonautics. This indicates that phenomena such as liberalisation, commercialisation, and openness to international cooperation affected only a small portion of the overall rocket and space field. The transformation of the sector's core assets had yet to fully unfold.

Conclusions

Summarising Glavkosmos' activities, Baklanov emphasised its achievements in both foreign policy and commercial ventures. In terms of foreign policy, he highlighted the signing of ten intergovernmental agreements by 1992, facilitated with Glavkosmos' participation. These agreements focused on cooperation in the exploration and use of outer space for peaceful purposes and covered the development of technologies in fields such as materials science, biology, medicine, remote sensing of Earth, and astrophysical research of the solar system, among others.

In the commercial sphere, Baklanov noted the establishment of contacts with foreign space firms and organisations. He stated that "the first commercial contracts for launching payloads of foreign customers on Soviet launch vehicles were realised". According to Baklanov, this activity "paved the way for domestic launch vehicle manufacturers to enter the international market for launch services. Additionally, Glavkosmos commercialised the organisation of flights for foreign cosmonauts and experiments on technology and materials aboard manned stations, generating $280 million from commercial contracts since 1985".[31]

In essence, Glavkosmos ushered the Soviet Union into the commercial space market, signalling a clear intent to capitalise on the financial

[30] Economic crisis will force cuts in Soviet Space budget. *Space news*, August 20–26, 1990.
[31] Baklanov, Rogozin. (2013), 458–459.

opportunities presented by the newly accessible, once-forbidden space industry. However, as part of the forced and accelerated transition to market-based relations, the restructuring of this complex system to align with the country's evolving needs posed numerous challenges, driven by both objective and subjective factors.

It was long before the breakup of the Soviet Union and the subsequent period that the commercialisation of Soviet space assets took the direction towards the transition to the market economy. Therefore, it is worth reconsidering the general belief that the "selling off" of space hardware and technologies started in Russia in the transitional 1990s. At the same time, it would be erroneous to claim that since the mid-1980s there has been a complete opening of the civil space sector towards the international market. Rather, this process could be correctly characterised by the formula "to open something without opening everything".

This represented an initial, relatively positive endeavour—particularly when compared to the preceding era—aimed at aligning the unwieldy, highly conservative, and exceptionally secretive Soviet VPK, including the space program, with the demands of the rapidly evolving domestic political and economic landscapes, as well as the international market. However, it would have been overly optimistic to expect unqualified success from the efforts of Soviet neo-entrepreneurs, given the persistence of the planned economy in the 1980s, the partial nature of Gorbachev's reforms, systemic inertia, internal political resistance, and the cautious stance adopted by international partners. Those newly formed entrepreneurs had to start from scratch, learn the rules of international trade in real time, adapt to the requirements of the global market, and develop a legal framework for protection. These challenges were compounded by the extreme secrecy of the space sector, made even more sensitive by its close ties to the military. Glavkosmos' achievements in promoting Soviet space assets and technologies to the international market should be considered a partial success. Meanwhile, events had taken their course and, despite the efforts of the Soviet military and political elite, had spiralled out of their control.

CHAPTER 14

Conclusions

The early *perestroika* period was characterised by a dynamic progress of the Soviet space program. The International Halley's Comet Project, including the Soviet VEGA mission (1984–1986), epitomised the remarkable expansion of scientific research. In 1986, the Soviet Union launched Mir, its third orbital station within a decade, and in 1988, the single flight of the Energya-Buran transportation system integrated all available cutting-edge technologies. The election of Mikhail Gorbachev as General Secretary of the CPSU in March 1985 and the introduction of his *new thinking* policy inevitably had a profound impact on the space sector. Similar to his predecessors, Gorbachev viewed space as an instrument of foreign policy, albeit one that underwent significant changes in its application. The priority of national and ideological prestige, as well as military-strategic importance, gradually gave way to the rhetoric of the *new thinking,* which promoted international cooperation in space endeavours and emphasised their exclusively peaceful nature. Over time, the space issue gradually receded from the political agenda as the political, economic, and social crises that Gorbachev faced in the latter half of his leadership deepened.

During the early *perestroika* period, the space sector continued to experience both the legacy of previous political approaches and the influence of Gorbachev's reforms. Funds continued to be spent for the use of space to

O. Dubrovina, *The Soviet Space Program and International Collaboration during Perestroika, 1985–1991*, Palgrave Studies in the History of Science and Technology,
https://doi.org/10.1007/978-3-032-03157-0_14

promote the advantages of the Communist system (thus, the last "free ticket" to the Mir station was given to the Afghan cosmonaut Abdul Ahad Momand in August 1988), albeit to a lesser extent than in the previous years. The climate of closeness and secrecy was ultimately not fully overcome, and the decades-old approach to relations with foreign partners persisted, although at a lower level. The development of civil space exploration continued within the framework of the VPK, which exerted a decisive influence on the entire program.

The decentralisation of management, conversion, and liberalisation at the end of the 1980s had contradictory effects on the development of the space sector. Moreover, the budget allocation for space activities was reduced each year. In 1990 for instance, the space budget was cut by 37% compared to the previous year. Despite *perestroika* team's initial intentions to reform the field plagued by significant shortcomings, these efforts ultimately resulted in failures and a further deterioration of the condition of the space sector. *Glasnost*, with its multifaceted aspects, fostered lively discussions about the deficiencies of the space sector and potential solutions. However, it also diverted public attention from its actual needs and fuelled resentment over expenditure for its maintenance and development.

Cornered by this deep crisis, the space sector management were forced to urgently seek a way out of the deadlock. Two primary solutions emerged: strengthening of the international scientific cooperation with international partners and the broadening of commercialisation. What had once appeared to be a generous gesture became a necessity, and the previously forbidden exchange of classified technologies transformed into a lucrative commodity.

A notable example of the Soviet effort to overcome the financial and technological crisis was the active development of cooperation between Soviet scientific institutions and ESA in space exploration. During this period, the Soviet Union's willingness and readiness for full-scale cooperation on a renewed basis aligned with the needs of ESA, which, due to the grounding of the USA Space Shuttle and Ariane's failures, was compelled to seek alternative opportunities for conducting scientific and technological experiments in orbit. Naturally, this rapprochement was facilitated by the broader international context and the transition from confrontation to dialogue initiated by Gorbachev.

This alignment of timing and circumstances facilitated the realisation of successful projects such as *Bion*, as well as the less successful *Phobos*, which revealed the complexities of collaboration at political, technical, and personal levels. The actual implementation of these missions depended on several key factors: a favourable foreign policy environment, the interest and support of political authorities, and the dedicated efforts of scientists and engineers, who remained equally committed to achieving common objectives and resolving emerging challenges. These projects also served as a compelling illustration of the intricate relationship between political power and science, highlighting both the potential and the limitations of space diplomacy.

Another solution to the collapsing space industry was the gradual transition of the still classified space sector, which had been a cornerstone of the strategic security system, to market relations. Considering all the inhibiting factors, in the second half of the 1980s, an exceptionally bold attempt was made to adapt the Soviet rocket and space industry to quickly enter the international market for launch and satellite services. At the same time, a strict condition for this restructuring was the preservation of the state's control over the commercialisation process within the space sector. Although this effort cannot be deemed entirely unsuccessful, it was for several reasons ultimately doomed to fail.

The space policies implemented during *perestroika* by the Soviet government and military-industrial establishment can be examined from both a narrow, space industry-specific perspective and a broader, historical one.

From the space industry-specific perspective, Soviet political and industrial leadership simultaneously had to urgently carry out two fundamentally contradictory tasks: ensuring national security (as it was then understood) and salvaging a rapidly disintegrating economy. While the concept of state security imposed a policy of secrecy and isolation, the need to survive in a collapsing economy required the opposite—openness, data sharing, and adherence to rules established by the international community. The Soviet leadership attempted to reconcile these two objectives in its approach to international space activities. However, in the absence of a clear distinction between military and civilian functions, as well as the duplication of managerial and operational resources, achieving these goals proved challenging. The sharp cuts in government funding, the gradual erosion of competencies, the emergence of institutions attempting to operate autonomously from the central authority (such as Glavkosmos), the establishment of direct engagements with foreign partners bypassing Glavkosmos, and the intensification of internal competition in international cooperation—these factors collectively precipitated the collapse of the predominantly vertical power structure. Consequently, this led to diminished state control over the commercialisation of the space sector.

To assess the processes that have occurred in cosmonautics within a broader historical perspective, it is useful to start with a seminal work in the political history of space exploration: Walter McDougall's *...The Heaven and the Earth*. According to McDougall, the American space program competed directly with the Soviet one, in part by imitating the Soviet planned economy model. In doing so, the United States effectively built its own military-industrial complex under the banner of fighting socialism. McDougall contends that as long as the USSR maintained its planned economy and strict political regime, "the Soviet state could make up in tyranny and disregard for deleterious side effects what its centralised technocracy lost in creativity".[1]

Considering the paradigm proposed by McDougall, during the *perestroika* period Soviet leadership attempted the opposite approach—transitioning from a planned economy to a more commercial model. As discussed in this book, cooperation with Western businesses was seen as essential for attracting new resources and rescuing the space sector from the internal economic pressures that threatened to destroy it. Yet, this transition had profound consequences. On the one hand, adopting capitalist business practices in cosmonautics—alongside democratic reforms such as freedom of speech, elected management, market competition, international partnerships, and decentralised control—undermined the authority and legitimacy of state structures, accelerating centrifugal forces within the Soviet VPK. On the other hand, the commercialisation of Soviet space activities—which essentially meant their transformation into consumer commodities and services—clashed with the traditional role of the space program as a source of national pride and a symbol of progress and a great future. This shift led one segment of the population to become disillusioned with space exploration, cultivating a "nostalgia for the future", while another segment grew increasingly frustrated, perceiving investments in space infrastructure as futile and financially unjustifiable. Thus, the liberalisation and commercialisation of Soviet cosmonautics, pursued during *perestroika* as a survival strategy, ultimately had detrimental effects on national identity and the collective sense of technological and ideological superiority.

[1] McDougall (1997), 444.

It was during *perestroika* that the systemic flaws of the Soviet space program first became apparent—flaws that would only deepen during the subsequent transition period. Efforts to address these issues by introducing elements of a market economy and liberal democratic reforms, inspired by Western models, failed to deliver the expected results. The trajectory of the Soviet space program closely mirrored the broader contradictions that shaped the Soviet Union/Russia's transition, particularly the stark gap between the idealised visions of capitalism and liberal democracy and the far more complex realities on the ground.

Returning to the central theme of this book—that cosmonautics reflected broader socio-political and economic shifts in the late Soviet period—a more meaningful conclusion can be drawn about the contradictions of the transition era. The inherent inconsistencies that had accumulated within the Soviet political and economic system, along with the rapid collapse of its main pillars—the Communist Party, the planned economy, and the ideological superstructure—in the final years of the USSR, combined with the sudden imposition of liberal-democratic and capitalist values on a society already strained by half a century of socio-economic burdens, created a problem so complex that Gorbachev's leadership could not resolve it. These unsettled issues were ultimately inherited by the newly formed governments of the former Soviet republics.

But that is another story to tell.

Index[1]

[1] Note: Page numbers followed by 'n' refer to notes.

O. Dubrovina, *The Soviet Space Program and International Collaboration during Perestroika, 1985–1991*, Palgrave Studies in the History of Science and Technology,
https://doi.org/10.1007/978-3-032-03157-0

FSC
www.fsc.org
MIX
Papier | Fördert
gute Waldnutzung
FSC® C083411

Kolibri 360 GmbH
Ferdinand-Jühlke-Straße 7
99095 Erfurt, Deutschland
produktsicherheit@kolibri360.de